Biodegradable Long Acting Injectables and Implants

Farshad Ramazani
Editor

Biodegradable Long Acting Injectables and Implants

An Industry View

Springer

Editor
Farshad Ramazani
Technical Research & Development
Novartis (Switzerland)
Basel, Switzerland

ISBN 978-3-031-72301-8 ISBN 978-3-031-72302-5 (eBook)
https://doi.org/10.1007/978-3-031-72302-5

This Springer imprint is published by the registered company Springer Nature Switzerland AG
The registered company address is: Gewerbestrasse 11, 6330 Cham, Switzerland

Preface

Parenteral drug delivery systems refer to the administration of drugs through routes other than the digestive tract, such as injections and surgical insertion of depots. Compared to oral administration, parenteral delivery offers advantages in terms of drug bioavailability, as it bypasses the drug absorption step, gastrointestinal enzyme degradation, and liver first-pass effect. However, for the management of chronic diseases frequent administration is often required leading to decreased patient compliance. To overcome these limitations, controlled release strategies have been developed to extend systemic drug exposure following a single dose. Long-acting injectables and implants (LAIs) first emerged around 70 years ago by developing benzathine penicillin G, an antibiotic. Several terms have been used to describe LAIs including sustained release, delayed release, prolonged release, modified release, extended release, controlled release, and depot formulation.

LAI is typically administered through intramuscular (IM) or subcutaneous (SC) routes, allowing for controlled release of the drug from its depot over an extended period (ranging from a few days to several months). There is a growing interest in the development of long-acting options for local drug delivery, alongside systemic administration through IM or SC injections. This has led to the exploration of alternative administration methods, such as intra-ocular, intra-articular, and peri-tendon delivery. As a result, we can expect more drug products based on LAI formulations to enter the clinical stage in the future. Despite the numerous benefits in terms of adherence, efficacy, and safety approximately 50 brand name medicines that utilize various biodegradable long-acting technologies have received FDA approval so far. The manufacturing process for LAI involves multiple steps, each presenting its own unique challenges. Consequently, pharmaceutical companies consider LAI formulations to be complex and difficult to develop. Most of the existing literature on this topic is authored by scientists who never worked in the industry and therefore have limited knowledge of large-scale industrial manufacturing and challenges. As a response, we aim to address the challenges associated with the development of biodegradable LAI from an industrial perspective.

In this book, we explore the drug substance characteristics that can be specifically modified for the development of LAI. We also delve into the manufacturing and characterization of biodegradable polymers with a strong track record of biocompatibility, known to prolong the release of drugs. Additionally, we outline manufacturing strategies for various LAI, such as polymeric particulate systems, implants, *in-situ* forming depots, suspensions of drug crystals, and oil-based formulations of lipophilic prodrugs. We then review the clinical applications of LAI in different disease areas. In addition to formulation and manufacturing, we discuss in detail the most important challenge in development of LAI for local drug delivery which is biocompatibility and local tolerability. We conclude the book by providing prospects for LAI.

Considering the wide range of subjects covered, I think this book will be highly relevant to professionals in both academia and pharmaceutical companies worldwide. Whether you're a project leader, a researcher, a scientist specializing in formulation and process development (with focus on LAI), an analytical scientist, a QC/QA professional, a regulatory staff member, or a manufacturing leader, this book is your roadmap to LAI success.

I would like to express my gratitude to all the authors for their significant contributions to each chapter. I also extend my thanks to Susanne Dathe and Reece Menzies from Springer Nature Group for their unwavering support. Finally, I am deeply grateful to my mother, Nimtaneh Ramazani, my wife, Samaneh Validi, and my daughter, Lara Ramazani, for their tremendous emotional support.

Wishing you an enjoyable read!

Basel, Switzerland Farshad Ramazani

Contents

About the Author

Farshad Ramazani is a Senior Expert Science and Technology at Novartis Pharma AG in Basel, Switzerland, specializing in parenteral drug development. He has been instrumental in developing several patient-centric long-acting injectable formulations, now in various clinical trial phases. Dr. Ramazani regularly publishes research and review articles in leading pharmaceutical journals and is a co-inventor on three patents on long-acting injectables. He received the TRD Culture Award in 2021 and 2023, and the Innovation Award of the Year in 2016 at Novartis. In 2010, he was among the top four candidates nationally to receive a PhD scholarship from the Iranian Health Ministry. He completed his PhD in 2015 at Utrecht University in the Netherlands, focusing on polymeric micro and nanoparticles for local and targeted drug delivery.

Chapter 1
Drug Design for the Development of Long-Acting Injectables

Thomas Ullrich

List of Abbreviations

ADME	Absorption/distribution/metabolism/excretion
Clog*P*	Calculated octanol-water partition coefficient
DMPK	Drug Metabolism and pharmacokinetics
eTPP	Early target product profile
FDA	Food and Drug Administration
HIV	Human immunodeficiency virus
LAIs	Long-acting injectables
LAR	Long-acting release
log*P*	Octanol-water partition coefficient
PD	Pharmacodynamic
PK	Pharmacokinetic
PLGA	Poly(lactic-co-glycolic) acid
PSD	Particle size distribution
R&D	Research and development
RoA	Route of administration
Sol	Solubility

1.1 Introduction

The past decades have seen an astonishing surge in development of biodegradable, long-acting drug products including crystalline suspensions, gels, implants, or polymer-based nano- and microparticles in order to extend drug release beyond conventional solution-based systems (Bari 2010; Gonzalez et al. 2020). Most

T. Ullrich (✉)
Global Discovery Chemistry, Biomedical Research, Novartis Pharma AG, Basel, Switzerland
e-mail: thomas.ullrich@novartis.com

F. Ramazani (ed.), *Biodegradable Long Acting Injectables and Implants*,
https://doi.org/10.1007/978-3-031-72302-5_1

approaches in the past were targeting strategies to tailor long-acting formulation principles to molecules that had already been optimized for potency and often also for oral bioavailability. Such strategies, however, required the development and characterization of a large number of formulation candidates through extensive and time-consuming experiments, thus putting a significant burden on LAI development (O'Brien et al. 2021). Our group has envisioned a different approach, i.e., to optimize drug substances not only for pharmacodynamic and pharmacokinetic properties but also for their compatibility—or even synergistic attributes—with selected formulation principles of choice. In that manner, a developmental candidate would not require such extensive formulation work anymore as it already features physical characteristics that make it amenable to a preferred sustained-release matrix. In cases where this is not successful, a thoroughly validated prodrug design can also be applied. In this chapter, we summarize the key aspects of LAI drug design and properties for consideration by the academic and industrial community, and even for future implementation in machine learning models to accelerate the design of, e.g., polymer-based LAIs (Bannigan et al. 2023).

1.2 Oral Drugs Versus LAI Drugs: How Are They Different?

In contrast to peptides, therapeutic proteins, or antibodies, low-molecular weight ("small") synthetic drug molecules can be easily modified by chemical processes to tailor their physical and chemical ("physicochemical") characteristics and make them amenable to LAI development. Such characteristics include chemical and metabolic stability, solubility in various excipients, and compatibility or suitability for reproducible manufacturing. At the very beginning of a drug discovery campaign, the research team usually defines an early target product profile ("eTPP") which must include the preferred route of administration (RoA). To this end, several different disciplines weigh in to provide scientific and technical insights, e.g., molecular biology, medicinal chemistry, technical development, drug metabolism and pharmacokinetics ("DMPK"), preclinical safety, and translational medicine. A locally delivered drug, for example, will likely require a relatively small dose to minimize the injectable or implantable volume/size and therefore alleviate the burden for manufacturers, healthcare providers, and patients alike. Such small doses, however, will only result in targeted tissue concentrations over a prolonged period if they are highly potent, i.e., engage their biological target at low local concentrations. This contrasts with systemic drugs that can be administered in substantially higher doses, e.g., via oral delivery. Moreover, different drug product applications will require different release mechanisms and therefore come with certain limitations on injectable volume and dosage form. Consequently, research teams apply distinct screening strategies and adapt their preclinical flowcharts to front-load certain assays and models that are of high importance for timely decision-making. Eventually, it is the medicinal chemist's task to engineer the most appropriate physicochemical and absorption, distribution, metabolism, and excretion (ADME) profile to achieve the expected in vivo performance of a drug candidate (Bassand et al. 2022).

Toward the end of the last century, the pharmaceutical industry suffered from a rather high attrition rate from oral clinical candidates to market launch, often because of inadequate compound quality (Campbell et al. 2018). Through a revolutionary period of progress in molecular biology, assay miniaturization, and computational power—to predict and quickly measure physicochemical properties—along with the help of large datasets analyzing the statistics of marketed and failed drugs, chemists have begun to improve that quality dramatically. Lipinski's famous Rule of Five (Lipinski et al. 2001) and the subsequently published metrics of optimization such as ligand efficiency (Kuntz et al. 1999) and molecular obesity (Hann 2011) have been instrumental in guiding the medicinal chemists toward compound characteristics, thereby standing a better chance of achieving sufficiently high absorption after oral administration. Advances in predictive computational power also led to an enhanced understanding of ADME design and, consequently, better anticipation of human dose and pharmacokinetics and, eventually, better translation into clinical trials. At present, research teams are perfectly capable of rationalizing oral bioavailability as a pivotal interplay of aqueous solubility, permeability, drug disposition/distribution, and first-pass hepatic metabolism (Aungst 2017).

The design of LAI small molecules is based on fundamentally different principles. The oral drug disposition depends on the molecule's ability to dissolve well in the gastrointestinal tract, which then permeates through the gut wall and escapes to the liver during first-pass metabolism/excretion. On the other hand, there are new prerequisites for an injectable drug: low solubility in the injection site's tissue compartment(s), excellent compatibility with a well-tolerated and biodegradable depot solution/suspension/carrier, and functional groups that can serve as anchor points for prodrug generation (Bassand et al. 2022). Furthermore, LAI substances aiming at prolonged residence times in the systemic circulation after absorption from the injected depot need to be metabolically very stable, while locally acting drugs (depot injection to the site of disease/injury) are often required to metabolize quickly once reaching the bloodstream, mostly to avoid systemic side effects.

1.3 Design of Drug Substances for Crystalline Suspensions

Micro- or nanocrystalline suspensions can offer very effective sustained release without any pharmaceutical carrier and usually represent a natural first choice of modality to develop injectable depots, owing to a straightforward technical development. In this modality, drug release essentially depends on the dissolution rate of the crystal particles. A slow dissolution rate can be engineered by making the drug substance poorly soluble in the biophase and by formulating it as particles with a sufficiently large size to provide an overall low depot surface. How do medicinal chemists design poor solubility (in water-based media representing body fluids)? The two key properties are high crystal lattice energy—manifesting in very high melting temperatures—and a limited ability to form hydrates in an aqueous environment. Rigid and flat molecules tend to pack very densely in crystal lattices and form strong intermolecular *van der Waals* bonds, π–π stacking, or hydrogen bonding interactions (Fagerberg et al. 2010; Wassvik et al. 2008). Moreover, high lipophilicity (the ability of a chemical compound to dissolve in nonpolar solvents) leads to poor hydration (Bergström et al. 2007). Others have also established a correlation between the number of aromatic rings in a molecule and its physicochemical profile; even within a defined range of lipophilicity, aqueous solubility is negatively correlated with the so-called aromatic ring count (Ritchie and Macdonald 2009). Finally, many locally delivered drugs, which must exert their biological activity in a confined physiological compartment—importantly without leading to undesirably high systemic exposure for safety considerations—should have in-built predetermined "breaking points." These are often metabolically labile molecular substructures or functional groups that are cleaved by typically enzymatic processes. Smart design of such "soft drugs" allows for rapid clearance of the parent drug from the systemic circulation, leaving behind an ideally biologically inactive (and non-toxic) metabolite that is gradually eliminated from the body.

The interplay of crystal lattice-limited and lipophilicity-limited solubility of crystalline drug substances becomes apparent when comparing 11 clinical/marketed LAI drugs (Fig. 1.1), which are all categorized as "poorly soluble in water" or "practically insoluble in water" in common databases (www.pharmacircle.com, www.drugbank.ca, www.chembk.com). Melting points (onset temperatures of melting processes) are plotted against calculated log*P* values. Interestingly, a successful

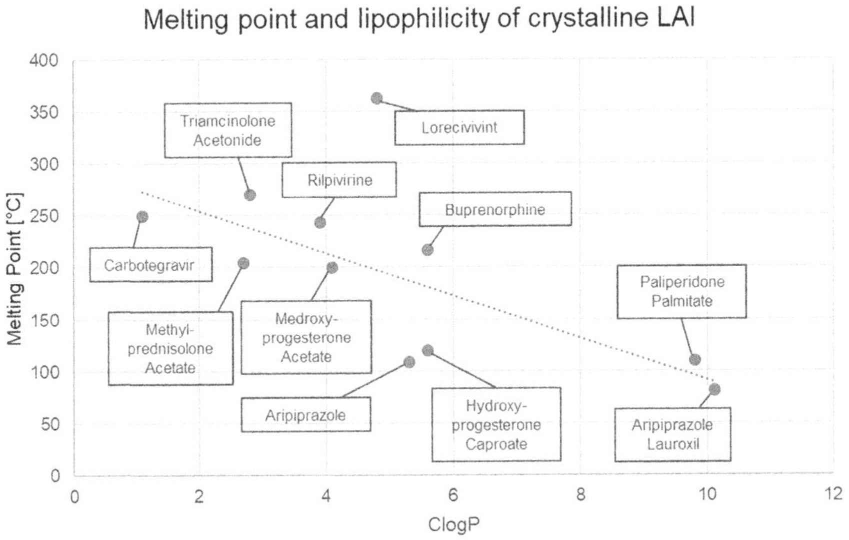

Fig. 1.1 Physicochemical properties of clinical/marketed crystalline LAI suspensions. Experimental melting points and calculated log*P* values taken from www.pharmacircle.com, www.drugbank.ca, and www.chembk.com

LAI physicochemical profile seems to be driven by either high log*P* or high melting point, and those two properties are largely anti-correlated. Very high-melting compounds, such as carbotegravir or triamcinolone acetonide, are rather mildly lipophilic, whereas the fatty acid prodrugs (see Sect. 7), such as paliperidone palmitate or aripiprazole lauroxil, have relatively low melting points (therefore low crystal lattice energy). However, due to the nature of their prodrug structure, they show unprecedentedly high lipophilicity. Both profiles lead to an overall very poor aqueous solubility and enable drug release from up to several weeks to months after subcutaneous or intramuscular injection. The chondrogenic clinical drug candidate lorecivivint/SM04690 (Deshmukh et al. 2018) sticks out with an extraordinarily high melting point (363 °C) and—combined with a high calculated octanol-water partition coefficient (Clog*P*) (4.8)—features quite unique characteristics that are thought to be responsible for a reported-release duration of up to 6 months in joint tissues and a clinical dosing interval of two intra-articular injections per year, thus showing a remarkable competitive advantage over existing local therapies.

Lorecivivint (structure shown in Fig. 1.2) is an outstanding example of a drug molecule designed to be injected into arthritic knee joints. The combination of high melting point and high lipophilicity leads to a physicochemical profile that renders the substance practically insoluble in water and enables crystal suspension formulations. Thus, lorecivivint microcrystalline suspension dissolves very slowly in articular joints after injection and the residence time in cartilage up to several months in rat models has been reported. Upon dissolution, lorecivivint distributes within joint and bone tissues but results in only negligible systemic exposure over the entire treatment period, leading to high safety margins.

In contrast to drug products aiming at local drug disposition (proximal to the administration site), depot injections to intramuscular or subcutaneous tissue serve

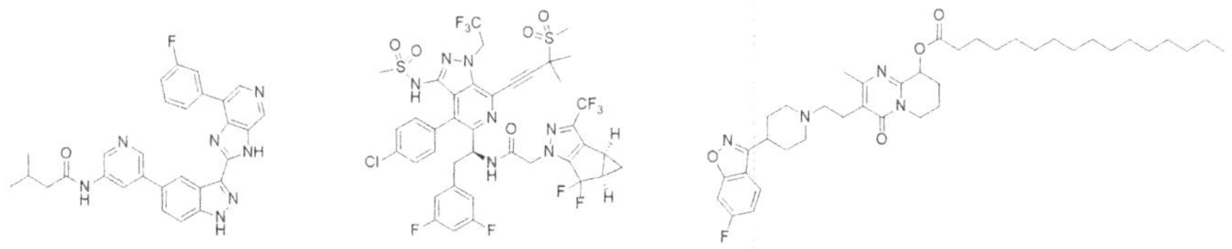

Lorecivivint[a]
MW = 505.6
Aromatic ring count = 6
logP = 6.4
Sol (pH 6.8): <0.001 mg/mL
Melting point: 356°C

Lenacapavir[b]
MW = 968.3
Aromatic ring count = 5
logP = 7.9
Sol (pH 6.8): insoluble
Melting point: n.d.

Paliperidone Palmitate[c]
MW = 664.9
Aromatic ring count = 3
logP: 9.21
Sol (pH 6.8): insoluble
Melting point: 114-116°C

Fig. 1.2 Characteristics of lorecivivint (for intra-articular injection as microcrystal suspension), lenacapavir (for subcutaneous injection), and paliperidone palmitate (for intramuscular injection as nanocrystal suspension). The low solubility of lorecivivint is achieved by high lipophilicity, high crystal lattice energy, and high aromatic ring count. Lenacapavir has an unusually high molecular weight and is also practically insoluble in water. Paliperidone palmitate owes its very low solubility to the high lipophilicity of the fatty acid ester prodrug. [a]Experimental data measured at Novartis. [b]Data taken from www.pharmacircle.com. [c]Data taken from www.labnetwork.com

as a reservoir for extended systemic drug exposure. Lenacapavir, an antiviral agent, has been reported to support a 6-month dosing interval after a single subcutaneous injection, owing to an exceptionally long plasma half-life in humans up to 175 days (Begley et al. 2020). The drug was optimized for ultra-high potency against its target, human immunodeficiency virus (HIV-1) capsid protein, and for high metabolic stability. The molecular properties of lorecivivint, lenacapavir, and palmiperidone palmitate are summarized in Fig. 1.2.

Medicinal chemists can build in key molecular features, such as lipophilicity, 3D shape, or aromatic ring count to impact the physicochemical profile toward an optimal range for crystal formulation development. To achieve this, research teams need to invest early on in the characterization of solid-state properties, such as crystallinity, melting temperature, polymorphism, and chemical stability during milling processes. Thermodynamic solubility should be determined from a well-defined crystalline batch in a relevant aqueous medium or biological fluid to select promising candidates for the development of injectable micro- or nano-suspension depots (Nkanga et al. 2020).

In an endeavor to streamline the decision-enabling screening cascade for the preclinical evaluation of injectable crystalline suspensions, our group has devised an early flowchart in which iterative cycles of compound optimization have been built (Fig. 1.3). Medicinal chemists submit their newly synthesized molecules to a primary biological assay which, thanks to the state-of-the-art molecular and cell biology technology, readily delivers key data on drug potency in a short time at low costs. Since exquisite potency is one of the key deliverables of any given LAI discovery campaign to warrant a low injectable dose, this first screening assay will usually come with a high attrition rate. Compounds surviving such a rigid potency filter (the exact definition of which can be deduced from pharmacodynamic models and compared to drug

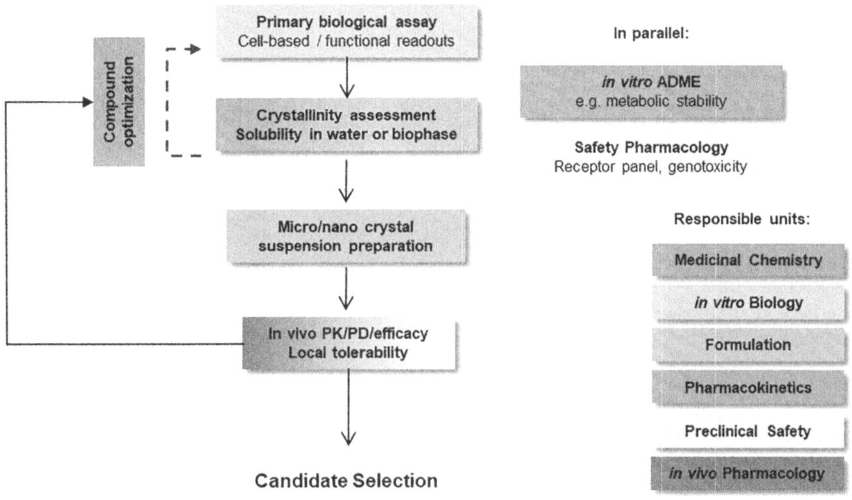

Fig. 1.3 Exemplary flowchart for the discovery of crystal suspension-based drug candidates

products already tested in clinical trials for the same indication) will be then tested for their physicochemical and in vitro ADME properties. Only stable crystal forms—devoid of early signs of complex polymorphism—with sufficiently high melting points and low solubility in aqueous buffer systems (as a surrogate for bio-relevant fluids if the latter are not readily available) will be allowed to pass. Exact thresholds for aqueous solubility are project-specific and depend on the minimum effective unbound drug concentrations that must be reached in the circulation or at the target tissue over the desired period of time. If none of the compounds fulfill such pre-defined criteria, the project will undergo another iteration of compound synthesis. Qualifying molecules can then be processed to micro- or nanocrystal suspensions, which may take up to 100 mg of substance for first milling experiments and can therefore be performed only on a small number of selected compounds. Stable crystal suspensions with a defined particle size distribution (PSD) are sent to the first line of in vivo experiments in rodents by which the pharmacokinetic profile after a single injection (usually three different doses) is determined. The monitoring period depends on the expected dissolution and elimination rates in the organism and may extend to 2 weeks or longer. While blood sampling can be done in a longitudinal fashion to provide multiple data points to measure systemic drug levels over time, local tissue sampling may be needed in local drug delivery projects to evaluate drug disposition and/or target engagement in tissues surrounding the injection site. Those sampling procedures, sometimes accompanied by tissue histology, are often terminal and can be performed only at the end of the in-life study phase. Ideally, they should be coupled with appropriate pharmacodynamic (PD) or even efficacy readouts to maximize the information gained from such long-lasting animal experiments. Likewise, the team can capitalize on the in vivo work to get insights into the local tolerability of the suspension formulation. Foreign-body reactions are often mild in nature and reversible over 1–2 days but under certain circumstances (compromised injection tissue, large microparticles, on-target biological effects of test compound) may lead to a more pronounced immune reaction that can be well classified by histology, on top of macroscopic signs of inflammation. Failure to achieve the targeted PK/PD profile in the absence of severe adverse histopathological findings at the injection sites will inevitably lead to another round of chemical optimization.

1.4 Drug Encapsulation in Depot Carriers

While crystal suspensions offer an "in-built" sustained-release profile, not every drug can be designed toward stable injectable suspension formulation of crystalline material. Many potent drug candidates with favorable ADME profiles cannot be derivatized to high-melting, crystalline forms or analogs—strongly lipophilic drugs often evade crystallization for the lack of intermolecular interactions to build up crystal lattices. Thus, precious time and material are wasted in unsuccessful crystallization screening campaigns. Furthermore, a complex polymorphism landscape may pose insurmountable hurdles to successful development projects. Efforts to predict crystal characteristics by computational methods are ongoing but not universally reliable and applicable yet.

If the drug substance itself cannot be turned into a slow-releasing crystal form, it may be encapsulated in a molecular carrier that protects it from fast dissolution, distribution, and elimination from the biophase. Injectable depots consisting of drug substance and carrier are commonly formulated as molecular or solid dispersions and act as effective permeation barriers, delaying drug release in a controlled fashion. Here, it is the carrier, rather than the drug substance, that fulfills the functional requirements of the LAI, and experts in macromolecular (bio)chemistry, polymer chemistry and material sciences are needed for the creation of smart depots. Yet, medicinal chemists also have the crucial task of designing molecules that are compatible with established or innovative LAI technologies (e.g., encapsulation in polymer-based microspheres, oil/lipid-based formulations, or liquid-crystal in situ gels). Ideally, such compatibility design goes in parallel to chemical lead optimization in the research phase, so that the end of the optimization phase should be marked by a clinical-ready drug product consisting of the best possible combination of drug substance and depot formulation. A molecule that is to be released from the depot carrier via passive diffusion and/or breakdown of the carrier should show low partitioning from the depot phase to the water-affine tissue compartments. It also must be considered that several manufacturing processes, e.g., oil-in-water emulsification for the preparation of poly(lactic-co-glycolic) acid (PLGA) microparticles, may require sufficient drug solubility in specific solvents (Nkanga et al. 2020).

Successful examples of polymer-based LAI drugs (Fig. 1.4) include risperidone PLGA microspheres (Risperdal Consta), the first long-acting injectable atypical antipsychotic to reach the pharmacy shelves (Rainer 2008). From a chemical property perspective, the compound does not readily lend itself to a prodrug strategy (for the lack of free functional groups such as OH or NH), but its high solubility in certain organic solvents (e.g., methylene chloride) facilitates encapsulation in polymeric carriers. Other established small-molecule products developed as PLGA microspheres are naltrexone (Vivitrol) that is used for treating opioid and alcohol dependence (Syed and Keating 2013) and triamcinolone acetonide (Zilretta), a corticosteroid for intra-articular injections (Paik et al. 2019).

These three drugs do not share any common structural elements, so one cannot assume a chemical design principle underlying the carrier-based formulation approach. With respect to their protonation state under physiological conditions,

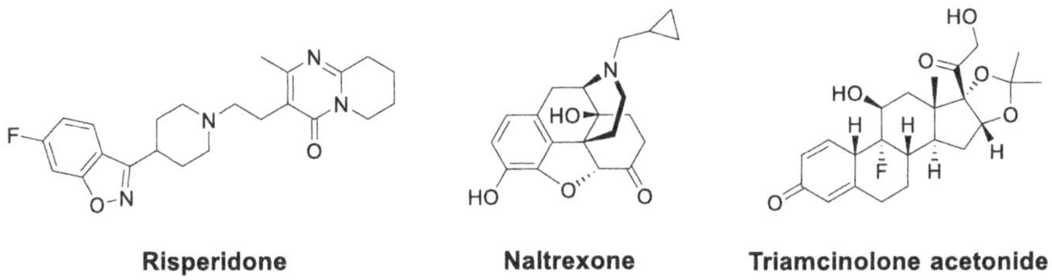

Fig. 1.4 Exemplary small molecules developed as PLGA microspheres for LAI applications

they could hardly be more diverse. Risperidone is a highly basic molecule; naltrexone is a zwitter-ionic; and triamcinolone acetonide is nonionizable and therefore is not charged in an aqueous environment. What all three compounds have in common, however, is a very limited solubility in water paralleled by high solubility in organic excipients used in polymer manufacturing. Research teams embarking on a polymer-based micro- or nanoparticle formulation strategy must therefore focus on such a physical profile of their drug candidates to achieve success in technical development.

Naturally, the carrier-based approach comes with a certain limitation of deliverable dose over the release period. Depending on the exact site of administration, LAI formulations are capped by a maximum tolerable injection volume. It also plays a role in whether the target tissue is compromised by acute or chronic inflammation processes that may further limit the injectable burden. With a given technical limit of a carrier's drug loading capacity, the amount of drug that can be administered per injection volume is quite restricted, and practically only highly bioactive substances that can be administered in low doses are amenable for such a strategy.

1.5 Prodrugs

The limitations described in the previous sections can be mitigated by using prodrug-based sustained-release concepts. Indeed, long-lasting injectable depots of small molecule prodrugs have been met with great clinical success. Figure 1.5 shows exemplary structures of corticosteroids, antipsychotic, and antiviral drugs featuring hydroxyl or amino groups that were tethered to appropriate, cleavable prodrug moieties. The widely studied corticosteroids, notoriously suffering from low therapeutic safety margins, have been modified to prodrugs for multiple therapeutic purposes (Hobson 2023), usually taking advantage of their C21 hydroxyl group to introduce carboxylic acids via ester bonds. Activation occurs through enzymatic ester cleavage effected by tissue esterases (for example: methylprednisolone acetate, Scheme 1.1).

In contrast to short carboxylic acid esters such as acetates, the introduction of more lipophilic carboxylic acids ("fatty acids") renders the drug much less soluble in biological fluids, prevents rapid uptake into the systemic circulation, and builds the foundation of long, sustained-release after injection. For instance, hydroxyprogesterone caproate (Proluton, Makena) (Boelig 2023) features a six-carbon-containing (C6) carboxylic acid ("caproic acid"); paliperidone palmitate (Invega) (Owen 2010) is a C16 carboxylic acid ("palmitic acid"). Indeed, the aqueous solubility of paliperidone palmitate in its crystalline form is so low—usually below limits of quantification—that exact figures have not been reported. Databases usually refer to descriptors like "sparingly soluble in water" or "extremely low water solubility."

The same concept has also been applied to noncorticosteroid drugs. Recently, antiviral therapies have been facing new horizons using LAI concepts. Long-acting

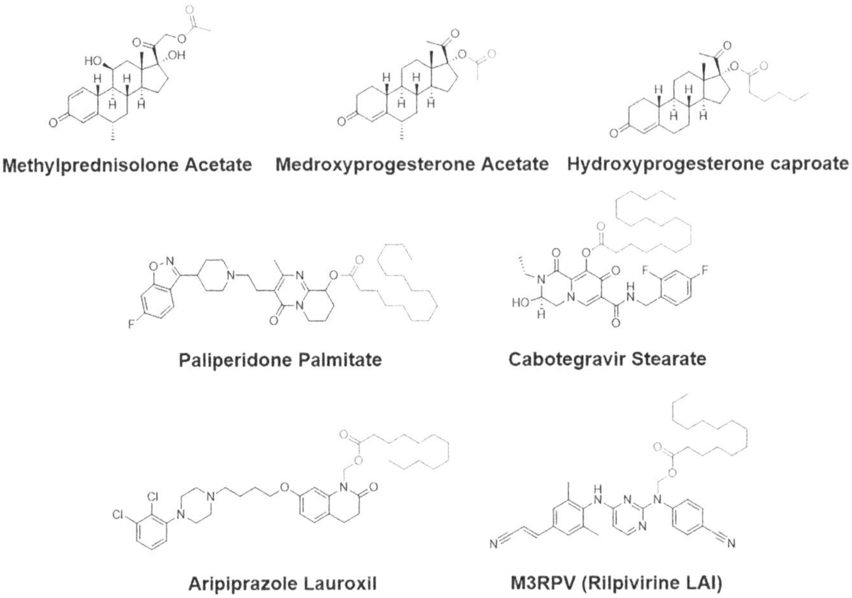

Fig. 1.5 Examples of LAI prodrugs based on chemical modification of hydroxy or amino groups present in the drug substance. The cleavable prodrug moieties are marked in red color

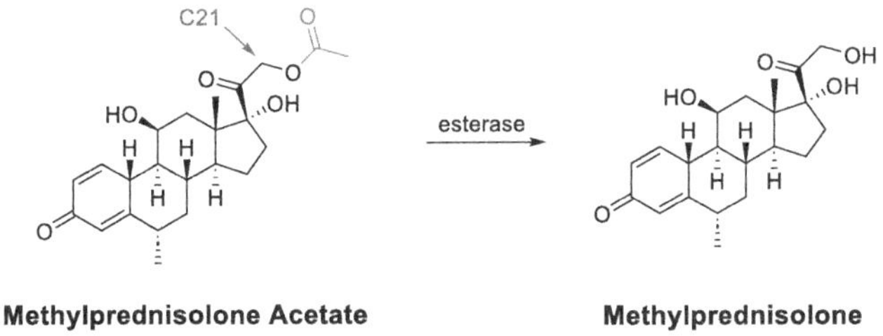

Scheme 1.1 Activation of methylprednisolone by esterases

cabotegravir was derivatized as a C18 carboxylic acid ("stearic acid") and formulated as nanocrystals to reach therapeutically active plasma concentrations in preclinical models for up to a year, after a single intramuscular injection (Kulkarni et al. 2020). Similar data have been reported for a long-acting prodrug of rilpivirine (Hilaire et al. 2019).

Rilpivirine LAI (M3RPV) and the antipsychotic drug aripiprazole lauroxil (Aristada) (Turncliff et al. 2014) are drug substances that do not feature hydroxyl groups that can be readily converted into esters. Here, medicinal chemists have taken advantage of accessible amino (NH) groups that were modified as acyloxymethyl prodrugs, a concept exemplified 40 years ago for allopurinol (Bundgaard 1985). Thus, the fatty acid, which is required to increase lipophilicity and reduce

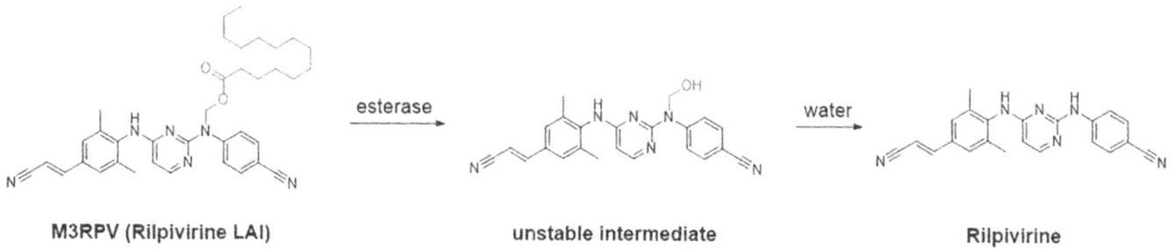

Scheme 1.2 Suggested activation of rilpivirine from its prodrug form

solubility, is tethered to the parent drug via a hemiaminal bridge leading to a two-step prodrug degradation (e.g., rilpivirine, Scheme 1.2). In the first step, the ester is hydrolyzed by esterases, resulting in an unstable hydroxymethyl intermediate; then, in a fast nonenzymatic step, the parent drug is released (Santos et al. 2022).

A big advantage of the prodrug approach is the ability not only to tune the prodrug release kinetics but also to offer different formulation options by a smart choice of the prodrug moiety. For example, paliperidone palmitate is delivered as a nanocrystal suspension of ester prodrug via intramuscular injection and is hydrolyzed to paliperidone with a mean half-life of 24–49 days in patients, depending on the dose (Owen 2010).

1.6 Salts

Ionizable drug molecules can form salts with acids or bases, which usually results in a significant change in physicochemical (aggregation state, polymorphism, solubility, lipophilicity, hygroscopicity, chemical stability/shelf-life) and ADME properties (dissolution, distribution, onset and termination of effect, drug interactions). Thus, salts can have a great impact on clinical dosage form and even extend the scope of different RoA (Gupta et al. 2018).

For salt formation, the appropriate choice of counterion from a wide range of clinically validated and Food and Drug Administration (FDA)-approved species is of critical importance (Wiedmann and Naqwi 2016). Many counterions have been selected in the past to address distinct liabilities of the original drug, e.g., to enhance the dissolution rate of poorly soluble, weakly acidic or basic, oral drugs. For LAI development, usually the opposite needs to be achieved, i.e., turning a highly soluble molecule into a sustained-release formulation by pairing it with a counterion that confers low solubility. Olanzapine (Fig. 1.6), an atypical neuroleptic drug, was originally produced as an oral treatment for schizophrenic patients. For a sustained therapeutic effect, the free base was paired with pamoic acid as counterion, resulting in a highly crystalline, low-solubility salt form that was approved as a microcrystal suspension for intramuscular injection (Chue and Chue 2012).

The selection of pamoic acid may be called a serendipitous case of success, as it resulted from a salt-screening effort encompassing highly lipophilic,

Fig. 1.6 Chemical structure of olanzapine pamoate: blue—olanzapine free-base, double-protonated; red—pamoate counterion

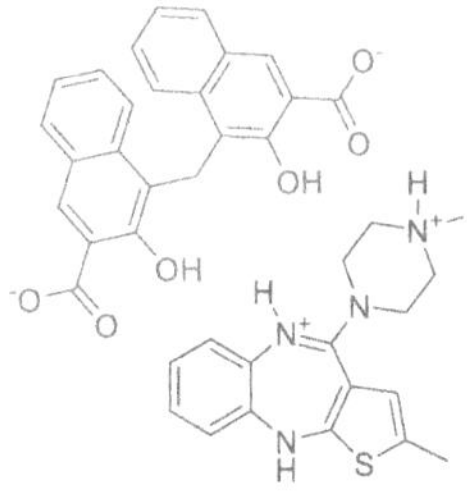

Olanzapine Pamoate

pharmaceutically safe carboxylic acids—rather than a precise design of physicochemical properties. It is noteworthy that appropriate salt forms for sustained-release formulations are not commonly easy to achieve—on top of high crystallinity and low solubility, their stability is of crucial importance, as salt dissociation of drug substance and counterion can adversely affect the performance of the drug product (Nie et al. 2017). Despite substantial investment in predictive models, there is no reliable way to accurately design salt properties based on the counterion used (Gupta et al. 2018).

A smart application of the salt concept has been successfully applied to modulate the physicochemical properties and stability of the charged, large hydrophilic large molecules including peptides and proteins. The drug substance is paired with hydrophobic counterions to result in neutral, water-insoluble complexes that can be formulated via numerous encapsulation techniques ("hydrophobic ion pairing") (Al Musaimi et al. 2022; Ristroph and Prud'homme 2019). A prominent example developed by Novartis is pasireotide pamoate (Signifor long-acting release (LAR)), a synthetic long-acting cyclic hexapeptide with somatostatin-like activity to treat Cushing's disease (McKeage 2013). The formulation process (loading onto PLGA microparticles) was greatly aided by the generation of a pamoate salt complex showing improved stability against sterilization through γ-irradiation.

1.7 Summary and Conclusions

Pharmaceutical research and development can rely on well-established principles and guidelines to design small molecules for long-acting depots, with and without pharmaceutical carriers. Lipophilicity and crystallinity determine drug solubility in water and organic excipients and are therefore prime optimization metrics. Beyond such physicochemical design parameters, metabolic stability plays a crucial role when distinguishing locally from systemically active drugs to lead to either desired fast elimination or long half-life after reaching the circulation. In the event of a prodrug approach, cleavage kinetics must also be considered and added to the complexity of preclinical characterization and prediction of pharmacokinetics in patients suffering from a broad diversity of diseases. Importantly, early chemistry optimization flowcharts are complemented by adequate physicochemical assays to characterize solid states and thermodynamic solubility in customized media. One of the most fundamental optimization principles that are common to all LAI approaches is to aim for the highest-possible drug potency in relevant biological assays to minimize the burden of injectable dose and volume, both contributing to the safety and tolerability of the final product. Those design principles, combined with recent technological advances in formulation sciences, are about to generate a multitude of opportunities for smart delivery principles that may soon outpace conventional methodologies and enable accelerated LAI drug discovery as an interplay of chemical and pharmaceutical expertise.

References

Al Musaimi O, Lombardi L, Williams DR, Albericio F. Strategies for improving peptide stability and delivery. Pharmaceuticals. 2022;15(10):1283. https://doi.org/10.3390/ph15101283.

Aungst BJ. Optimizing oral bioavailability in drug discovery: an overview of design and testing strategies and formulation options. J Pharm Sci. 2017;106(4):921–9. https://doi.org/10.1016/j.xphs.2016.12.002.

Bannigan P, Bao Z, Hickman RJ. Machine learning models to accelerate the design of polymeric long-acting injectables. Nat Commun. 2023;14:35. https://doi.org/10.1038/s41467-022-35343-w.

Bari H. A prolonged release parenteral drug delivery system—an overview. Int J Pharm Sci Rev Res. 2010;3:1.

Bassand C, Villois A, Gianola L, Laue G, Ramazani F, Riebesehl B, Sanchez-Felix M, Sedo K, Ullrich T, Duvnjak RM. Smart design of patient-centric long-acting products: from preclinical to marketed pipeline trends and opportunities. Expert Opin Drug Deliv. 2022;19(10):1265–83. https://doi.org/10.1080/17425247.2022.2106213.

Begley R, Lutz J, Rhee M, Dvory-Sobol H, Chiu A, West SK, Corpus J, Ling J, German P. Lenacapavir sustained delivery formulation supports 6-month dosing interval. In: Poster presented at 23rd international AIDS conference virtual, 6–10 July 2020.

Bergström CAS, Wassvik CM, Johansson K, Hubatsch I. Poorly soluble marketed drugs display solvation limited solubility. J Med Chem. 2007;50:5858–62. https://doi.org/10.1016/j.ejps.2019.104967.

Boelig RC. Obstetric indications for progestin therapy. Obstet Gynecol Clin N Am. 2023;50(1):101–7. https://doi.org/10.1016/j.ogc.2022.10.004.

Bundgaard H. Formation of prodrugs of amines, amides, ureides, and imides. Methods Enzymol. 1985;112:347–59. https://doi.org/10.1016/s0076-6879(85)12029-x.

Campbell IB, Macdonald SJF, Procopiou PA. Medicinal chemistry in drug discovery in big pharma: past, present and future. Drug Discov Today. 2018;23(2):219–34. https://doi.org/10.1016/j.drudis.2017.10.007.

Chue P, Chue J. A review of olanzapine pamoate. Expert Opin Pharmacother. 2012;13(11):1661–70. https://doi.org/10.1517/14656566.2012.686169.

Deshmukh V, Hu H, Barroga C, Bossard C, Kc S, Dellamary L, Stewart J, Chiu K, Ibanez M, Pedraza M, Seo T, Do L, Cho S, Cahiwat J, Tam B, Tambiah JRS, Hood J, Lane NE, Yazici Y. A small-molecule inhibitor of the Wnt pathway (SM04690) as a potential disease modifying agent for the treatment of osteoarthritis of the knee. Osteoarthr Cartil. 2018;26(1):18–27. https://doi.org/10.1016/j.joca.2017.08.015.

Fagerberg JH, Tsinman O, Sun N, Tsinman K, Avdeef A, Bergström CA. Dissolution rate and apparent solubility of poorly soluble drugs in biorelevant dissolution media. Mol Pharm. 2010;7(5):1419–30. https://doi.org/10.1021/mp100049m.

Gonzalez AZ, Jermain SV, Roberts EB, Schwarzwalder GM. The drug substances of long-acting injectables: examples from hormone, CNS, and antiviral therapies. Med Chem Rev. 2020;55:383. (Chapter 16)
Gupta D, Bhatia D, Dave V, Sutariya V, Varghese GS. Salts of therapeutic agents: chemical, physicochemical, and biological considerations. Molecules. 2018;23(7):1719. https://doi.org/10.3390/molecules23071719.
Hann MH. Molecular obesity, potency and other addictions in drug discovery. MedChemComm. 2011;2:349–55. https://doi.org/10.1039/C1MD00017A.
Hilaire JR, Bade AN, Sillman B, Gautam N, Herskovitz J, Dyavar Shetty BL, Wojtkiewicz MS, Szlachetka A, Lamberty BG, Sravanam S, Fox HS, Alnouti Y, Dash PK, McMillan JM, Edagwa BJ, Gendelman HE. Creation of a long-acting rilpivirine prodrug nanoformulation. J Control Release. 2019;311–312:201–11. https://doi.org/10.1016/j.jconrel.2019.09.001.
Hobson A. Prodrugs. In: The medicinal chemistry of glucocorticoid receptor modulators. Cham: Springer Nature Switzerland; 2023. https://doi.org/10.1007/978-3-031-28732-9_4.
Kulkarni TA, Bade AN, Sillman B, Shetty BLD, Wojtkiewicz MS, Gautam N, Hilaire JR, Sravanam S, Szlachetka A, Lamberty BG, Morsey BM, Fox HS, Alnouti Y, McMillan JM, Mosley RL, Meza J, Domanico PL, Yue TY, Moore G, Edagwa BJ, Gendelman HE. A year-long extended release nanoformulated cabotegravir prodrug. Nat Mater. 2020;19(8):910–20. https://doi.org/10.1038/s41563-020-0674-z.
Kuntz ID, Chen K, Sharp KA, Kollman PA. The maximal affinity of ligands. Proc Natl Acad Sci. 1999;96(18):9997–10002. https://doi.org/10.1073/pnas.96.18.9997.
Lipinski CA, Lombardo F, Dominy BW, Feeney PJ. Experimental and computational approaches to estimate solubility and permeability in drug discovery and development settings. Adv Drug Deliv Rev. 2001;46(1-3):3–26. https://doi.org/10.1016/s0169-409x(00)00129-0.
McKeage K. Pasireotide: a review of its use in Cushing's disease. Drugs. 2013;73(6):563–74. https://doi.org/10.1007/s40265-013-0052-0.
Nie H, Byrn SR, Zhou QT. Stability of pharmaceutical salts in solid oral dosage forms. Drug Dev Ind Pharm. 2017;43(8):1215–28. https://doi.org/10.1080/03639045.2017.1304960.
Nkanga CI, Fisch A, Rad-Malekshahi M, Romic MD, Kittel B, Ullrich T, Wang J, Krause RWM, Adler S, Lammers T, Hennink WE, Ramazani F. Clinically established biodegradable long acting injectables: an industry perspective. Adv Drug Deliv Rev. 2020;167:19–46. https://doi.org/10.1016/j.addr.2020.11.008.
O'Brien MN, Jiang W, Wang Y, Loffredo DM. Challenges and opportunities in the development of complex generic long-acting injectable drug products. J Control Release. 2021;336:144–58. https://doi.org/10.1016/j.jconrel.2021.06.017.
Owen RT. Paliperidone palmitate injection: its efficacy, safety and tolerability in schizophrenia. Drugs Today. 2010;46(7):463–71. https://doi.org/10.1358/dot.2010.46.7.1514647.
Paik J, Duggan ST, Keam SJ. Triamcinolone acetonide extended-release: a review in osteoarthritis pain of the knee. Drugs. 2019;79:455–62. https://doi.org/10.1007/s40265-019-01083-3.
Rainer MK. Risperidone long-acting injection: a review of its long term safety and efficacy. Neuropsychiatr Dis Treat. 2008;4(5):919–27. https://doi.org/10.2147/ndt.s3311.
Ristroph KD, Prud'homme RK. Hydrophobic ion pairing: encapsulating small molecules, peptides, and proteins into nanocarriers. Nanoscale Adv. 2019;1(11):4207–37. https://doi.org/10.1039/c9na00308h.
Ritchie TJ, Macdonald SJF. The impact of aromatic ring count on compound developability – are too many aromatic rings a liability in drug design? Drug Discov Today. 2009;14:1011–20. https://doi.org/10.1016/j.drudis.2009.07.014.
Santos SS, Gonzaga RV, Scarim CB, Giarolla J, Primi MC, Chin CM, Ferreira EI. Drug/lead compound hydroxymethylation as a simple approach to enhance pharmacodynamic and pharmacokinetic properties. Front Chem. 2022;9:734983. https://doi.org/10.3389/fchem.2021.734983.
Syed YY, Keating GM. Extended-release intramuscular naltrexone (VIVITROL®): a review of its use in the prevention of relapse to opioid dependence in detoxified patients. CNS Drugs. 2013;27:851–61. https://doi.org/10.1007/s40263-013-0110-x.

Turncliff R, Hard M, Du Y, Risinger R, Ehrich EW. Relative bioavailability and safety of aripiprazole lauroxil, a novel once-monthly, long-acting injectable atypical antipsychotic, following deltoid and gluteal administration in adult subjects with schizophrenia. Schizophr Res. 2014;159(2–3):404–10. https://doi.org/10.1016/j.schres.2014.09.021.

Wassvik CM, Holmén AG, Draheim R, Artursson P, Bergström CA. Molecular characteristics for solid-state limited solubility. J Med Chem. 2008;51(10):3035–9. https://doi.org/10.1021/jm701587d.

Wiedmann TS, Naqwi A. Pharmaceutical salts: theory, use in solid dosage forms and in situ preparation in an aerosol. Asian J Pharm Sci. 2016;11(6):722–34. https://doi.org/10.1016/j.ajps.2016.07.002.

Chapter 2
Biodegradable Polymers for Development of LAIs

Seán McMahon and Patrick Duffy

List of Abbreviations

API	Active pharmaceutical ingredient
ANDA	Abbreviated New Drug Application
ASTM	American Society for Testing and Materials
BCNU	1,3-bis[2-chloroethyl]-1-nitrosourea
C-NMR	Carbon nuclear magnetic resonance
CAS	Chemical Abstract Services
CFR	Code of Federal Regulations
CPP	Central precocious puberty
CRCG	Center for Research on Complex Generics
DETOSU	3,9-Diethylidene-2,4,8,10-tetraoxaspiro[5.5]undecane
DMSO	Dimethyl sulfoxide
DSC	Differential scanning calorimetry
FDA	Food and Drug Administration
GA	Glycolide
GC	Gas chromatography
GMP	Good Manufacturing Practices
GPa	Gigapascal
GPC	Gel permeation chromatography
HME	Hot melt extrusion
H-NMR	Proton nuclear magnetic resonance
HOSs	Higher order structures
IID	Inactive Ingredient Database
IR	Infrared spectroscopy
ISO	International Standards Organisation

S. McMahon (✉) · P. Duffy (✉)
Ashland Specialties Ireland Limited, National Science Park, Building V, Petitswood, Mullingar, Westmeath, Ireland
e-mail: sean.mcmahon@ashland.com; patrick.duffy@ashland.com

F. Ramazani (ed.), *Biodegradable Long Acting Injectables and Implants*,
https://doi.org/10.1007/978-3-031-72302-5_2

kDa	Kilo Dalton
LA	Lactide
LAI	Long-acting injectable
LAR	Long-acting release
MALS	Multiangle light scattering
mg	Milligram
Mn	Number average molecular weight
MPa	Megapascal
Mw	Weight average molecular weight
NCA	N-carboxyanhydrides
NCR	New chemical entity
NMP	N-methyl-2-pyrrolidone
P(3-HB-co-4-HB)	Poly-3-hydroxybutyrate-co-4-hydroxybutyrate
PACA	Poly(alkyl cyanoacrylates)
PBAE	Poly(β-amino esters)
PCL	Polycaprolactone
PCPP-SA	Poly[(carboxyphenoxypropane)-(sebacic acid)]
PDi	Polydispersity index
PDLLA	Poly D,L-lactide
PEA	Poly ester amide
PEG	Poly ethylene glycol
PGA	Poly glycolide
pH	Potential of hydrogen
PHA	Polyhydroxyalkanoates
PHB	Poly-3-hydroxybutyrate
PHBH	Polyhydroxybutyrate-co-hexanoate
PHBV	Poly-3-hydroxybutyrate-co-valerate
PLCL	Polylactide-co-caprolactone
PLGA	Poly D,L-lactide-co-glycolide
PLLA	Poly L-lactide
POE	Polyorthoester
PPE	Polyphosphoester
Ppm	Parts per million
Q1	Qualitative
Q2	Quantitative
Rc	Randomness
ROP	Ring-opening polymerization
SEC	Size exclusion chromatography
SR	Sustained release
SRS	Substance Registration System
T_g	Glass transition temperature
T_m	Melt temperature
USA	United States of America
USP	United States Pharmacopeia
UV-Vis	Ultraviolet–Visible spectroscopy

2.1 Introduction

Classification of Biodegradable Polymers

Biodegradable polymers are a class of polymer which undergo chemical breakdown in physiological environments into biocompatible oligomer or monomer units recognized as metabolic species for clearance from the body. Degradation is usually driven by hydrolysis or enzymes and by-products should be non-toxic and non-immunogenic and should be eliminated from the body through natural metabolic pathways. Biodegradable polymers can be found naturally or produced synthetically and are typically classified as such.

Classification of bioresorbable or biodegradable polymers (Makadia and Siegel 2011) can be subdivided into two major categories including (1) synthetic bioresorbable polymers such as polymers of α-hydroxy acids and (2) naturally occurring polymers such as hyaluronan, chitosan, and inorganic materials.

History of Biodegradable Polymers for Medical and Pharmaceutical Applications

Synthesis of aliphatic polyesters from polycondensation of diols with dicarboxylic acids was first reported by Wallace Hume Carothers and co-workers in the 1930s (Carothers et al. 1932); however, the materials initially synthesized presented poor physical properties and low melting points due to difficulty obtaining high molecular weights which delayed widespread usage. Over many years, technology and understanding developed both in terms of achieving more useful degradable polymer properties and their potential value-added applications (Okada 2002). Almost 20 years later, a patent (2005) detailing the development of devices for tissue repair using absorbable polymers (PLA and PGA) from Smith and Nephew Inc. was published in 1955.

In 1966, Kulkarni et al.'s (1966) paper on "polylactic acid for surgical implants" detailed how ring-opening polymerization (ROP) could be used to synthesize suitably high molar mass polymers that could be cast into films or spun into fibers while also indicating that the material could be degraded by the body and provide a safe immunological response. Then in 1967, Schmitt and Polistina patented (Schmitt and Polistina 1964) their method for production of polyglycolide (PGA)-based sutures which 4 years later led to the marketing of Dexon (Frazza and Schmitt 1971). Ethicon modified this approach by incorporating the cyclic dimer lactide leading to the creation of Vicryl which entered the market in 1974 (J&J 1974). Vicryl was produced from the material polyglactin 910 which is a copolymer of glycolide (90%) and lactide (10%).

In tandem with this work, pharmaceutical applications were emerging. The first patent on microencapsulation (Ranger 1974) appeared in 1957. The first patent on "polylactide-drug mixtures" in this area of drug delivery using lactide/glycolide polymers was assigned to Boswell and Scribner from E.I. du Pont de Nemours and filed in November 1973 (Boswell et al. 1970). In the 1980s, Swiss pharmaceutical company Debiopharm started working on microsphere formulations of triptorelin using the methodology from Southern Research Institute and Syntex research team which was developed for nafarelin acetate (2017). In 1986 (Blasi 2019), a pharmaceutical application segment was commercially realized with the first long-acting lactide glycolide microparticle product Decapeptyl from Debiopharm (Tice 2023). Since then, medical applications for biodegradable polymers for human and animals have expanded greatly in both medical device and pharmaceutical drug applications.

While microspheres were first to market, there were two additional depot formulation formats using lactide/glycolide-type polymers in development in the in situ depot and solid implant categories. In 1987, Richard Dunn et al. developed an in situ depot technology (Dadey 2013) at Southern Research Institute which was licensed to Vipont (became Atrix in 1989 and QLT in 2004). This subsequently formed the Atrigel technology which was used to create Atridox approved in 1998 and Eligard approved in 2002. ICI Pharma was working on PLG implants for 1 month release of goserelin which led to the launch of Zoladex in 1990.

Lactide/glycolide polymers were also utilized in nanoparticle formulations. A di-block structure of methoxy-poly(ethylene glycol)-block-poly D,L-lactide (mPEG-PDLLA) was used by Samyang in a paclitaxel formulation to create self-assembling nanoparticles in the range of 20–50 nm. This led to the approval of Genexol PM in Korean market in 2007 for breast and small cell cancer (Pillai 2014) (Fig. 2.1).

2.2 LAI Market Products Based on Biodegradable Polymers

Products in the LAI space can be grouped into three categories as (1) particle systems including microspheres, microparticles, and nanoparticles; (2) in situ depots; and (3) solid implants (Tables 2.1 and 2.2).

2.3 Biodegradable Polymers and Synthesis

Natural Polymers

Natural polymers, often termed biopolymers, are those derived from biological sources such as microbials, plants, animals, or marine sources. Their synthesis generally involves enzyme-catalyzed, chain growth polymerization reactions of activated monomers, which are typically formed within cells by metabolic processes. Biomass or agro-waste-derived polymers make up most natural biodegradable polymers used for pharmaceutical applications and can be split between polysaccharides and proteins (Samir et al. 2022). Biopolymers can also be derived or extracted from naturally occurring or genetically modified organisms.

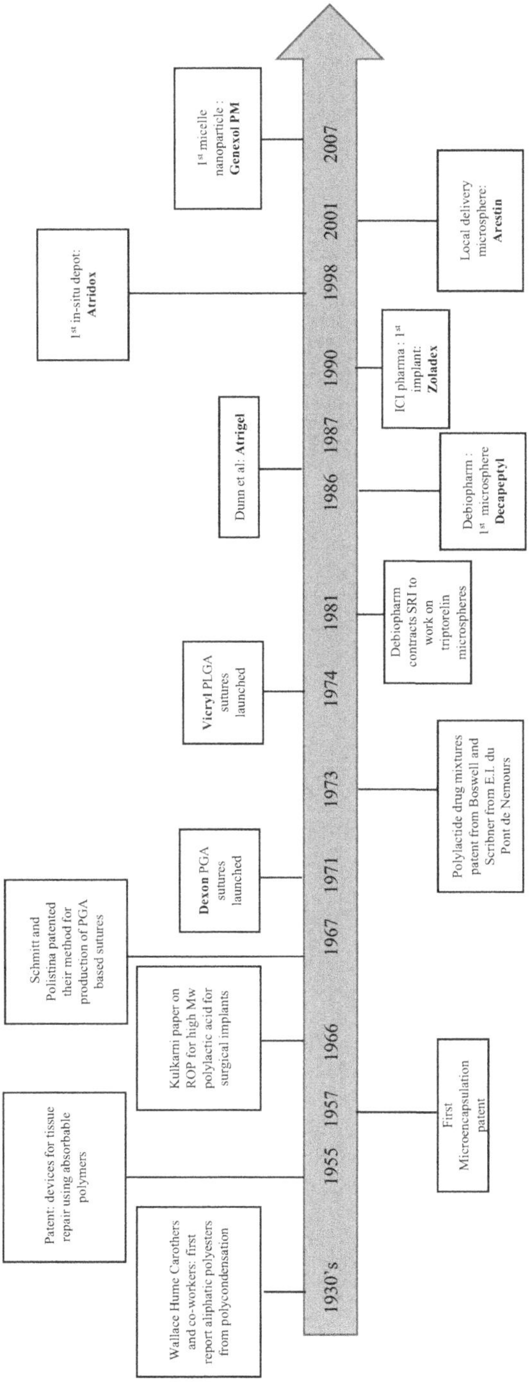

Fig. 2.1 Timeline of development in polymerization, drug delivery, and medical devices of lactide/glycolide copolymers

Table 2.1 Approved lactide/glycolide marketed drug products

Drug product	Active ingredient	Strength	Dosage form, route of administration, indications	Polymer	Company name
Microparticle technologies (microspheres, microparticles)					
Lupron®/Lupron Depot®/Lupron Depot-PED® (2024)	Leuprolide acetate	3.75 mg and 11.25 mg 7.5 mg, (1 month) 22.5 mg, (3-month) 30 mg, (4-month) 45 mg, (6-month)	Microspheres, Intramuscular, Endometriosis, uterine fibroids, advanced prostate cancer, central precocious puberty (CPP)	3.75 mg: PLGA 7.5 mg: PLGA 11.25 mg: PLGA 22.5 mg: PLA 30 mg: PLA 45 mg: PLA	Takeda Pharmaceutical Company Limited
Sandostatin LAR® (2024)	Octreotide acetate	10 mg, 20 mg, or 30 mg every 4 weeks	Microspheres, subcutaneous, indicated for acromegaly	Glucose star polymer PLGA 10 mg: 188.8 mg 20 mg: 377.6 mg 30 mg: 566.4 mg	Novartis
Trelstar® (2024)	Triptorelin pamoate	3.75 mg every 4 weeks 11.25 mg every 12 weeks 22.5 mg every 24 weeks	Microspheres, intramuscular, indicated for advanced prostate cancer	3.75 mg: 138 mg PLGA 11.25 mg: 120 mg PLGA 22.5 mg: 183 mg PLGA	Debiopharm, AbbVie
Vivitrol® (2024)	Naltrexone	380 mg every 4-weeks	Microspheres, intramuscular, indicated for the treatment of alcohol dependence	PLGA, 75:25, 377 mg of naltrexone per gram of microspheres	Alkermes, Inc.
Arestin® (2024)	Minocycline HCl	1 mg, variable dosing frequency	Microspheres, periodontal, indicated for periodontitis	PLGA	Bausch Health Companies Inc./Orapharma Inc.
Risperdal Consta® (2024)	Risperidone	12.5 mg, 25 mg, 37.5 mg, or 50 mg, every 2 weeks	Microspheres, intramuscular, indicated for schizophrenia and bipolar disorder	12.5 mg, 25 mg, 37.5 mg, or 50 mg, PLGA 75:25, 381 mg risperidone per gram microspheres	Alkermes, Inc., Janssen, Division of Ortho-McNeil-Janssen Pharmaceuticals, Inc.

Bydureon® (2024)	Exenatide	2 mg, every 7 days	Microspheres, subcutaneous, type 2 diabetes	37.2 mg of PLGA, 50:50 per dose	AstraZeneca
Signifor® LAR (2024)	Pasireotide pamoate	10 mg, 20 mg, 30 mg, 40 mg, or 60 mg, every 28-days	Microspheres, subcutaneous, acromegaly	Mixture of PLGAs: PLGA 50–60/40–50 PLGA 50:50	Novartis
Decapeptyl® SR	Triptorelin acetate	3.75 mg, 11.2 mg, 22.5 mg,	Microspheres, intramuscular, indicated for lowering sexual hormones, prostate cancer, endometriosis, fertility treatment, IVF	PLGA	Ferring Pharmaceuticals GmbH, Debiopharm, Ipsen
Zilretta® (2024)	Triamcinolone acetonide	32 mg	Microspheres, intraarticular, indicated for osteoarthritis pain of the knee	PLGA, 75:25	Flexion Therapeutics, Inc./ Pacira BioSciences
Smart Shot® B12	Vitamin B12	6 mg, every 4 months and 8 months	Microparticles, injection, indicated in deficiency in animals	PLA	Stockguard Labs (Virbac)
Nanoparticle: Polymeric Micelle Technology					
Genexol® PM injection	Paclitaxel	30 mg, 100 mg/vial	Polymeric micelle, lyophilized injection, indicated for the treatment of breast and ovarian cancer	30 mg, 100 mg, mPEG-PDLLA	Samyang Holdings Biopharmaceuticals Division, Korea
Nanoxel® M injection	Docetaxel	20 mg, 80 mg/vial	Polymeric micelle, lyophilized injection, indicated for the treatment of breast cancer, NSCLC, esophageal cancer, ovarian cancer, H&N cancer, gastric cancer, and prostate cancer	20 mg, 80 mg, mPEG-PDLLA	Samyang Holdings Biopharmaceuticals Division, Korea

(continued)

Table 2.1 (continued)

Drug product	Active ingredient	Strength	Dosage form, route of administration, indications	Polymer	Company name
In situ forming depot gels					
Atridox®	Doxycycline hyclate	50 mg for 7-days	Atrigel delivery system, subgingival application, indicated for periodontitis	36.7% PLA with 63.3% NMP	Atrix Laboratories, Inc. USA
Eligard®	Leuprolide acetate	7.5 mg, every 1 months 22.5 mg, every 3 months 30 mg, every 4 months 45 mg, every 6 months	Atrigel delivery system, subcutaneous injectable suspension, indicated for advanced prostate cancer	7.5 mg, 50:50, PLGA acid end + NMP; 22.5 mg, 75:25 PLGA ester end + NMP. 30 mg, 75:25 PLGA ester end + NMP; 45 mg, 85:15 PLGA ester end + NMP	Tolmar Therapeutics Inc., USA
Sublocade®	Buprenorphine	100 mg/0.5 ml and 300 mg/1.5 ml for 1 month	Atrigel delivery system, subcutaneous injection, indicated for opioid use	100 mg and 300 mg, 50:50, PLGA + NMP	Indivior Inc., USA
Perseris®	Risperidone	90 mg or 120 mg once monthly	Atrigel delivery system, subcutaneous injection, indicated for the treatment of schizophrenia in adults	90 mg and 120 mg, 80:20, PLGH + NMP	Indivior Inc., USA
Longrange®	Eprinomectin	50 mg for 5 months	Theraphase delivery system, parasite prevention	75:25, PLGA + NMP	Boehringer Ingelheim Animal Health USA Inc.
Fensolvi® (2024)	Leuprolide acetate	45 mg	In situ depot Atrigel delivery system, subcutaneous injection, central precocious puberty	PLGA 8515 (165 mg)	Tolmar Inc.
CAMCEVI®	Leuprolide mesylate	42 mg leuprolide equivalent to 48 mg leuprolide mesylate	In situ depot, subcutaneous, advanced prostate cancer	PDLLA (184 mg)	Foresee Pharmaceuticals Co. Ltd. (2024)

Uzedy® (2024)	Risperidone	50 mg, 75 mg, 100 mg, 125 mg, 150 mg, 200 mg, 250 mg	BEPO system in situ depot, subcutaneous, antipsychotic indicated for schizophrenia	mPEG-PDLLA (15% w/w), PDLLA-PEG-PDLLA (10% w/w)	MedinCell/Teva
Implantable systems					
Zoladex®	Goserelin acetate	3.6 mg or 10.8 mg every 28-days	Implant, subcutaneous, indicated for prostate cancer	PLGA	TerSera Therapeutics LLC, USA, AstraZeneca
Ozurdex®	Dexamethasone	0.7 mg (700 mcg), variable dosing frequency	Implant, intravitreal injection, indicated for uveitis affecting the posterior segment of the eye, diabetic macular edema	Mixture of PLGAs	Allergan, Inc., USA/ AbbVie
Suprefact® Depot	Buserelin acetate	6.3 mg, every 2 months and 9.45 mg, every 3 months	Implant, subcutaneous injection, indicated for prostate cancer	6.3 mg and 9.45 mg, 75:25, PLGA	Cheplapharm Arzneimittel GmbH, Germany
Durysta®	Bimatoprost	10 µg, 4– 6 months	Implant, intracameral, indicated for glaucoma or ocular hypertension	PDLLA-PLGA-PDLLA, acid end	Allergan, Inc., USA/ AbbVie
Propel® Mini Sinus Implant	Mometasone furoate	370 µg, every 1 month	Implant, indicated for treatment following sinus surgery	PLGA	Intersect ENT, Inc., USA
Revalor XS (2024)	Trenbolone/estradiol	200 mg trenbolone acetate and 40 mg of estradiol	Implant, growth promotion cattle, subcutaneous into the ear	PLGA	Merck Animal Health
Sinuva™ (2024)	Mometasone Furoate	1350 mcg/implant	Sinus implant, indicated for the treatment of nasal polyps following ethmoid sinus surgery, intranasal	PLGA	Intersect ENT (Subsidiary of Medtronic)
Scenesse®	Afamelanotide	16 mg, every 2 months	implant, subcutaneous injection, indicated for erythropoietic protoporphyria (light intolerance)	PLGA (15.3–19.5 mg)	Clinuvel Inc.

(continued)

Table 2.1 (continued)

Drug product	Active ingredient	Strength	Dosage form, route of administration, indications	Polymer	Company name
Discontinued					
Nutropin® Depot	Somatropin (rDNA origin)-recombinant human growth hormone (rhGH)	13.5 mg, 18 mg, or 22.5 mg, Once or twice monthly depends on the patient's body weight	Microspheres, subcutaneous, growth hormone deficiency	13.5 mg, 18 mg, or 22.5 mg: PLGA	Genentech, Inc.

Abbreviation: *PLGA*, poly lactide-co-glycolide or poly lactic-co-glycolic acid; *PDLLA/PLA*, poly D,L-lactide or poly lactic acid; *NMP*, N-methyl-2-pyrrolidone; *mPEG*, methyl ether-polyethylene glycol; *HPMC*, hydroxypropyl methylcellulose

Table 2.2 Long-acting systems using alternative carriers

Drug product	Active ingredient	Dosage form, route of administration, Indications	Technology/carrier	Company name
Zynrelef	Bupivacaine and meloxicam (29.25 mg/mL and 0.88 mg/mL)	Extended-release solution, postoperative pain from small- to medium-sized surgical wounds	DETOSU/triethylene glycol/triethylene glycol polyglycolide copolymer, 730 mg, triacetin 293 mg, dimethyl sulfoxide (117 mg) and maleic acid	Heron Therapeutics (2024)
SUSTOL (2016)	Granisetron (10 mg/0.4 mL)	Extended-release injection, indicated for prevention of nausea and vomiting	DETOSU/triethylene glycol/triethylene glycol polyglycolide copolymer 392 mg, polyethylene glycol monomethyl ether, NF, 98 mg	Heron Therapeutics
BRIXARDI (2024) weekly	Buprenorphine (50 mg/mL)	Weekly subcutaneous	10% w/w, anhydrous ethanol and soybean phosphatidylcholine/ glycerol dioleate in a weight ratio 50/50 to final volume	Braeburn Inc.
BRIXARDI monthly	Buprenorphine (356 mg/mL)	Monthly, subcutaneous	30% w/w N-methyl pyrrolidine and soybean phosphatidylcholine/ glycerol dioleate in weight ratio 40/60 to final volume	Braeburn Inc.
Abilify Maintena (2024)	Aripiprazole	Intramuscular, extended-release suspension	NanoCrystal technology	Alkermes

In general, natural biodegradable polymers are readily and abundantly available in nature. In addition, (Wright and Burgess 2012) they can have an affinity for cell- or tissue-specific binding, are typically hydrophilic, and can be water soluble. However, this hydrophilicity often means a high rate of water uptake and consequently faster rates of drug release. Enzymatic degradation is typical for natural biopolymers in vivo and as such the rate of degradation is tissue specific which makes drug release challenging to predict and control (Wright and Burgess 2012). Another drawback of these natural polymers from a pharmaceutical application standpoint is the relatively difficult, unscalable, and costly nature of their manufacturing processes. Production often involves multiple steps and extensive purifications making control of physicochemical properties much more difficult compared with synthetic polymers. The resulting batch-to-batch variance in key chemical properties such as composition and molecular weight can result in variable performance such as inconsistent processing and formulation, and difficult to reproduce degradation and drug release kinetics. Various naturally derived biodegradable polymers have shown success in preclinical trials. In more recent years, there have been significant developments surrounding bacterially produced biodegradable polymers, scale-up, and control which may provide valuable candidate materials for consideration in future developments.

Polysaccharides

Polysaccharides are biopolymers composed of many monosaccharides (sugars) connected by glycosidic linkages generated during condensation (Baranwal et al. 2022). Of the polysaccharides, cellulose, alginate, dextran, hyaluronic acid, and chitosan have been the most utilized in a drug delivery context (Idrees et al. 2020; George et al. 2019). Chitosan is produced commercially by partial deacetylation of chitin, which is present in the skeleton of marine crustaceans (Kou et al. 2021; Sanjanwala et al. 2022). Due to their cationic charge, chitosan nanoparticles have received considerable interest as a delivery vehicle for negatively charged nucleic acid therapeutics (Wright and Burgess 2012; Prabaharan and Mano 2005). Alginates and hyaluronic acid are anionic linear polysaccharides. The former is typically extracted from seaweed as alginic acid before being converted into sodium alginate, while the latter is found in the extracellular matrix of vertebrate epithelial, neural, and connective tissues (Tønnesen and Karlsen 2002; Sze et al. 2016). Alginate can be easily functionalized because of its free hydroxyl and carboxyl groups, while hyaluronic acid contains surface receptors which allow for targeted drug delivery. Both are typically utilized in hydrogel formats (Huang and Huang 2018). Dextran is a complex, branched glucan and is synthesized by lactic acid bacteria or their enzymes in the presence of sucrose (Díaz-Montes 2021). Dextran nanoparticles have more recently become a focus area for controlled-release and targeted drug delivery research, due to their ease of surface modification for selectivity for specific target organs, tissues, and cells (Huang and Huang 2019).

Proteins

Polypeptides or proteins are biopolymers capable of forming amorphous three-dimensional structures stabilized mainly by noncovalent interactions. The functional properties of these materials are highly dependent on structural heterogeneity, thermal sensitivity, and hydrophilic behavior of proteins. For drug delivery purposes, collagen, gelatin, and albumin are perhaps the most utilized animal and vegetable-derived proteins. Collagen, the major component of skin, bones, and connective tissue, is a fibrous protein consisting of three intertwined polypeptide chains, whereas gelatin is obtained by controlled hydrolysis of the fibrous, insoluble collagen. Collagens that are currently in use by the medical field are mostly animal derived (bovine, porcine) and purified, but resulting batch-to-batch variabilities and potential immunogenic responses have prompted huge efforts to develop novel methods for the manufacture of recombinant collagens (Wang et al. 2017). Collagen has a hydrophilic hydrogel-like nature due to enormous substance of essential, acidic, and hydroxylated amino corrosive buildups and has innate ability to gel with temperature change resulting in its use as an injectable hydrogel for sustained-release drug delivery (An et al. 2016). However, this hydrogel nature also results in poor dimensional stability, substantial swelling in vivo, and consequently unpredictable drug release kinetics. Albumin is a family of globular proteins most typically obtained from bovine or human serum. Albumin nanoparticles have received significant attention as chemotherapeutic delivery vehicle due to their ability to accumulate in the tumor sites (Kratz and Elsadek 2012).

Microorganism-Derived/Microbial Biopolymers

Microorganism-derived or microbial biopolymers are secondary metabolites generated by microbes and plants. The microbial polyesters are produced by biosynthetic function of a microorganism and readily biodegraded by microorganisms and within the body of higher animals, including humans. From a pharmaceutical standpoint, polyhydroxyalkanoates (PHAs) are the commonly used example. PHAs are polyesters formed in nature and produced directly by bacterial metabolism. Stored as inclusion bodies in bacteria, PHAs are generated and aggregated intracellularly, serving as both a source of energy and a carbon store (Muhamad et al. 2017; Jendrossek 2009). Despite their microbial origin, PHAs are thermoplastics similar to other polyesters and can be melt-processed and shaped accordingly. PHAs are degraded in vivo enzymatically or by surface erosion via chemical hydrolysis and as such offer a promising solution to combat the burst release and associated acidic environment generated after degradation of more commonly used synthetic polyesters such as poly(lactic acid) (Martin and Williams 2003). PHAs are synthesized directly via fermentation of a carbon substrate inside a microorganism. Depending on the microorganisms and the cultivation conditions, homo- or copolyesters with different hydroxylalkanoic acids can be obtained (Fuessl et al. 2012). The most prominent commercially available PHAs are poly-3-hydroxybutyrate (PHB), poly-3-hydroxybutyrate-*co*-4-hydroxybutyrate (P(3-HB-*co*-4-HB)), poly-3-hydroxybutyrate-*co*-valerate (PHBV), and polyhydroxybutyrate-*co*-hexanoate (PHBH). The primary drawbacks and limitations of PHA for pharmaceutical use is the costly and complicated nature of their production process which has been estimated to be three to four times more expensive than common synthetic plastics such as polyethylene and polypropylene (Kourmentza et al. 2017). Processing methods require substantial fine-tuning and are very polymer specific resulting in batch-to-batch variability and inconsistent material properties (Bugnicourt et al. 2014).

Synthetic Polymers

Synthetic biodegradable polymers are produced chemically by conventional polymerization processes, similar to regular plastics. As a consequence of these well-established production processes, high-quality materials can be robustly produced at industrial scales under GMP conditions. Crucially, polymer characteristics such as composition, structure, and chain length can be tightly controlled and tailored during synthesis to allow for highly predictable mechanical and physical properties. From a drug delivery context, this allows for more reproducible degradation rates and drug release profiles. Moreover, functional groups can be easily incorporated into synthetic polymers to allow for surface modification of chemical moieties for improved functionality as a drug delivery vehicle (Wright and Burgess 2012; Duncan 2007). This combats a typical disadvantage of synthetic polymers in that, in their native form, they generally cannot attach with receptor binding ligands on cells, which hinders their ability for targeted, site-specific drug delivery. Also, synthetic polymers are inherently less biocompatible compared with natural biopolymers and can have associated toxicity problems. Up to now, only polyesters and polyanhydrides have been used in commercially available and approved products (Wright and Burgess 2012). However, a vast and increasing number of novel synthetic biodegradable polymers have demonstrated success in preclinical studies. The most researched and utilized synthetic biodegradable polymers are discussed below.

Aliphatic Poly(esters)

Aliphatic polyesters are the most extensively utilized class of biodegradable polymers within polymeric LAIs. These polymers contain an aliphatic ester bond in their backbone, whose linkages are hydrolyzed in aqueous environments. Although various types of aliphatic polyesters exist (such as the naturally derived polyhydroxyalkanoates or poly(alkene dicarboxylate)s, the major industry focus is on poly-α-hydroxy acids or poly-α-hydroxyesters, whose repeat units are shown in Fig. 2.2.

Despite being somewhat hydrophobic, poly-α-hydroxyesters degrade via hydrolysis of the ester linkages through bulk erosion (Gopferich 1997). There is also literature to suggest that enzymatic degradation can play a part in the biodegradation of aliphatic polyesters, particularly at the latter stages of degradation as erosion and

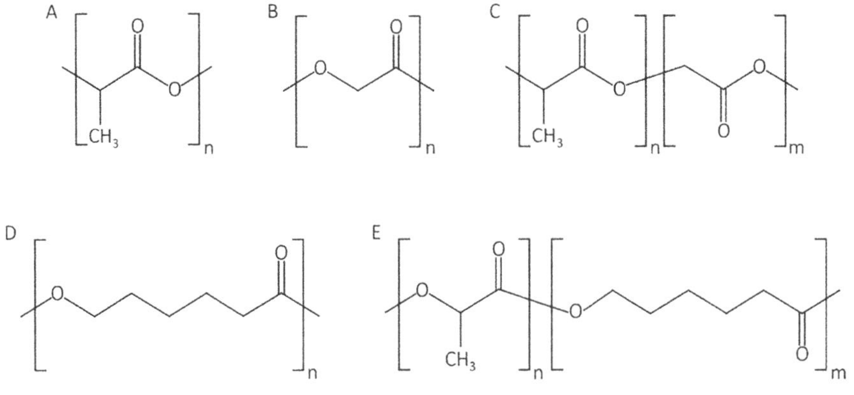

Fig. 2.2 Structure of (**a**) polylactide, (**b**) polyglycolide, (**c**) polylactide-co-glycolide, (**d**) polycaprolactone, and (**e**) polylactide-co-caprolactone repeating units

physical fragmentation of the polymer occurs and particularly for "rubbery" polymers with low glass transition temperatures (Holland et al. 1986).

Aliphatic polyesters can be synthesized via direct polycondensation of respective difunctional acids (lactic and/or glycolic acid); however, this typically involves high temperatures, long reaction times, and the prevalence for side reactions. The resulting polymers typically have a lower molecular weight, broad polydispersity, and weak mechanical properties, unsuitable for most medical applications (Löfgren et al. 1995; Athanasiou et al. 1998). For pharmaceutical applications, these aliphatic polyesters are most commonly and effectively synthesized via ring-opening polymerization (ROP) of their respective cyclic ester monomers, known as lactones (lactide, glycolide, ε-caprolactone). ROP allows for milder, controlled reaction conditions from which high-molecular-weight polymers can be easily prepared with close control of monomer sequence and chain length in much shorter reaction times and with reduced impact of side reactions (Jérôme and Lecomte 2008; Okada 2002). For medical applications, ROP in bulk or melt is often favored over solution polymerization because of the elimination of potential residual presence of organic solvents. ROP of lactide, glycolide, and caprolactone requires a suitable catalyst to proceed in reasonable conditions and to obtain polymers with controlled properties. Many cationic, anionic, and coordinative initiators or catalysts have been reported, but organometallic compounds are most typically utilized in commercial processes (Jérôme and Lecomte 2008). For pharmaceutical and industrial applications, tin (II) 2-ethylhexanoate or stannous octoate ($Sn(Oct)_2$) is almost extensively used by commercial manufacturers. $Sn(Oct)_2$ is widely and readily available, easy to handle, and soluble in various organic solvents and in melt monomers. Moreover, it is highly reactive and allows for the creation of high-molecular-weight polyesters in minutes to hours in bulk at temperatures in the range of 140–180 °C (Dechy-Cabaret et al. 2004). Despite its FDA approval, the toxicity associated with tin compounds is a drawback of $Sn(Oct)_2$ and residues must be tightly controlled. Other zinc or magnesium-based catalysts can also be used, and are receiving great interest as potential non-toxic, more biocompatible alternatives to tin. ROP using $Sn(Oct)_2$ requires an active hydrogen compound as a co-initiator. $Sn(Oct)_2$ reacts with the co-initiators' free hydroxyl to form a tin alkoxide active center, followed by a coordination-insertion ring-opening polymerization mechanism, although there exists some debate on this topic (Storey and Sherman 2002; Kowalski et al. 1998). Thus, calculated addition of co-initiator to the polymerization reaction is the most rational approach to controlling molecular weight based on monomer-to-initiator molar ratio. In practical sense, by reducing the initiator amount you can increase the molecular weight since fewer initiators will generate fewer chains of longer length. For controlled-release applications, the molecular weight of aliphatic polyesters is usually in the low to medium range, often less than 50 kDa with few cases above 100 kDa.

With ROP, the co-initiator utilized determines the polymer end group chemistry from primary reactions. Monofunctional alcohols such as butanol, hexanol, decanol, and dodecanol allow for the formation of ester-terminated polymers (i.e., a capped end group with an ester on one end of the polymer chain and a hydroxyl group on the other end). Alternatively, α-hydroxy acids such as glycolic acid, lactic acid, and oligomeric lactic acid or water can be used to form acid-terminated

Table 2.3 Co-initiators used with $Sn(Oct)_2$ for ROP of cyclic lactones/lactides to create linear copolymers

Co-initiator utilized in ROP	Resulting end groups
Monofunctional alcohol	Ester and hydroxyl
Diol	Hydroxyl and hydroxyl
α-Hydroxyl acid	Carboxylic acid and hydroxyl
Polyethylene glycol methyl ether (mPEG)	PEG and hydroxyl

Table 2.4 Properties of synthetic poly-α-hydroxyesters with data from Kapoor et al. (2015), Nair and Laurencin (2007), Van de Velde and Kiekens (2002), Zhang et al. (2021)

Polymer	T_g (°C)	T_m (°C)	Tensile strength (MPa)	Tensile modulus (GPa)	Elongation at break (%)	Degradation time (months)
PGA	35–45	>220	60–100	6–7	1.5–20	6–12
PLLA	55–65	>170	45–70	2–4	3–10	>24
PDLLA	45–55	Amorphous	15–30	<2	2–10	12–16
PLGA (50–85% DLA)	40–55	Amorphous	40–55	1–2	2–10	1–6
PCL	−60	60	20–34	0.2–0.35	>700	>24
PLCL (70% LLA)	10–20	135–155	17–22	0.02–0.04	300–700	>24

polymers (one end has a carboxylic acid group and the other end has a hydroxyl group). Similarly, polyethylene glycol (PEG) can also be used to initiate polymerization, either via a singular hydroxyl group in the case of PEG methyl ether which results in the creation of a di-block copolymer (Azhari et al. 2023) or via both hydroxyl groups on conventional PEG to form tri-block copolymers. Dextrose, glucose, or other compounds with several hydroxyl groups allow for the creation of star-shaped or branched polyesters. A summary of the most utilized co-initiators and their resulting end group chemistry can be found in Table 2.3.

Moreover, the different monomers within this family can be used in copolymerization or blending strategies to modulate final material properties. The type and number of monomers used in synthesis significantly influences the physical, chemical, crystalline, mechanical, and degradation properties of the resulting polyester. This versatility, coupled with their thermoplastic and biocompatible nature, is what makes the aliphatic polyesters attractive for pharmaceutical applications. The intrinsic material properties of each of the most popular poly-α-hydroxyesters will now be discussed separately. A summary can be found in Table 2.4. Other cyclic lactones such as p-dioxanone and trimethylene carbonate can be used in ROP to create polydioxanone (PDO) and poly(trimethylene carbonate) (PTMC), respectively; however, these are more suited to medical devices and therefore will not be discussed in detail.

Poly(Glycolic Acid)/Polyglycolide (PGA)

PGA is the simplest linear aliphatic polyester. It is a rigid, thermoplastic polymer with relatively high crystallinity (45–55%), a T_m of more than 220 °C, and a T_g of 35–45 °C (Nair and Laurencin 2007). PGA has a tensile strength of 60–100 MPa, a Young's modulus of 6–7 GPa, elongation at break of 1.5–20%, and a degradation time frame of between 6 and 12 months (Middleton and Tipton 2000; Van de Velde and Kiekens 2002) depending on its molecular weight. PGA's use in pharmaceutical applications is limited by two main factors: poor solubility in most organic solvents and unsuitably long degradation times yielding high rate of acidic by-products. Consequently, copolymerization with other biodegradable monomers such as caprolactone and lactide or blending with other polymers is typically utilized.

Poly(Lactic Acid)/Polylactide (PLA)

PLA is derived from lactic acid, a chiral molecule which exists as two optically active stereoisomers, namely L (−) and D (+) enantiomers. The mixture of both D and L results in the meso equimolar D,L lactic acid (Gupta et al. 2007a, b). For medical applications, PLA is typically synthesized via ROP of lactide cyclic dimers. Polymerization of L-lactide or D-lactide results in isostatic, semi-crystalline polymers named poly(L-lactide) (PLLA) or poly(D-lactide) (PDLA), while polymerization of the monomer D,L-lactide results in atactic amorphous polymer poly(D,L-lactide) (PDLLA) (Vert 2002). PLLA has a T_g of 55–65 °C, a T_m in excess of 170 °C, a tensile strength of 45–70 MPa, a Young's modulus around 2–4 GPa, and a strain at break of approx. 3–10% (Mochizuki and Hirami 1997; Vouyiouka and Papaspyrides 2012). Alternatively, amorphous PDLLA has no distinct melting point, presents a T_g between 45–55 °C, and is mechanically less stiff with a greatly reduced tensile strength (15–30 MPa) and modulus (<2 GPa) compared to PLLA. The presence of additional methyl groups in PLA coupled with the steric shielding effect of these side groups causes it to be more hydrophobic and resistant to hydrolysis than PGA (Mochizuki and Hirami 1997). The degradation time to complete mass loss for PLLA and PDLLA has been reported as greater than 2 years and between 12 and 16 months, respectively (Middleton and Tipton 2000; Ikada and Tsuji 2000a, b).

Poly(Lactic-Co-Glycolic Acid)/Polylactide-Co-Glycolide (PLGA)

PLGA is by the far the most dominant aliphatic polyester utilized for controlled-release applications. Although PLGA can be synthesized to contain either enantiomeric form of lactide (D+ or L-), it is almost exclusively produced using racemic D,L-lactide as poly(D,L-lactide-co-glycolide) for drug delivery purposes. The most utilized compositions contain 50, 75, or 85%mol D,L-lactide. Mechanical properties of these compositions have been reported as ranging from 40 to 55 MPa for tensile strength, 1–2 GPa for modulus, and 2–10% for strain at break (Van de Velde and Kiekens 2002). Biodegradation is also greatly influenced by the copolymer ratio. As mentioned, the presence of methyl side groups in lactide backbone makes PLA polymers more hydrophobic in comparison to glycolide containing chains. Thus, PLGA copolymers that are rich in lactide are more hydrophobic, absorb less water, and thus degrade via hydrolysis at a slower rate. Degradation ranges from a few weeks to several months depending on lactide content (Gentile et al. 2014).

Another important factor which influences PLGA properties is the microstructure or sequence distribution of lactide and glycolide units along the polymer chain backbone. This is often referred to as "randomness" or "blockiness." ROP naturally results in a somewhat randomized distribution of monomer units; however because glycolide has a lower melting point and higher reactivity than lactide, polyesters made with higher glycolide content (>40%mol) have the tendency to form glycolate blocks (GGGG) in the polymer chain. These glycolate blocks and their length are not easily controlled. In theory, there are many contributing variables while the most significant process parameters are usually considered as temperature, rate of monomer addition, stirring, reaction duration, catalyst type/amount, and general processing setup. For example, higher temperatures and longer reaction time frames will tend to create more randomized structures whereas lower temperatures over shorter periods limit transesterification and a reduction in randomness. As the length of the polymer chain grows, the resulting chain that forms increases steric hindrance at the hydroxyl reaction site (Lan et al. 2004) which means that higher molecular weight and longer reactions times will also impact this outcome. A "blocky" PLGA polymer with a low degree of randomness and large regions of glycolate blocks has the potential to form crystalline domains which can decrease their solubility in the typically used organic solvents affecting processing. Moreover, the G-G linkages are hydrolyzed faster than G-L which impacts degradation behavior.

Polycaprolactone (PCL)

PCL is a semi-crystalline linear polyester which is soluble in a wide range of organic solvents. PCL is highly processable due to its low T_m of 60 °C, low T_g of −60 °C, and high thermal stability. These thermal properties make PCL a semi-rigid, ductile material at room temperature (Vroman and Tighzert 2009). PCL has a relatively low tensile strength and Young's modulus of between 20–34 MPa and 206–345 MPa, respectively, but an extremely high elongation at break, reported well in excess of 700%. In physiological conditions, PCL degradation can take up to 2–3 years and is characterized by a slow decrease in molecular weight without deformation, followed by the metabolization of its by-products such as caproic acid (Pitt et al. 1987). Due to its slow degradation and high permeability to many drugs, PCL is typically investigated as a long-term, ultra-slow-release drug delivery vehicle (Sinha et al. 2004). Modification of PCL with other polymers (such as block copolymer synthesis or blending with PLGA and PLA) has led to new use polymers (Dash and Konkimalla 2012). PCL-PEG block copolymers create an amphiphilic structure used in self-assembling micelles for targeted delivery of chemotherapeutics (Li and Tan 2014; Kamaly et al. 2016).

Polylactide-Co-Caprolactone (PLCL)

Copolymerization of lactide and caprolactone monomers yields polylactide-co-caprolactone (PLCL). PLCLs with a CL content from 15 to 30% yield elastomeric performance with T_g's higher than 20 °C, whereas those with a CL content from 5 to 15%, although stiffer, may still be applied as ductile materials (Fernandez et al. 2014). Typically, PLCL copolymers contain below 50% caprolactone as higher fractions yield a sticky and difficult to handle form (Fernandez et al. 2012). Copolymerization with caprolactone interrupts the formation of long crystalline regions of lactide in the polymer backbone and thus also provides an increased susceptibility to hydrolysis and an acceleration to the overall degradation rate. Similar to PLGA, PLCL is available as D-, L- and D, L isomers. D- and L-PLCL can have some degree of crystallinity as well as a high capacity for crystallization during degradation, while D, L-PLCL or PDLCL is completely amorphous at all compositions (Fernandez et al. 2014). Consequently, PDLCL has a lower mechanical strength, greater elongation, and a faster degradation rate when compared with PLCLs from both enantiomeric forms of LA, containing the same ratio of LA:CL (Fernandez et al. 2014). Compared to PLA and PLGA, PLCL copolymers are far less utilized for controlled-release applications; however, these polymers are being used in more and more medical device applications that seek some mechanical flexibility (Table 2.5).

Poly(anhydrides)

Polyanhydrides are a class of biodegradable polymers which consist of aliphatic anhydride bonds that connect repeat units of the polymer backbone chain. These anhydride bonds are hydrolytically labile, resulting in two easily metabolized carboxylic acid groups, making these polymers biodegradable and biocompatible (Saltzman 2001). Despite the water-labile anhydride bonds, the polymer itself is hydrophobic which impedes water absorption into the polymer matrix and results in a surface erosion degradation mechanism. This contrasts with the bulk eroding polyesters and in theory can provide a more constant zero-order drug release profile. However, due to their chemical nature (a lot of highly reactive functional groups for hydrolytic degradation), polyanhydrides degrade very quickly (from a few days up to weeks) (Goepferich 1999). As a result, polyanhydrides are more limited to shorter-term delivery. In addition, their hydrolytic instability demands storage in moisture-free, freezer-like conditions. Polyanhydrides can be synthetized using various polymerization methods such as melt condensation, ROP, interfacial condensation, dehydrochlorination, or solution polymerization. Melt condensation is most common as it results in high-molecular-weight polymers and is typically approached at high temperatures under a vacuum to react dicarboxylic acid monomers with an excess of acetic anhydride. The best established polyanhydride is poly[(carboxyphenoxypropane)-(sebacic acid)] (PCPP-SA) which is a copolymer of sebacic acid and 1,3-bis(p-carboxyphenoxy) propane. PSA is a linear aliphatic polymer that degrades within days, whereas PCPP is a hydrophobic, aromatic polymer which degrades much slower. Thus, varying degradation rates can be achieved

Table 2.5 Solubility of poly-α-hydroxyesters in different organic solvents (X = insoluble, PS = partially soluble, S = fully soluble)

	Acetone	Ethyl acetate	Chloroform	Methylene chloride	Tetrahydrofuran	N-methyl-2-pyrrolidone	Hexafluoro-isopropanol	Dimethyl formamide
PGA	X	X	X	X	X	X	PS	X
PLLA	X	X	S	S	X	PS	S	X
PDLLA	S	S	S	S	S	S	S	S
PLGA (75% DLA)	S	S	S	S	S	S	S	S
PLGA (50% DLA)	PS	PS	S	S	PS	S	S	S
PCL	S	S	S	S	S	S	S	S

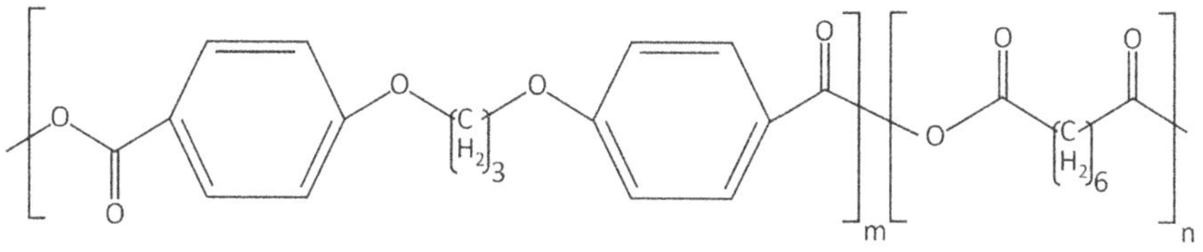

Fig. 2.3 Structure of poly[(carboxyphenoxypropane)-(sebacic acid)] (PCPP-SA)

by altering the ratios of these two components. PCPP-SA (Fig. 2.3) is used in the commercially available Gliadel® wafer, a localized implant for the treatment of brain cancer. The polymer acts as a delivery vehicle for controlled release of the chemotherapeutic agent carmustine (1,3-bis[2-chloroethyl]-1-nitrosourea [BCNU]). In addition, a polyanhydride copolymer of 1:1 sebacic acid and erucic acid dimer is used in Septacin®, another localized implant containing gentamicin sulfate for sustained delivery of the antibiotic to infected bone for the treatment of osteomyelitis (Li et al. 2002).

Poly(orthoesters)

POEs have been under development since 1970, primarily by Heller, and over the last 50 years have evolved through four generations, designated as POE I, POE II, POE III, and POE IV. POE I is obtained by transesterification of an α, ω-diol with 2,2-diethoxytetrahydrofuran but autocatalytically hydrolyzes uncontrollably in an aqueous environment. POE II was developed to alleviate this and is prepared by polyaddition of a α, ω-diol and the diketene acetal 3,9-diethylidene-2,4,8,10-tetraoxaspiro [5.5] undecane (DETOSU). The resulting polymer is more hydrophobic and requires an acidic compound to initiate degradation. POE III is prepared by transesterification using a triol and an orthoester. In this case, the efforts to limit the backbone hydrophobicity ultimately resulted in lengthy, complex, and poorly reproducible synthesis. Finally, POE IV came as further development of POE II which resolves the slow degradability by incorporating short sequences of glycolide and/or lactide in the polymer chain backbone. POEs are more hydrophobic than poly(α-esters) and polyanhydrides, owing to their lack of carbonyl functionality, with the hydrolysis of the orthoester linkage being comparable to that of an anhydride (Park et al. 2005). The less acidic microenvironment is a potential advantage in POE application compared to polyesters for larger molecules. In addition, POEs degrade via a surface erosion mechanism due to hydrophobic hydrocarbon–ether ring which, like polyanhydrides, has been shown to provide a more zero-order-like drug release profile. Despite these promising properties, POEs' commercial application has been limited by the long and arduous development process to arrive at the relatively recent POE IVs as well as the complex and challenging nature of their synthesis. Nevertheless, POE IVs have been utilized in various successful clinical trials and are currently the main excipient used in ZYNRELEF®, a bupivacaine and meloxicam extended-release solution for instillation to produce postsurgical analgesia.

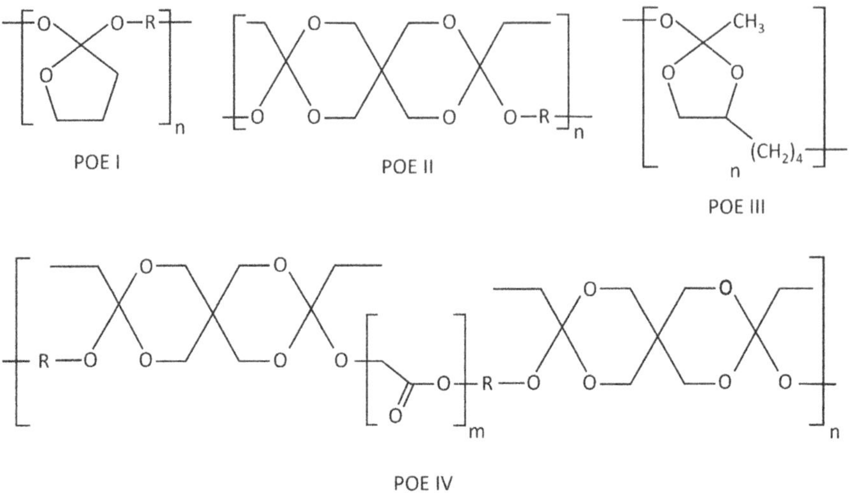

Fig. 2.4 Structure of the four generations of poly(orthoesters)

Biochronomer® is the proprietary brand name of the POE IV used in ZYNRELEF® and is manufactured by A.P. Pharma (Redwood, California) who also produce a di-block copolymer of POE IV and PEG marketed as Bioerodimer® (Fig. 2.4).

Other Emerging Biodegradable Polymers

As discussed, most commercial polymeric LAI products utilize aliphatic polyesters with some cases using polyanhydrides and polyorthoesters. Beyond these, there are also other polymers in R&D worth considering.

Poly(amino acids), for example, are linear synthetic biodegradable polymers made from amino acid repeating units. Their degradation rates are dependent on the hydrophilicity of the respective amino acids present in the polymer structure (Martin et al. 1971; Marck et al. 1977). The development of controlled ROP of α-amino acid *N*-carboxyanhydrides (NCAs) in the past two decades has enabled the synthesis of a large quantity of polyamino acid materials with predictable molecular weights and narrow polydispersities (Deming 1998). With properties that mimic proteins, poly-amino acids have shown promise as gene delivery vectors in various preclinical studies. These properties include increasing the solubility and stability of drug attachments, drug encapsulation, drug targeting, bypassing multidrug resistance factors, minimal stimulation of the immune system, and low toxicity. Examples include functional poly(L-glutamic acid) and poly(L-lysine), poly(L-arginine), and polysarcosine. Positively charged poly(L-lysine) has garnered attention as a delivery vehicle for negatively charged nucleic acids, but its cationic nature also poses a toxicity concern (Tiera et al. 2011). Tyrosine-derived polymers are a type of pseudo-poly(amino acids) without peptide-like backbones where amino acids are linked through side groups which have shown great promise due to their excellent biocompatibility, degradability, and functional as well as engineering properties (Jain et al. 2023).

Another class of biodegradable polymer which has promise as a nucleic acid delivery vehicle is poly(phosphoester) (PPE) due to its structural similarity (Zhao et al. 2003). Its unique structure consists of phosphorus atoms attached to either an oxygen or carbon atom, allowing for the conjugation of a range of side chains. PPEs can be synthesized by ROP, condensation, or addition polymerizations, and degrade in vivo by hydrolysis and enzymatic digestion, producing non-toxic by-products (Ulery et al. 2011).

Poly[(organo)phosphazenes] are inorganic/organic hybrid polymers consisting of alternating nitrogen and phosphorus atoms and organic substituents linked to the phosphorus atoms as side groups. Polydichlorophosphazene is the primary precursor and although it is extremely hydrolytically unstable, it can be readily substituted with nucleophilic substituents to give a wide range of stable poly[(organo)phosphazenes] with an extremely diverse set of properties (Allcock 2003). The hydrolytic stability can be tuned (depending on the properties of the side substituents and their hydrophilicities) to produce a wide spectrum of materials with differing degradation rates. Polydichlorophosphazene is typically synthesized by thermal ROP of hexachlorophosphazene but results in limited to no control over molecular weight (Teasdale and Brüggemann 2013). The development of a living polymerization route to polyphosphazenes was a key advancement allowing for polymers with controlled molecular weights and narrow polydispersities (Allcock et al. 1996). Poly[(organo)phosphazenes] have shown good potential as drug and gene delivery vehicles in numerous in vitro and in vivo studies but use in a commercially approved market product remains far away (Teasdale and Brüggemann 2013).

Poly(ester amide)s (PEAs) combine the thermal and mechanical properties of polyamides with the biocompatibility and biodegradability of polyesters. The amine/ester bond ratio and microstructure can be tuned to modulate the physical degradation properties of PEAs. Generally, a higher ester content favors an increased degradation rate, whereas a higher amide content favors stability against degradation as well as thermal stability (Rodríguez-Galán et al. 2011). A lower degradation rate and lower pH drop compared to polyesters are a primary benefit of PEAs as a drug delivery vehicle. PEAs can be synthesized by ROP of cyclic monomers such as lactams, lactones, or morpholino-2,5-diones or polycondensation of linear monomers such as diols and diamide diester or dicarboxylic acids and amino acids. Poly(β-amino esters) (PBAEs), synthesized via the Michael addition reaction, allow for biodegradable polymers with various backbones and have received much interest as gene delivery vehicles (Siegwart et al. 2011; Anderson et al. 2003; Green et al. 2009; Akinc et al. 2003).

A final example is poly(alkyl cyanoacrylates) (PACAs), a family of biodegradable and biocompatible polymers produced from acrylic monomers that rapidly polymerize in the presence of water due to the presence of two highly reactive electron-withdrawing groups on the alkyl cyanoacrylate monomers (Graf et al. 2009). PACAs have been most successfully utilized in drug delivery as passively targeting stealth nanoparticles. Synthesis can be from alkyl cyanoacetate either via a condensation reaction leading to oligomers or via a thermal depolymerization reaction of these oligomers (Nicolas and Couvreur 2009). To allow for complete degradation and elimination of monomers, molecular weights should be between 5 and 10 kDa (Nicolas and Couvreur 2009).

2.4 Regulatory Status of Lactide/Glycolide Polymers and Requirements for Use

Given that lactide glycolide copolymers are the most utilized polymeric technology for long-acting injectable market products, we have completed a deeper review with respect to this family of polymers.

Established Safety

Synthetic aliphatic polyesters have a well-established safety profile. PLGA or poly(lactic-co-glycolic acid) copolymers are used in many Food and Drug Administration (FDA)-approved therapeutic devices with acceptable biodegradability and biocompatibility (Crotts 1998; Chen et al. 2017; Gunatillake and Adhikari 2003; Ikada and Tsuji 2000a, b; Allahyari and Mohit 2016; Choi and Han 2018). Under typical use, PLGA is anticipated to undergo hydrolytic degradation and cleavage of backbone ester linkages to finally break down into monomeric units of lactic and glycolic acid. These degradation by-products are natural metabolites which are cleared by normal pathways (tricarboxylic acid cycle) in the body and excreted as water and carbon dioxide. This has been demonstrated both in vivo and in vitro for various drug types and proteins with different polymer ratios (Amann et al. 2010; Ramchandani and Robinson 1998).

FDA Status

The FDA maintains an inactive ingredient database (IID). Per 21 CFR 210.3 (b)(8), an inactive ingredient is any component of a drug product other than the active ingredient (2024a). The IID provides information on inactive ingredients present in FDA-approved drug products. For new developments, once the inactive ingredient appears in a new approved drug product for a specified route of administration then that ingredient is no longer considered new and may receive a less extensive review than the next time it is included in a new drug formulation. This IID also lists the highest amount of the excipient per unit dose within each dosage form that it is used in. Lactide/glycolide polymers are listed on the FDA's inactive ingredient database (2024b) under various dosage forms with different polymer compositions (50:50, 75:25, 100:0), end groups, and molecular weights. Routes of administration include intravitreal, subcutaneous, intramuscular, intraarticular, periodontal, dental, and

Table 2.6 IID Database for lactide/glycolide containing polymers. Note: results compressed per material. Full IID can be consulted (2024c). IM Intramuscular, IV Intravenous, SC: Subcutaneous

Inactive ingredient	Route	Dosage form	CAS number	UNII	Maximum potency per unit dose	Maximum daily exposure (MDE)
IID Database results for lactide and glycolide containing registrations						
DL-Lactide and glycolide (50:50) copolymer 12000 acid	Endosinusial, intravitreal, SC	Implant, injection, solution	26780507	WE369X5600	1.35 mg, 116 mg	533 mg
DL-lactide and glycolide (75:25) copolymer 20000 acid	IM	Injection, powder, for suspension	26780507	58X445TH30		145 mg, 99 mg
Polylactide	IM, IV, periodontal	Injection, system	26680104	459TN2L5F5	264.8 mg, 0.37 mg/mg	99 mg
Polylactide (14000 MW)	IM	Injection, powder, for suspension		9J7G894K2M		188 mg
Polyglactin	Dental, IM, SC	Powder, pellet injection, lyophilized, solution, suspension, implant,	26780507	NA	132.4 mg, 28.32%w/v, 25.2 mg, 211.5 mg, 132.6 mg	NA, 33 mg, 138 mg, 304 mg, 4 mg
Detosu/ triethylene Glycol/ triethylene Glycol polyglycolide copolymer	SC	injection		JC2Q67KRW7		392 mg

intravenous, and these polymers have been approved in many drug products as well as medical devices (Table 2.6).

UNII stands for Unique Ingredient Identifier and is part of the joint FDA Substance Registration System (SRS)/United States Pharmacopeia (USP). This is a 10-character alphanumeric identifier linked to a substances structure or descriptive information which is contained in domestically marketed foods, drugs, and medical devices. CAS stands for Chemical Abstracts Service which is a division of the American Chemical Society and assigns a unique CAS registry number to chemical substances (Table 2.7).

Table 2.7 Unique Ingredient Identifier (UNII) codes per FDA website connected with lactide/glycolide polymers (2024)

Ingredient description	UNII	CAS Registry
Polylactide (0.4 DL/G)	7B01B0FN51	26680-10-4
DL-Lactide and glycolide (50:50) copolymer 12000 ethyl ester	7SVE964630	1354955-03-5
DL-Lactide and glycolide (50:50) copolymer 12000 acid	WE369X5600	26780-50-7
Lauryl polylactide (14000 MW)	AOA7A18GO2	202832-99-3
Polylactide (14000 MW)	9J7G894K2M	26680-10-4
DL-Lactide and glycolide (75:25) copolymer 140000 acid	1SIA8062RS	26780-50-7
DL-Lactide and glycolide (50:50) copolymer 15000 acid	C329FPO0UC	26780-50-7
DL-Lactide and glycolide (75:25) copolymer 20000 acid	58X445TH30	26780-50-7
DL-Lactide and glycolide (50:50) copolymer 22000 lauryl ester	RIU41B2XDV	197526-99-1
DL-Lactide and glycolide (85:15) copolymer 23000 acid	93PPD1S477	26780-50-7
Polylactide (23000 MW)	E14P5SD3J4	26680-10-4
DL-Lactide and glycolide (80:20) copolymer 27000 acid	K6S9F6373P	26780-50-7
DL-Lactide and glycolide (50:50) copolymer 30000 acid	U5MPH5YWJ9	26780-50-7
Polyglactin 370	USA36U2P3T	385434-35-5
DL-Lactide and glycolide (50:50) copolymer 46000 acid	LQ35R50VR1	26780-50-7
DL-Lactide and glycolide (50:50) copolymer 63000 acid	QD923V64H3	26780-50-7
DL-Lactide and glycolide (50:50) copolymer 64000 lauryl ester	YK5TFZ8PC3	197526-99-1
DL-Lactide and glycolide (55:45) copolymer 70000 acid	HFB0YWNRAP	26780-50-7
DL-Lactide and glycolide (50:50) copolymer 7500 ethyl ester	M5MB55C36J	1354955-03-5
Polyglactin 910	51L8Z461QZ	385434-35-5
DL-Lactide and glycolide (75:25) copolymer 9500 methyl ester	I8DU4QS1I2	731852-64-5
DL-Lactide and glycolide (75:25) copolymer 9500 acid	SJN700Q87A	26780-50-7
Polylactide	459TN2L5F5	26680-10-4
Polylactide, L-	W92NDP347A	26811-96-1
Poliglecaprone 25	5W4V3X9K92	41706-81-4
Polyglyconate	KOF6AD0GWS	75734-93-9
Poliglecaprone 90	UX47P67NE5	41706-81-4
Polyglycolic acid	H1IL6F7KB8	26009-03-0

Toxicology

Various studies of PLGA-based devices suggest that generally mild local tissue reactions at the site of application may occur (Dailey et al. 2006). PLGA has been shown to be extremely safe as a material and deemed to have insignificant human health and environmental impacts (Conaway 2011).

Regulatory Standards Guiding Polymer Specifications for Medical Use

There are a range of ASTM and ISO standards that guide requirements in terms of specifications for bioresorbable polymers. Prominent standards are summarized below.

ASTM Standards

- ASTM-F2313 › Standard Specification for Poly(glycolide) and Poly(glycolide-co-lactide) Resins for Surgical Implants with Mole Fractions Greater Than or Equal to 70% Glycolide
- ASTM-F2579 › Standard Specification for Amorphous Poly(lactide) and Poly(lactide-co-glycolide) Resins for Surgical Implants
- ASTM F1925-22: Standard Specification for Semi-Crystalline Poly(lactide) Polymer and Copolymer Resins for Surgical Implants

ISO Standards

- ISO 5425:2023: Specifications for use of poly(lactic acid) based filament in additive manufacturing applications
- ISO 15814:1999(en): Implants for surgery—Copolymers and blends based on polylactide—In vitro degradation testing
- ISO 13781:2017: Implants for surgery. Homopolymers, copolymers and blends on poly(lactide) In vitro degradation testing

Monographs and Development

Beyond those listed standards, there are monographs either existing or in development as detailed below:

Chinese Pharmacopoeia Commission has three existing monographs which detail specification requirements per grade.

- Poly(lactide-co-glycolide) (50 50) (for injection)
- Poly(lactide-co-glycolide) (75 25) (for injection)
- Poly(lactide-co-glycolide) (85 15) (for injection)

At the time of writing this chapter, we note the proposed 2020–2025 Complex Excipients Expert Committee Work Plan (2023) to develop a set of USP monographs for lactide/glycolide copolymers. Based on the FDA IID listings and communications with the FDA compendial office, USP has finalized the following prioritized list of PLGA polymers for monograph development (Table 2.8).

Table 2.8 USP's prioritized LG polymer list (proposed) for monograph development (2022)

#	LG polymer	Approved/Proposed Monograph Title by USP	UNII#
1	Poly (DL-lactic-co-glycolic acid), (50:50;12000 MW)	DL-Lactide and glycolide (50:50) copolymer 12000 acid	WE369X5600
2	Poly (DL-lactic-co-glycolic acid), ethyl ester terminated, (50:50;12000 MW)	DL-Lactide and glycolide (50:50) copolymer 12000 ethyl ester	7SVE964630
3	Poly (DL-lactic-co-glycolic acid), (75:25;20000 MW)	DL-Lactide and glycolide (75:25) copolymer 20000 acid DL-Lactic acid and glycolic acid (75:25) copolymer 20000	58X445TH30
4	Poly (DL-lactic-co-glycolic acid), (50:50;46000 MW)	DL-Lactide and glycolide (50:50) copolymer 46000 acid	LQ35R50VR1
5	Poly (DL-lactic-co-glycolic acid), (50:50;63000 MW)	DL-Lactide and glycolide (50:50) copolymer 63000 acid	QD923V64H3
6	Polylactide	To be determined	459TN2L5F5

2.5 Polymer Properties and Analytical Testing

Ultimately, the physicochemical properties of polymers determine their functional performance. For successful application as controlled-release drug delivery vehicles, physicochemical properties should be well understood and selected to match the intended goal in application. As such, analytical testing is vital to the understanding and use of biodegradable polymers. Characterization and quality control testing is discussed below and has been primarily done in the context of the aliphatic polyesters as they represent the vast majority of biodegradable polymers in commercial use today.

Polymer Identification, Composition, and Structure

Proton nuclear magnetic resonance (^{1}H-NMR) spectroscopy is perhaps the most useful technique to properly characterize and understand the chemical structure of biodegradable polymers. The proton environment in the polymer molecule can be determined by its resonance frequency, with each proton having its own unique chemical shift δ. This is typically expressed in parts per million (ppm) by frequency. The polymer is dissolved in a suitable deuterated solvent with a reference of known frequency. Chemical structures can be identified from a combination of chemical shift data and spin–spin splitting, derived from proton–proton interaction. Thus, NMR is used to study polymer stereochemistry and monomer composition and sequencing.

For aliphatic polyesters, polymer samples are typically dissolved in deuterated chloroform and the spectrum of the polymer solution is recorded from—1 to 15 ppm relative to tetramethylsilane. The spectrum generated during this analysis is then compared against that of a reference standard material to confirm the identity. ^{1}H-NMR spectroscopy is also utilized to determine the molar ratio of monomers in copolyesters with analysis conducted on the same spectrum obtained in the identification test. This is done by integration of the peaks related to specific hydrogens for each repeating co-monomer unit in the polymer which can then be used to determine the relative amount (molar ratio) of each unit that is present. Generally, an NMR instrument with a frequency of at least 400 MHz is utilized to achieve a suitable resolution for reliable and precise integration of peaks in the spectrum. For PLGA copolymers, the signals at approx. 4.8 ppm correspond to the two protons from the glycolide repeating unit (methylene group). The signals at approx. 5.2 ppm (methine group) and 1.6 ppm (methyl group) correspond to the protons from the lactide repeating unit. The molar ratio of lactide glycolide copolymers is calculated by comparing the integrated areas at 5.2 and 4.8 ppm. Typical acceptance criteria are ±3% of the desired ratio (i.e., 50:50, 75:25, 85:15, etc.).

Carbon-13 NMR spectroscopy (^{13}C-NMR) is also used to determine polymer chemical structure and, for PLGA copolymers, it is required to accurately distinguish the microstructural arrangement or distribution of the monomer units in the polymer chain backbone, i.e., the "blockiness." This analysis is best determined in DMSO-d6 at elevated temperatures (>40 °C). The blockiness (Rc) is determined by dividing the peak intensity of the glycolide-glycolide carbonyl group (located at approx. 166.36 ppm) by the peak intensity of the glycolide-lactide carbonyl group (located at approx. 166.44 ppm). The higher Rc value indicates higher blockiness, i.e., higher heterogeneity in the distribution of lactide and glycolide monomers. Average glycolide block length (L_G) is the metric most reported and is equal to the Rc + 1. ^{1}H-NMR can also be utilized to provide an indication of glycolide block length by comparing the peak patterns in the GA pentad structure at 4.8 ppm. The higher intensity of these peaks relative to the rest of the CH_2 region indicates a higher glycolide block length.

$$\mathrm{Rc} = \frac{I_{\mathrm{GG}}}{I_{\mathrm{GL}}} \qquad L_{\mathrm{G}} = \frac{I_{\mathrm{GG}}}{I_{\mathrm{GL}}} + 1$$

Both ^{1}H-NMR and ^{13}C-NMR can also be used to clearly identify the polymer's end groups. For example, polyesters which have ester-capped end groups can easily be identified by ^{1}H-NMR by its distinct chemical shifts relative to the lactide and glycolide repeat units. These NMR methods follow the basic outline and principles described in USP <761> and <1761>.

Infrared spectroscopy (IR), Raman spectroscopy, ultraviolet–visible spectroscopy (UV–Vis), and specific or optical rotation are also commonly utilized to identify and determine the structure and composition of biodegradable polymers. Fourier-transform infrared spectroscopy (FTIR) is commonly used as a "fingerprint" identification test for aliphatic polyesters as well as many others for pharmaceutical excipients. However, for quality control purposes, manufacturers primarily rely on NMR spectroscopy to confirm and determine the chemical nature and composition of supplied biodegradable polymers.

Molar Mass: Molecular Weight and Inherent Viscosity

Average Molecular Weight and Polydispersity

Polymer molecular weight is the molecular mass of a polymer chain and is a result of polymerization and overall molecular weight of the repeating unit. Molecular weight is a central aspect of polymer properties.

Molecular weights of polymers are usually expressed as average molecular weights because synthetic polymers typically consist of a distribution of polymer chains with different molecular weights. Gel permeation or size-exclusion chromatography (GPC/SEC) is the most common technique used to the measure the number average molecular weight (Mn), weight average molecular weight (Mw), and polydispersity (PDI) of biodegradable polymers. For aliphatic polyesters supplied commercially, polymer samples are prepared by dissolving in tetrahydrofuran or chloroform which is used as the mobile phase and contains an appropriate stabilizer. One to four columns of appropriate pore sizes are utilized with temperatures typically set at 30–40 °C for columns and detector. Columns are calibrated relative to polystyrene (PS) or poly(methyl methacrylate) (PMMA). Mw, Mn, and PDI data are collected using refractive index (RI) peak values.

While conventional GPC/SEC methods work well to measure these characteristics, it should be noted that it has limitations when it comes to operating principles and instrument-to-instrument variability, especially when the same technique is done using instruments from different manufacturers. Conventional GPC methods provide results which are relative to test parameters and conditions such as the solvent used as mobile phase, the type of columns, the column/detector temperature, and, crucially, the standards it is calibrated against. The accuracy of measured molecular weights depends on the degree of similarity between the calibration polymer and the polymer of interest. This accuracy can be improved by a universal calibration method utilizing Mark-Houwink parameters which are specific to the exact polymer being examined. It is for these reasons that aliphatic polyesters are generally not sold on the basis of their average molecular weights. Specifications relating to molar mass are instead represented by inherent viscosity as will be outlined below.

Alternatively, multiangle light scattering (MALS) analysis can be used to determine absolute Mw, while membrane osmometry or ^{1}H-NMR can be used to determine absolute Mn (comparing the relative integration of the repeating units to the end groups).

GPC/SEC methodologies can also be utilized to characterize branched or multiarm polymers and determine the degree of branching. Glucose-initiated PLGA for example has been used in Sandostatin® LAR Depot (octreotide acetate) and is a 5-arm *star*-shaped polymer. This analysis is conducted on GPC equipped with quadruple detection systems (GPC-4D) which provides information on the branched polymers concentration, absolute molecular weight, hydrodynamic radius, and intrinsic viscosity compared to a linear counterpart (Hadar et al. 2019). This allows for determination of the branch units per molecule or "degree of branching."

Inherent Viscosity

Inherent viscosity (IV) is an alternative representative measure of polymer molar mass which is considered more practical, simple, and very reliable. IV is the increased resistance to flow of a polymer solution relative to the pure solvent used to dissolve the polymer. Higher-molecular-weight polymers increase the viscosity of liquids in which they are dissolved. The IVs of commercially available aliphatic polyesters are most typically determined in a 0c or I type Ubbelohde capillary viscometer at 25 or 30 °C relative to chloroform at concentrations of 0.1 or 0.5%w/v and reported in deciliters per gram (dl/g). If polymer solubility dictates, other solvents such as hexafluoro-2-propanol, benzene, tetrahydrofuran, *N*, *N*-dimethylformamide, methanol, acetone, and ethyl acetate can also be utilized. The flow times of the solvent and polymer solution are measured and the IV (η_{inh}) is calculated accordingly, as per the formula below:

$$\eta_{\text{inh}} = \ln(\eta_{\text{rel}}) / (C \times F)$$

η_{rel} = relative viscosity.
C = concentration of the *Sample solution* (g/mL).
F = unit conversion factor from g/mL to g/dL, 100.

In general, this method follows USP General Chapter <911> Method 1. Hagenbach correction is used when the flow time is less than the minimum flow time listed in national and international norms per capillary size number. Manufacturers supplying aliphatic polyesters offer polymers with a wide range of inherent viscosities but for drug delivery purposes they are typically in the range of 0.1–1.5 dl/g, in chloroform at 25 °C. Specifications and acceptance criteria are outlined on a polymer-to-polymer basis and vary depending on manufacturer. Tolerances to inherent viscosity are generally agreed based on end user requirements versus a safe range of performance (Fig. 2.5).

Thermal Properties

Thermal properties such as melting temperature (T_m) and the glass transition temperature (T_g) of polymers greatly influence polymer functional performance. These temperatures are determined using differential scanning calorimetry (DSC) and are directly related to crystallinity which can also be determined via DSC. Below the T_g, the polymer is in a glassy state; i.e., it has limited mobility and low diffusion rates. Above the T_g, the polymer is in a rubbery state, which facilitates higher mass transfer rates of water and drug molecules throughout the matrix (Liechty et al. 2010). The type of polymer and chemistry of its repeat unit dictate whether it is amorphous or semi-crystalline. Amorphous polymers have no distinct melting point and exhibit reduced mechanical properties and degradation rates compared to crystalline or

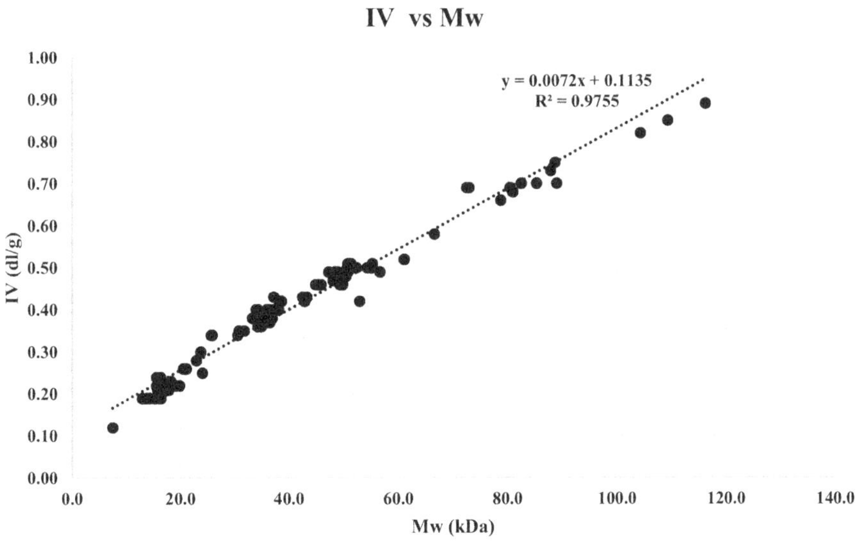

Fig. 2.5 Relationship between weight average molecular weight (Mw) from GPC (THF, 30 °C, PS standards) and inherent viscosity (IV) from U-tube viscometry ($CHCl_3$, 25 °C, c = 0.1%w/v) for a range of PLGA 75:25 copolymers of both acid and ester-capped end groups. Data provided by Ashland LLC from their viatel™ biodegradable polymer portfolio

semi-crystalline polymers which have a clearly defined T_m. T_g is also affected by the molecular weight of the polymer and by chain cross-linking. T_g increases with increasing molecular weight and cross-linking.

Thermogravimetric analysis (TGA) is another analytical technique used to determine the thermal stability of polymers. TGA can clearly indicate the temperature at which polymer degradation or decomposition will onset (Td), a critical metric for successful downstream processing of thermoplastics. TGA can also provide an indication on the presence or absence of residual monomers.

Residual Monomers

Post-synthesis or post-extraction, biodegradable polymers often contain some degree of unreacted residual monomers. These monomers can act as impurities, impacting polymer processing and performance and thus must be tightly controlled and minimized, if not eliminated altogether via purification methods. In the case of the aliphatic polyesters, ROP of lactide is an equilibrium reaction; i.e., a plateau is eventually observed once that the polymerization has reached the ring-chain equilibrium state where the amount of remaining monomer corresponds to the equilibrium monomer concentration (Masutani and Kimura 2014). This polymerization and depolymerization equilibrium always results in some degree of unreacted residual monomers, typically in the range of 1–5 wt.%.

Acceptable limits or specifications vary between different manufacturers and differing standard bodies. ASTM have standards for surgical grade PLA and PLGA resins and specify less than 2%, with 3% being optional if accepted by the customer (2018). USP are developing monographs and are proposing limits of 1 and 2% depending on the type of polymer. Ester-terminated PLGA polymers often contain lower levels of residual monomers compared to acid-terminated counterparts due to the efficiency of their respective initiating molecules. Many pharmaceutical applications require very low levels of <0.5 or even 0.1%wt.

The content of unreacted residual monomers remaining in biodegradable polymers is best determined via gas chromatography (GC), equipped with a flame ionization detector (FID) with instrument conditions and procedures as per USP <621>. Standards are typically prepared by dissolving known amounts of the residual monomers to be determined and the internal standard in a suitable solvent, followed by addition of isopropanol. Polymer samples are dissolved containing a known amount of internal standard and the polymer is then precipitated in isopropanol and then centrifuged. An aliquot from the vials with sample preparation and standard preparation is injected in replicate onto the column, and separation of the peaks to be determined is achieved using a gradient temperature program. The residual monomer content is calculated using the formula for internal standard calibration methods.

NMR and FTIR can also be used to measure the amount of unreacted residual monomers but are considered less accurate and with higher limits of detection compared to GC methodologies. In the case of NMR, it is determined by comparing the area of signals which correspond to the monomer vs. their respective polymer signals.

Residual Solvents

Biodegradable polymer synthesis or purification is often carried out in the presence of organic solvents. Despite intensive removal techniques, solvent residues are generally not entirely eliminated and inevitably remain in the polymers produced from these methods. These residues can be toxic and inflammatory in vivo and can also impact polymer processing, and thus, like residual monomers, must be tightly monitored and controlled. Acceptable residual levels of a particular solvent are reflective of toxicity, with a maximum acceptable limit consistent with ICH Q3C. The detection limit for the chosen analytic method must be adequate to ensure compliance with the applicable ICH guideline and the determined residual(s) and applied concentration limit(s) must be reported. To minimize potential for toxic interaction of solvent combinations, a cumulative total residual solvent content is often limited to 1000 ppm regardless of whether individual solvent toxicity limits could be higher than 1000 ppm.

Residual solvent content is determined with headspace GC using a standard addition calibration method, as per USP general chapter <467>. Unspiked samples are prepared by dissolving the polymer sample in a suitable solvent in headspace vials. DMF and DMSO are most used in the case of the aliphatic polyesters. Spiked samples are prepared by dissolving the polymer sample in standard addition solution in headspace vials. The standard addition solution contains defined concentrations of the residual solvents to be determined in the test solvent. The GC device is typically equipped with an FID detector and a column suitable for separation of the residual solvents to be determined. An aliquot from the headspace of all vials with standard solution and unspiked and spiked samples is injected once onto the GC column, and separation of the peaks to be determined is achieved using an appropriate temperature gradient. The residual solvent content is calculated for each residual solvent to be determined using the calculation formula for standard addition calibration methods.

Residual Moisture/Water

Polyesters degrade via hydrolysis and should thus contain very low levels of residual moisture. Most standards and manufacturers specify limits of ≤0.5%wt. up to ≤1% wt. The water content in biodegradable polymers is commonly measured by gravimetric and Karl Fischer (KF) titration. In the case of aliphatic polyesters, it is best determined via a coulometric KF technique equipped with the oven as per USP <921> Method I, Method Ic.

Acid Content

For aliphatic polyesters, acid number can also be used as a secondary form of identification test and provides some information relating to the polymer end group. As discussed above, polyesters are typically terminated with either a carboxylic acid or ester end group. Ester-terminated polymers present a lower acid number value compared to their acid-terminated counterparts. Acid number is usually a higher value for low-molecular-weight acid-terminated polymers versus higher molecular weights comparisons. In this regard, we can say acid number is inversely proportional to polymer molecular weight. Acid number is determined via potentiometric titration as per USP <541> and measured in mgKOH/g. The polymer is dissolved in a suitable combination of solvents (e.g., chloroform/methanol/THF (80/10/10 v/v/v)) and titrated with methanolic potassium hydroxide (KOH) to a phenolphthalein endpoint.

Tin (Sn) Content

As has been discussed, tin-based catalyst stannous octoate ($Sn(Oct)_2$) is almost exclusively used by manufacturers of aliphatic polyesters who supply to the pharmaceutical marketplace. However, tin, like other metal catalyst residues, stays in the human body after polymer resorption and can have a toxicological impact when present in sufficiently large amounts. Residual tin in commercially used medical polymers is regulated (Stjerndahl et al. 2007) and general guidelines appear to set limits at $\leq$200 ppm. Moreover, the International Conference on Harmonization (ICH) allows for a risk analysis approach within the context of the product's intended application and should consider the element itself, its oxidation state (if applicable), the element's concentration within the polymer, and the final pharmaceutical product's total amount of polymer and its anticipated rate of degradation (2018). Sn content is most accurately determined via inductively coupled plasma-mass spectroscopy (ICP-MS) as per USP <730>. Samples are typically prepared by microwave digestion using nitric acid and/or hydrochloric acid. Atomic absorption spectroscopy (USP <852>) has also long been used to determine Sn content. In addition, since the polymer supplier can provide accurate information regarding the amount of added catalyst, reporting of actual added catalyst can be substituted for direct elemental testing, where approved.

Elemental Impurities

In the case of biodegradable polymers, elemental impurities can include catalysts which are of course added intentionally or environmental contaminants which may occur naturally or be introduced inadvertently (e.g., by interactions with processing equipment or packaging). The ICH provides Guidelines for Elemental Impurities to limit patient exposure to certain elements above specified limits during their treatment. USP <232> and <233> outline the limits and procedures, respectively, for elemental impurities in drug products. Polymers used for controlled-release injectables should meet the limits for parenteral products outlined in USP 232. Testing is conducted via inductively coupled plasma-optical emission spectroscopy (ICP-OES) or inductively coupled plasma-mass spectroscopy (ICP-MS).

Sulfated Ash/Residue on Ignition

Residue on ignition, also known as sulfated ash content, is a measure of the amount of residual substance not volatilized from a sample when it is ignited in a crucible in the presence of sulfuric acid. For aliphatic polyesters, testing should be conducted as per USP <281> and acceptance criteria are typically not more than 0.1 or maximum 0.2%.

Bioburden/Microbial Count (TAMC, TYMC)

Bioburden is a measure of the number of viable cell colonies (aerobic, anaerobic, and spore cells) per gram of polymer material. For aliphatic polyesters, bioburden testing is conducted to determine the total yeast and mold count (TYMC) and total aerobic microbial count (TAMC). This is performed according to USP <61>. Limits can somewhat vary from suppliers but should not exceed 2000 colony-forming units (CFU) per gram, as per USP <1111> which details the acceptance criteria for non-sterile pharmaceutical products.

Endotoxin Limit

Endotoxins are toxic components found in the cell walls of gram-negative bacteria which can cause fever (pyrogenic response) and other adverse effects when introduced into the bloodstream. Since gram-negative bacteria are ubiquitous in nature, endotoxins are likely present in most raw materials and parenteral products at some level. For aliphatic polyesters, testing for endotoxin is performed according to the USP <85> Method A (Gel clot method) (Table 2.9).

Table 2.9 Key characteristics/properties of commercially available aliphatic polyesters and their associated primary analytical methods

Characteristic/property	Primary test method
Polymer identification/structure	^{1}H-NMR spectroscopy
Copolymer molar ratio	
Glycolide block length	^{13}C-NMR spectroscopy
Average molecular weight and polydispersity	Gel permeation chromatography
Inherent viscosity	Ubbelohde capillary viscometry
Residual monomer content	Gas chromatography
Residual solvent content	
Water content	Karl Fischer titration
Melting point	Differential scanning calorimetry
Glass transition temperature	
Acid number	Potentiometric endpoint titration
Tin content	Inductively coupled plasma-mass spectroscopy
Elemental impurities	
Bioburden	USP <61>
Endotoxin	USP <85> method A (gel clot method)

2.6 Polymer Parameters and their Influence on Performance

Mechanism of Drug Release from Lactide/Glycolide Polymers

Mechanistic release should be considered holistically for the depot system rather than any single feature or component of the formulation. These drug depot systems are in a constant state of change over the long-acting profile which needs consideration as we interpret dissolution data.

Fredenberg et al. (2011) published an extremely detailed review of mechanisms of release including discussion on several features used to control the same. This work discussed four mechanisms:

- Diffusion through water-filled pores
- Diffusion through the polymer
- Osmotic pumping
- Erosion

The typical chain of events described is as follows:

- Delivery of depot resulting in water contact. Initially, drug present on depot surface is released (the initial burst release).
- Absorption of water into the depot and swelling (initial pores are usually too small to allow significant drug movement). This happens slowly at first and gradually accelerates over time.
- Water continues to act as a pore-forming agent. Over time, the number and size of pores become larger which facilitates increased diffusion (Webber et al. 1998; Mochizuki et al. 2008). Diffusion rates vary and provide the sustained-release profile.
- Hydrolysis which is the breakage of ester bonds in the polymer backbone proceeds and causes a reduction to molecular weight. At first, this reduction will usually not lead to mass loss as chain length/entanglement is sufficient to maintain structural integrity.
- At each chain breakage, additional -COOH or carboxylic acid end groups are formed which lead to increased acidity within the matrix. This also causes an autocatalysis phenomenon which can result in a faster degradation at the core of the implant vs. the surface like a tooth cavity effect (Li and McCarthy 1999).
- As molar mass reduces, the polymer becomes less hydrophobic while in tandem, more and more pores are forming, pores are merging, and water presence is rising. When molecular weights reach a sufficiently low point (approximately 1100 Daltons), fragments become more readily soluble and can themselves become water soluble and diffused (Park 1994).
- These dissolved polymer degradation by-products can drastically affect the overall depot. Firstly, they create a more acidic water profile when solubilized. Again, catalyzing the rate and process of degradation. They also lead to a reduced glass

transition temperature of the polymer (T_g) due to plasticization effect, increased osmolarity inside the matrix leading to water absorption.

- Erosion or mass loss of the polymer proceeds as those dissolved oligomers and monomers diffuse out of the matrix. This erosion process for PLGA is classified as bulk erosion.
- Once mass loss occurs, progression of physical degradation can occur quite rapidly since the structural integrity of implant is greatly reduced. In lab settings, we observe these implants crumbling under mild stresses.

All these interactions may be observed over the lifetime of a long-acting injectable. Depot changes can also be viewed through another lens. For example, the most practical being the release profile itself or changes to molar mass or changes to physical criteria. Swelling is one such criterion.

Swelling

Bode et al. (2019) published a comprehensive explanation of the hypothesized drug release and the orchestral role of swelling. Drug is initially trapped within a typically dense dry state polymeric matrix. Following interaction with aqueous fluids, drug present at or near the implant surface is released and water penetration begins. Swelling progression rate and ultimate magnitude will depend on hydrophobicity of the matrix. As small amounts of water enter, they cause a cascade of events including physical swelling, leeching of water-soluble fragments including monomers, porosity formation, chain dis-entanglement, polymer chain degradation, reduction in molecular weight, and increase in carboxylic acid end groups, among others. Those newly generated carboxylic end groups, reduced chain entanglement, and continued swelling lead to significant water content build-up within the implants. This swelled scenario facilitates improved drug mobility and due to concentration gradients, we observe drug diffusion out of the implants into the surrounding fluids. Over time, polymer molecular weights continue to decrease. Small oligomer chains generate sufficient mobility to diffuse out of the medium which is where we observe mass loss (Milosevic et al. 2018).

While the above theory is representative of general polymer swelling behavior and the relationship with release, it is also important to consider the impact from the delivery system. In the case of microparticles, surface area increases as particle size decreases which is usually correlated with faster surface release and swelling. In the case of implants, there is a much smaller overall exposed surface area. In situ depots present inversion-dependent porosity which is affected by many factors which are well detailed by Parent et al. (2013).

Polymer Storage

In the case of aliphatic polyesters such as lactide/glycolide/caprolactone copolymers, control of storage is important. These polymers are typically supplied with storage guidelines in controlled temperatures such as fridge (2–8 °C) or freezer (< − 15 °C). Handling of these materials is also critical where good practice suggests that the material would be stored in an airtight containment system, in the absence of moisture. Cold storage removal procedure recommendations are to take the polymer in its sealed packaging from cold storage, allow acclimatization to room temperature before opening, and re-seal it before restoring. Shelf life for these polymers is typically recommended in the range of 3–5 years from manufacture date when stored appropriately. Packaging systems should prevent moisture, create a gas barrier, and protect from light as good practices. Multi-layer systems are commonly used to improve protection and support GMP/ISO packaging requirements.

Polymer Features: Composition, Crystallinity, and Thermal Properties

Lactide acid (2-hydroxypropionic acid) exists as two enantiomers (L- and D-) which are identified by their effect on polarized light (Gupta et al. 2007a, b). For drug delivery applications, amorphous structures are usually used due to the programmable degradation profile and suitability versus pharmaceutical timelines. Semi-crystalline forms are more routinely used in medical devices (McMahon et al. 2018) for their load-bearing properties and longer degradation times. That said, there are exceptions, and it is extremely important to pay attention to crystallinity as well as thermal properties of these polymers. Polymerization of glycolide with a racemic mixture D,L-lactide yields amorphous poly D,L-lactide-co-glycolide (PLGA) which is by far the most commonly applied LAI delivery vehicle. For D,L-lactide/glycolide amorphous copolymers, we typically observe an increased degradation time frame when the D,L-lactide fraction is increased. As discussed, this is due to the presence of additional methyl groups in lactide repeat units which are not present in glycolide (Fig. 2.6).

Glass transition temperature and melt temperature are two basic values which play a significant role in potential processing and application of these polymers. Melt temperature presents an obvious phase change whereby materials change stage. Glass transition temperature is a fundamentally important concept for long-acting release systems and represents the point where polymer is converted from a glassy state to a rubbery state. In the glassy state, material chains have less flexibility to move, whereas in the rubbery state, chains become more flexible, and performance is drastically affected. In general, for PLGA polymers, T_g will be increased with higher molecular weights and in the case of amorphous copolymers of lactide/glycolide, the T_g's will tend to increase with increasing lactide fraction. PLGA T_g's

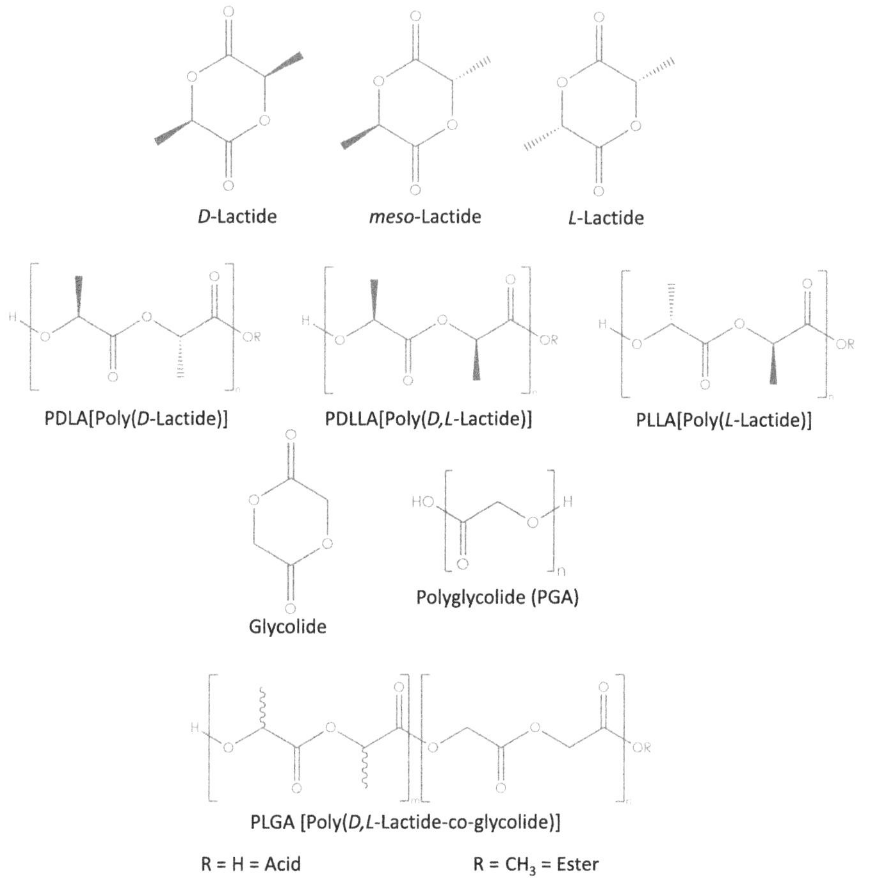

Fig. 2.6 Molecular structures and stereochemistry of the monomers lactide and glycolide, of the homopolymers polylactide and polyglycolide and the copolymer poly(D,L-lactide-co-glycolide)

range between 38and 60 °C broadly. Attention to T_g is also critical for any accelerated in vitro dissolution apparatus as release above T_g may not provide useful comparative data for expected performance.

Polymer Architecture

In terms of structure, we can broadly characterize these polymers as either linear or branched structures. In the linear cases, there are many different sub-structures including randomized, blocky, di-block, tri-block, or multi-block structures. The vast majority of PLGAs used in the market products today are linear randomized structures. Exceptions include di-block copolymers used in Genexol-PM (Lee et al. 2018) micelle design and the BEPO® (Roberge et al. 2020) in situ forming depots.

In the case of branched structures, it is possible to create low to very high degree of branching potential. An example of a commercial product using a branched structure is Sandostatin LAR which uses a glucose star-initiated PLGA polymer. Examples of more heavily branched polymers include dendritic structures. Linear and branched polymers can have very different properties in terms of processing, degradation, physical chemical properties, and sustained-release performance.

Monomer Distribution or Microstructure

During ROP, glycolide is far more reactive than lactide (Rodriguez-Galan et al. 2011), and this creates the capacity to form variation in terms of blockiness of PLGAs which was discussed in earlier sections of this chapter. In terms of degradation, segments of repeating glycolide units (G-G) are more reactive than lactide units (L-L) and, therefore, would expect an increased rate of hydrolysis if compared with an equal polymer containing lower G-G block length (Sun et al. 2022). Block length can also have a major impact on solubility of the polymer. For example, in the case of PLGA containing a large glycolide block length (PLGA 50:50), we observe reduced solvent solubility. This may not present a problem for solvent such as dichloromethane but is a major issue for use of ethyl acetate due to the weaker overall solubility profile. This will also impact that rate of solubility within a solvent where more random polymers will solubilize faster than their blocky counterparts (Fig. 2.7).

Molar Mass

Molar mass impacts on release can be considerable. Very low molecular weights may present as non-solid materials such as low-molecular-weight polycaprolactones. As molecular weight increases so too do physical-mechanical properties and

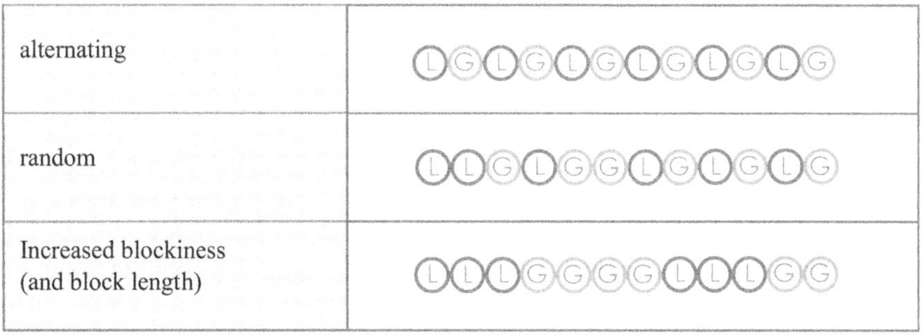

Fig. 2.7 Possible monomer distribution or microstructure in lactide glycolide polymers

chain entanglement. Lower-molecular-weight polymers show faster water update and degradation versus more hydrophobic higher-molecular-weight comparisons. Longer chains often present elevated mechanical properties due to chain entanglements and increased crystallinity for semi-crystalline or crystalline polymers. In theory, these longer chains would lead to longer lasting polymers due to increased hydrophobicity and entanglement-resistant moisture ingress and hydrolytic degradation starting from a higher baseline.

After a polymer implant or depot is delivered into the body, water penetration and uptake occur which lead to swelling. This swelling will cause reductions to mechanical strength and as it does so, we start to see breakage of polymer backbone. Typical breakage of backbone results in the formation of additional carboxylic acid end groups at the point of chain breakage which causes increased acidity. Molecular weight will start to decrease and ultimately, when the Mw has reached a critical low point, we will observe mass loss.

In real applications, teams will typically study a range of molar mass polymers in formulations. Larger Mw's typically lead to longer degradation time frame for the polymers. Practical considerations for this are also with respect to molar mass and how it affects viscosity in solvent solutions, shear stress in melt processing, and potential irradiation-induced damage to molar mass. In the case of solvent-based solutions such as Atrigel formulations, it is typical to work with relatively low molar mass polymers such that polymer weight concentrations can be increased without creating a viscosity too large to pass through associated needles. In the case of hot melt extrusion, higher molar mass polymers usually require higher processing temperatures, and due to thermal energy and shear stresses, these will often experience higher % loss to molar mass. In the case of irradiation using gamma or e-beam, higher molar mass polymers tend to experience more significant degradation from irradiation exposure versus their lower-molecular-weight counterparts (Fig. 2.8).

Polydispersity index (PDi) provides information on chain length deviation. Polymers having a broad range of variable chain lengths present a higher PDi and more standardized consistent chain lengths will have a lower PDi. A PDi of 1.0 indicates that all chains are the same length. Typical values for PDi for PLGA polymers fall in the range of 1.1–3.5 and results depend heavily on the configuration of the associated test, type of polymer, type of synthesis, and catalyst used.

End Groups

The most common two types of end groups in commercially used PLGAs are carboxylic acid or aliphatic ester with the other end of the chain terminated by a single hydroxyl group. Typical synthesis of these polymers uses either a hydroxyl-containing initiator to create acid-terminated or an aliphatic alcohol to create an ester termination. Outside of these, there are many types of end groups that can be

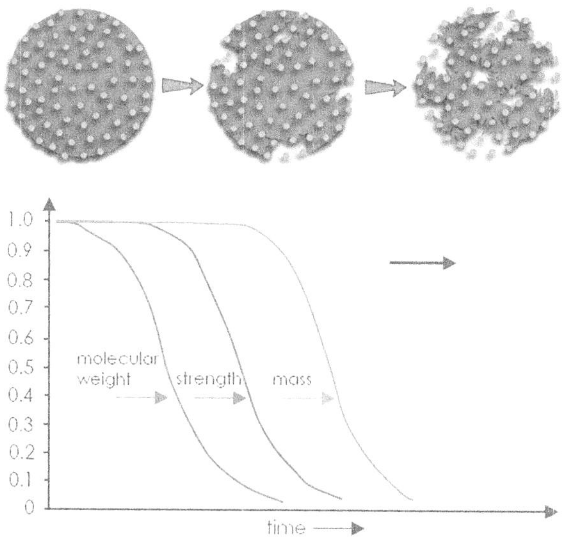

Fig. 2.8 Degradation scheme starting with molecular weight, followed by strength and mass loss

created. Some examples which exist in commercial PLGAs today include a hydroxyl-hydroxyl end group which is created using a Diol and used in the product Eligard (Sartor 2003). There are also products which contain polyethylene glycol-terminated polymers which contain a methyl ether PEG and hydroxyl termination on either side of the chain. Beyond these, there are more and more reported cases of functionalized PLGAs containing amine, maleimide, folate, or other terminations in development.

End group will cause impact on the performance of these polymers. Namely, acid-terminated PLGAs tend to be slightly more acidic than ester terminated. The rate of water ingress will also be impacted by the end group and hydrophilicity. For example, carboxylic acid will gather moisture faster than an ester termination and therefore lead to faster/greater swelling which in turn typically means faster release and degradation. In all LAI formulations, the drug molecule itself must be considered in terms of what or how it might behave within the system. For example, forces of attraction or repulsion with certain end groups may impact potential for encapsulation efficiency. So, while a conventional rule of degradation of PLGAs is true (degradation rate of ester<acid<PEG terminations), it may not deliver the maximized result in terms of target release especially if one end group tends to capture, bind, or encapsulate more of the drug. For this reason, it is important to consider multiple end group options in formulation screening (Table 2.10).

Table 2.10 Initiator and end group impact on relative hydrophobicity

Initiator	End groups	Relative effect
Monofunctional alcohol	Ester and hydroxyl	Most hydrophobic
Diol	Hydroxyl and hydroxyl	More hydrophobic
α-Hydroxyl acid	Carboxylic acid and hydroxyl	More hydrophilic
Polyethylene glycol methyl ether (mPEG)	PEG and hydroxyl	Most hydrophilic

Polymer Features and Impurities: Form, Monomer Residues, Water Content

Polymer form is not often discussed but can be an important design element to a successful development. The majority of suppliers today provide granules, flakes, foams, powders, or chunks. In the case of hot melt extrusion (HME), many teams will micronize the material before extrusion process. This can help better dry state mixing but can also contribute to a more reproducible process. Powder material has higher surface area and melts earlier in the extrusion barrel, therefore contributing to a lower overall shear but also requiring lower overall energy. In prior work, we observed reduced loss to molecular weight using powder form versus larger granule forms in HME. In solvent-based formulations, larger scale processes may, depending on the setup, prefer more dense material to avoid polymers floating to surfaces in mixing tanks. The reverse may also be true if systems are better configured to handle higher surface area powders to accelerate dissolution time frame with inclusion of baffles.

Residual monomer impacts have not been well documented in the past but can have a major impact on the performance of long-acting formulations. A comparison study of multiple PLGA suppliers (Walker et al. 2023) found that a lower residual monomer level in acid-terminated PLAs resulted in lower water uptake and slower overall degradation and erosion. Monomers are small units of acid such as lactic or glycolic acid. These contribute to increased acidity but also cause higher degree of swelling and porosity in the matrix after implantation. Their presence therefore contributes to faster drug release in the majority of cases. Note: some process techniques such as solvent-based emulsions to prepare microspheres can wash or remove the majority of monomers such that the final particles may not contain any. In other cases such as hot melt extrusion or in situ depots where the polymer and any monomers contained within the raw material are formulated into final product, significant impacts are observed. In the case of in situ depots, residual monomers in the PLGA raw materials contribute to reduced pH environment of the polymer and NMP solution. This in turn can accelerate degradation even in shelf life before the depot is deployed. The presence of the same monomers can act as accelerants to degradation of the polymer post-implantation but can also affect phase inversion (Parent et al.

2013) of the in situ depot. Water content should be minimized in these systems as they are hydrolytically degraded. Typical PLGA/PDLLA polymer specifications will range from ≤0.5 to ≤1.0% wt.

Tracking Changes to Polymer Chemistry Through Drug Development

There are various techniques used to produce long-acting injectable formulations which are addressed in other chapters of this book. In the context of lactide/glycolide/caprolactone copolymers, these processes can be broadly classified into solvent-based and melt-based formulation approaches.

A successful long-acting injectable development needs to pay close attention to the evolution of polymer characteristics from raw material through formulation process and during shelf-life studies. These stages include raw material storage, solvent- and melt-based formulation approach, terminal sterilization, and within final packaging during shelf-life studies. Close observation of these characteristics will help to explain changes in depot performance. The essential polymer criteria include paying close attention to changes in number average molecular weight (Mn), weight average molecular weight (Mw), polydispersity index, lactide/glycolide molar ratio, residual monomer content, glass transition temperature, moisture content, sequence distribution, or randomness (Table 2.11).

A typical example of this is molar mass changes during the development process, whereby molar mass is reduced by energy-intensive processes such as hot melt extrusion or downstream in gamma or electron beam irradiation. Conventional management approaches rely on minimizing the loss to molar mass to create a consistent change followed by compensating for the predicted molar mass reduction in the raw material specification. Losses to molar mass are often accompanied by reduction in T_g's.

Table 2.11 *Essential polymer characteristics to track during* development stages

Characteristic	Typical test method
Number average molecular weight (Mn), weight average molecular weight (Mw), polydispersity index	Size-exclusion chromatography (SEC)
Lactide/glycolide molar ratio	Proton nuclear magnetic resonance (H-NMR)
Residual monomer content	Gas chromatography (GC)
Glass transition temperature (T_g)	Differential scanning calorimetry (DSC)
Moisture content	Karl Fisher titration
Monomer sequence distribution or randomness	Carbon nuclear magnetic resonance (C-NMR)
pH changes (solutions)	pH meter
Acid number	Acid titration

In the case of in situ depot technologies, there are several formulations which combine PLGA or PDLLA polymers in solution state within N-methyl-2-pyrrolidone (NMP) solvent and store these systems for 2 years at fridge conditions. Since these polymers are hydrolytically degraded, minimizing moisture content within these formulations is critical to preserve molar mass and achieve an acceptable shelf life. When polymer chains break, additional carboxylic acid end groups typically form which can cause increased acidity. As chains reduce in size, oligomers and monomers form. Tracking pH of solutions can be a useful early indicator of such degradation activity.

2.7 Current Challenges and Future Perspective

Since the first bioresorbable medical device DEXON in 1971 and first long-acting drug product Decapeptyl in 1986, we have seen an explosion of commercial products taking advantage of degradable polymers across drug delivery, medical devices, dermal fillers, and tissue engineering scaffold. These successes have been enabled by the broad tunability of these biodegradable polymers such that the application and drug molecule needs can be tailored and accommodated for by intuitive programming of polymer chemistry for formulation format. That same tunability provides a template for a multitude of permeations in terms of polymer variants and this can be both a solution for innovators and a major hurdle to generics.

There are many challenges in this space which continue to hamper developments and delay products from market. The fact that these polymers are degradable allows formulation of depots that degrade over time; however, this same instability of the polymer makes it challenging to manage degradation through process and shelf life. For innovators developing LAIs, the time required to develop a formulation from concept to market authorization can be very long. This coupled with the risks of failure undermine developments. Continued improvement of in vitro dissolution release models, accelerated release methodologies, and in vitro/in vivo correlation data is needed. Furthermore, even when formulations show promise at small scale or early trials, scale-up issues are often encountered which create hurdles in creating large-scale repeatability of LAIs.

Indeed, the field of long-acting generics is famously complex (O'Brien et al. 2021) which has meant that innovator LAI systems have preserved value and sales in several instances beyond patent expiration including the well-documented Novartis' Sandostatin LAR case (Sagonowsky 2020; Skidmore et al. 2019). ANDA approval of a generic parenteral requires (21 CFR 314.94) demonstration that the PLGA or PDLLA used is qualitatively (Q1) and quantitatively (Q2) the same as that used in the reference listed drug. This challenge is notoriously difficult due to the inherent heterogeneity of PLGAs resulting from raw material input, manufacturing process, sterilization process, and product storage. In this regard, the US FDA and other regulatory agencies continue to publish guidance on and make efforts to support development of complex generics. The FDA's Competition Action Plan is such an effort. The Center for Research on Complex Generics (CRCG) aims to enhance collaboration with generic industry and further the FDA's mission of increasing access to safe effective generic products.

Standardization of polymers and testing is an important effort that is underway at USP (2024) with monograph developments as well as continued guidance on general chapter methods. These efforts help to establish more uniform criteria for polymer testing and specifications. The availability of high-quality polymers which can be quickly tuned or adjusted to meet the need is critical to successful LAI formulations. In this regard, excipient suppliers have a more prominent role in these developments and the authors would argue that experienced formulators will seek to engage closely with their suppliers of these polymers to increase probability of success.

Beyond the existing polymers that we see in the market today, there is room for new or modified chemistry solutions in this space. Lactide/glycolide copolymers have been successfully used to create long-acting small molecule and peptide formulations up to a maximum peptide size of 4187 Daltons (2024) in Bydureon's exenatide microparticle. There are still major issues formulating sensitive peptides and proteins with PLGA which is why we see no protein LAIs based on this technology. This has been a key challenge in the field for the past 30 years (Gonella et al. 2022). Issues exist with respect to the polymer and challenges with acidity build-up over time versus stability. These issues also present in manufacturing methods, solvent compatibility, thermal compatibility, and sterilization. These proteins are large macromolecules which can be affected by changes in their environment such as pH, shear, temperature, and solvent. Correlation between protein structure and function means that even slight changes to higher order structures (HOSs) can compromise efficacy of the product.

References

Abilify Maintena. Abilify maintena. accessdata.fda.gov. 2024. https://www.accessdata.fda.gov/drugsatfda_docs/label/2016/202971s008lbl.pdf. Accessed 25 Mar 2024.

Akinc A, Lynn DM, Anderson DG, Langer R. Parallel synthesis and biophysical characterization of a degradable polymer library for gene delivery. J Am Chem Soc. 2003;125:5316–23.

Allahyari M, Mohit E. Peptide/protein vaccine delivery system based on PLGA particles. Hum Vaccin Immunother. 2016;12:806–28.

Allcock HR. Chemistry and applications of polyphosphazenes. Hoboken, NJ: Wiley; 2003.

Allcock HR, Crane CA, Morrissey CT, Nelson JM, Reeves SD, Honeyman CH, Manners I. "Living" cationic polymerization of phosphoranimines as an ambient temperature route to polyphosphazenes with controlled molecular weights. Macromolecules. 1996;29:7740–7.

Amann LC, Gandal MJ, Lin R, Liang Y, Siegel SJ. In vitro-in vivo correlations of scalable PLGA-risperidone implants for the treatment of schizophrenia. Pharm Res. 2010;27:1730–7.

An B, Lin Y-S, Brodsky B. Collagen interactions: drug design and delivery. Adv Drug Deliv Rev. 2016;97:69–84, ISSN 0169-409X. https://doi.org/10.1016/j.addr.2015.11.013.

Anderson DG, Lynn DM, Langer R. Semi-automated synthesis and screening of a large library of degradable cationic polymers for gene delivery. Angew Chem Int Ed. 2003;42:3153–8.
Arestin. Arestin® (minocycline hydrochloride) microspheres, 1 MG. 2024. https://www.accessdata.fda.gov/drugsatfda_docs/label/2012/050781s019lbl.pdf. Accessed 18 Mar 2024.
ASTM International. Standard specification for amorphous poly(lactide) and poly(lactide-co-glycolide) resins for surgical implants. F2579. 2018. https://www.astm.org/f2579-18.html. Accessed 25 Mar 2024.
Athanasiou KA, Agrawal CM, Barber FA, Burkhart SS. Orthopaedic applications for PLA-PGA biodegradable polymers. Arthroscopy. 1998;14(7):726–37. https://doi.org/10.1016/s0749-8063(98)70099-4.
Azhari Z, Smith P, McMahon S, Wang W, Cameron RE. Modulating drug release from short poly(ethylene glycol) block initiated poly(L-lactide) di-block copolymers. Pharm Res. 2023;40:1697–707. https://doi.org/10.1007/s11095-022-03228-8.
Baranwal J, Barse B, Fais A, Delogu GL, Kumar A. Biopolymer: a sustainable material for food and medical applications. Polymers. 2022;14(5):983. https://doi.org/10.3390/polym14050983.
Blasi P. Poly(lactic acid)/poly (lactic-co-glycolic acid)-based microparticles: an overview. J Pharm Investig. 2019;49:337–46. https://doi.org/10.1007/s40005-019-00453-z.
Bode C, Kranz H, Fivez A, Siepmann F, Siepmann J. Often neglected: PLGA/PLA swelling orchestrates drug release: HME implants. J Control Release. 2019;306:97–107. https://doi.org/10.1016/j.jconrel.2019.05.039. Epub 2019 May 29
BRIXARDI. Brixadi prescribing information. 2024. https://braeburnrx.com/wp-content/uploads/2023/05/brixadi-prescribing-information.pdf. Accessed 25 Mar 2024.
Bugnicourt E, Cinelli P, Lazzeri A, Alvarez VA. Polyhydroxyalkanoate (PHA): review of synthesis, characteristics, processing and potential applications in packaging. Express Pol Lett. 2014;8:791–808. https://doi.org/10.3144/expresspolymlett.2014.82.
Bydureon. Bydureon® (exenatide extended release) for injectable ... 2024. https://www.accessdata.fda.gov/drugsatfda_docs/label/2018/022200s026lbl.pdf. Accessed 18 Mar 2024.
Carothers WH, Dorough GL, Natta FJ. Studies of polymerization and ring formation. X. the reversible polymerization of six-membered cyclic esters. J Am Chem Soc. 1932;54:761–72.
Chen X, Chen J, Li B, Yang X, Zeng R, Liu Y, Li T, Ho RJY, Shao J. PLGA-PEG-PLGA triblock-copolymeric micelles as oral drug delivery system: in vitro drug release and in vivo pharmacokinetics assessment. J Colloid Interface Sci. 2017;490:542–52.
Choi YH, Han HK. Nanomedicines: current status and future perspectives in aspect of drug delivery and pharmacokinetics. J Pharm Investig. 2018;48:43–60.
Conaway CC. Green chemistry, polymers of low concern, and animal testing and welfare. Toxicol Mech Methods. 2011;21(8):569–70.
Crotts G. Protein delivery from poly(lactic-co-glycolic acid) biodegradable microspheres: release kinetics and stability issues. J Microencapsul. 1998;15(6):699–713.
Dadey EJ. Modified-release drug delivery technology. 2nd ed. Taylor & Francis; 2013.
Dailey L, Jekel N, Fink L, Gessler T, Schmehl T, Wittmar M, Kissel T, Seeger W. Investigation of the proinflammatory potential of biodegradable nanoparticle drug delivery systems in the lung. Toxicol Appl Pharmacol. 2006;215:100–8.
Dash TK, Konkimalla VB. Polymeric modification and its implication in drug delivery: poly-ε-caprolactone (PCL) as a model polymer. Mol Pharm. 2012;9(9):2365–79. https://doi.org/10.1021/mp3001952.
Dechy-Cabaret O, Martin-Vaca B, Bourissou D. Controlled ring-opening polymerization of lactide and glycolide. Chem Rev. 2004;104(12):6147–76.
Deming TJ. Amino acid derived nickelacycles: intermediates in nickel-mediated polypeptide synthesis. J Am Chem Soc. 1998;120(17):4240–1. https://doi.org/10.1021/ja980313i.
Díaz-Montes E. Dextran: sources, structures, and properties. Polysaccharides. 2021;2:554–65. https://doi.org/10.3390/polysaccharides2030033.
Duncan R. Designing polymer conjugates as lysosomotropic nanomedicines. Biochem Soc Trans. 2007;35:56–60.

Fensolvi. Full prescribing information. accessdata.fda.gov. 2024. https://www.accessdata.fda.gov/drugsatfda_docs/label/2020/213150s000lbl.pdf. Accessed 18 Mar 2024.
Fernandez J, Etxeberria A, Ugartemendia J, Petisco S, Sarasua JR. Effects of chain microstructures on mechanical behaviour and aging of poly(L-lactide-co-e-caprolactone) biomedical thermoplastic elastomer. J Mech Behav Biomed Mater. 2012;12:29–38.
Fernandez J, Larranaga A, Etxeberria A, Wang W, Sarasua JR. A new generation of poly(lactide/ecaprolactone) polymeric biomaterials for application in the medical field. J Biomed Mater Res A. 2014;102(A):3573–84.
Foresee Pharmaceuticals. 2.1 recommended dosage 2.2 preparation and administration. 2024. https://www.accessdata.fda.gov/drugsatfda_docs/label/2021/211488s000lbl.pdf. Accessed 18 Mar 2024.
Frazza EJ, Schmitt EE. A new absorbable suture. J Biomed Mater Res. 1971;5(2):43–58. https://doi.org/10.1002/jbm.820050207.
Fredenberg S, Wahlgren M, Reslow M, Axelsson A. The mechanisms of drug release in poly (lactic-co-glycolic acid)-based drug delivery systems—a review. Int J Pharm. 2011;415(1–2):34–52. ISSN 0378-5173. https://doi.org/10.1016/j.ijpharm.2011.05.049.
Fuessl A, Yamamoto M, Schneller A. 5.03 - Opportunities in bio-based building blocks for polycondensates and vinyl polymers. In: Matyjaszewski K, Möller M, editors. Polymer science: a comprehensive reference. Elsevier; 2012. p. 49–70., ISBN 9780080878621. https://doi.org/10.1016/B978-0-444-53349-4.00132-1.
Gentile P, et al. An overview of poly(lactic-co-glycolic) acid (PLGA)-based biomaterials for bone tissue engineering. Int J Mol Sci. 2014;15(3):3640–59.
George A, Shah PA, Shrivastav PS. Natural biodegradable polymers-based nano-formulations for drug delivery: a review. Int J Pharm. 2019;561:244–64., ISSN 0378-5173. https://doi.org/10.1016/j.ijpharm.2019.03.011.
Goepferich A. Biodegradable polymers: polyanhydride. In: Mathiowitz E, editor. Encyclopedia of controlled drug delivery. New York: Wiley; 1999. p. 61–71.
Gonella A, Grizot S, Liu F, Noriega AL, Richard J. Long-acting injectable formulation technologies: challenges and opportunities for the delivery of fragile molecules. Expert Opin Drug Deliv. 2022;19(8):927–44. https://doi.org/10.1080/17425247.2022.2105318.
Gopferich A. Polymer bulk erosion. Macromolecules. 1997;30(9):2598–604.
Graf A, McDowell A, Rades T. Poly(alkylcyanoacrylate) nanoparticles for enhanced delivery of therapeutics – is there real potential? Expert Opin Drug Deliv. 2009;6:371–87.
Green JJ, Zugates GT, Langer R, Anderson DG. Poly (beta-amino esters): procedures for synthesis and gene delivery. Methods Mol Biol. 2009;480:53–63.
Gunatillake P, Adhikari A. Biodegradable synthetic polymers for tissue engineering. Eur Cells Mater. 2003;5(20):1–16.
Gupta B, Revagade N, Hilborn J. Poly(lactic acid) fiber: an overview. Prog Polym Sci. 2007a;32(4):455–82.
Gupta B, Revagade N, Hilborn J. Poly(lactic acid) fiber: an overview. Prog Polym Sci. 2007b;34:455–82.
Hadar J, Skidmore S, Garner J, Park H, Park K, Wang Y, Qin B, Jiang X. Characterization of branched poly(lactide-co-glycolide) polymers used in injectable, long-acting formulations. J Control Release. 2019;304 https://doi.org/10.1016/j.jconrel.2019.04.039.
Heron Therapeutics. Heron Therapeutics Announces U.S. FDA Approval of SUSTOL® (granisetron) Extended-release injection for the prevention of chemotherapy-induced nausea and vomiting. 2016. https://herontherapeutics.gcs-web.com/node/7956/pdf. Accessed 06 Mar 2024.
Heron Therapeutics. ZYNRELEF® highlights of prescribing information. 2024. https://zynrelef.com/prescribing-information.pdf. Accessed 06 Mar 2024.
Holland SJ, Tighe BJ, Gould PL. Polymers for biodegradable medical devices. 1. The potential of polyesters as controlled macromolecular release systems. J Control Release. 1986;4(3):155–80.
Huang G, Huang H. Application of hyaluronic acid as carriers in drug delivery. Drug Deliv. 2018;25(1):766–72. https://doi.org/10.1080/10717544.2018.1450910.

Huang S, Huang G. The dextrans as vehicles for gene and drug delivery. Futur Med Chem. 2019;11(13):1659–67. https://doi.org/10.4155/fmc-2018-0586.
Idrees H, Zaidi SZJ, Sabir A, Khan RU, Zhang X, Hassan SU. A review of biodegradable natural polymer-based nanoparticles for drug delivery applications. Nanomaterials (Basel, Switzerland). 2020;10(10):1970. https://doi.org/10.3390/nano10101970.
Ikada Y, Tsuji H. Biodegradable polyesters for medical and ecological applications. Macromol Rapid Commun. 2000a;21(3):117–32.
Ikada Y, Tsuji H. Biodegradable polyesters for medical and ecological applications. Macromol Rapid Commun. 2000b;21(3):34–6.
Jain S, John A, George CE, Johnson RP. Tyrosine-derived polymers as potential biomaterials: synthesis strategies, properties, and applications. Biomacromolecules. 2023;24(2):531–65. https://doi.org/10.1021/acs.biomac.2c01232.
Jendrossek D. Polyhydroxyalkanoate granules are complex subcellular organelles (carbonosomes). J Bacteriol. 2009;191(10):3195–202. https://doi.org/10.1128/JB.01723-08.
Jérôme C, Lecomte P. Recent advances in the synthesis of aliphatic polyesters by ring-opening polymerization. Adv Drug Deliv Rev. 2008;60(9):1056–76. https://doi.org/10.1016/j.addr.2008.02.008. ISSN 0169-409X. https://www.sciencedirect.com/science/article/pii/S0169409X08000562
Kamaly N, Yameen B, Wu J, Farokhzad OC. Degradable controlled-release polymers and polymeric nanoparticles: mechanisms of controlling drug release. Chem Rev. 2016;116(4):2602–63. https://doi.org/10.1021/acs.chemrev.5b00346.
Kapoor DN, et al. PLGA: a unique polymer for drug delivery. Ther Deliv. 2015;6(1):41–58.
Kou S, Peters LM, Mucalo MR. Chitosan: a review of sources and preparation methods. Int J Biol Macromol. 2021;169:85–94. https://doi.org/10.1016/j.ijbiomac.2020.12.005.
Kourmentza C, Placido J, Venetsaneas N, Burniol-Figols A, Varrone C, Gavala HN, et al. Recent advances and challenges towards sustainable polyhydroxyalkanoate (PHA) production. Bioengineering (Basel). 2017;4:E55. https://doi.org/10.3390/bioengineering4020055.
Kowalski A, Duda A, Penczek S. Kinetics and mechanism of cyclic esters polymerization initiated with tin(II) octoate, 1. Polymerization of ε-caprolactone. Macromol Rapid Commun. 1998;19(11):567–72.
Kratz F, Elsadek B. Clinical impact of serum proteins on drug delivery. J Control Release. 2012;161(2):429–45., ISSN 0168-3659. https://doi.org/10.1016/j.jconrel.2011.11.028.
Kulkarni RK, Pani KC, Neuman C, Leonard F. Polylactic acid for surgical implants. Arch Surg. 1966;93(5):839–43. https://doi.org/10.1001/archsurg.1966.01330050143023.
Lan P, Zhang YP, Gao QW, Shao HL, Hu XC. Studies on the synthesis and thermal properties of copoly (L-lactic acid/glycolic acid) by direct melt polycondensation. J Appl Polym Sci. 2004;92:2163–8.
Lee SW, Kim YM, Cho CH, Kim YT, Kim SM, Hur SY, Kim JH, Kim BG, Kim SC, Ryu HS, Kang SB. An open-label, randomized, parallel, phase II trial to evaluate the efficacy and safety of a cremophor-free polymeric micelle formulation of paclitaxel as first-line treatment for ovarian cancer: a Korean Gynecologic Oncology Group Study (KGOG-3021). Cancer Res Treat. 2018;50(1):195–203. https://doi.org/10.4143/crt.2016.376. Epub 2017 Mar 21. PMID: 28324920; PMCID: PMC5784626
Li S, McCarthy S. Further investigations on the hydrolytic degradation of poly (dl-lactide). Biomaterials. 1999;20:35–44.
Li Z, Tan BH. Towards the development of polycaprolactone based amphiphilic block copolymers: molecular design, self-assembly and biomedical applications. Mater Sci Eng C. 2014;45:620–34. https://doi.org/10.1016/j.msec.2014.06.003.
Li LC, Deng J, Stephens D. Polyanhydride implant for antibiotic delivery—from the bench to the clinic. Adv Drug Deliv Rev. 2002;54(7):963–86. https://doi.org/10.1016/S0169-409X(02)00053-4. ISSN 0169-409X. https://www.sciencedirect.com/science/article/pii/S0169409X02000534

Liechty WB, Kryscio DR, Slaughter BV, Peppas NA. Polymers for drug delivery systems. Annu Rev Chem Biomol Eng. 2010;1:149–73.

Löfgren A, Albertsson A-C, Dubois P, Jérôme R. Recent advances in ring-opening polymerization of lactones and related compounds. J Macromol Sci C. 1995;35(3):379–418. https://doi.org/10.1080/15321799508014594.

Lupron Depot. LUPRON DEPOT (leuprolide acetate for depot suspension). 2024. https://www.lupron.com/. Accessed 18 Mar 2024.

Makadia HK, Siegel SJ. Poly lactic-*co*-glycolic acid (PLGA) as biodegradable controlled drug delivery carrier. Polymers. 2011;3:1377–97. https://doi.org/10.3390/polym3031377.

Marck KW, Wildevuur CH, Sederel WL, Bantjes A, Feijen J. Biodegradability and tissue reaction of random copolymers of L-leucine, L-aspartic acid, and L-aspartic acid esters. J Biomed Mater Res. 1977;11:405–22.

Martin DP, Williams SF. Medical applications of poly-4-hydroxybutyrate: a strong flexible absorbable biomaterial. Biochem Eng J. 2003;16:97–105. https://doi.org/10.1016/s1369-703x(03)00040-8.

Martin EC, May PD, McMahon WA. Amino acid polymers for biomedical applications. I. Permeability properties of L-leucine DL-methionine copolymers. J Biomed Mater Res. 1971;5:53–62.

Masutani K, Kimura Y. In: Jiménez A, Peltzer M, Ruseckaite R, editors. Poly(lactic acid) science and technology: processing, properties, additives and applications, Chapter 1. The Royal Society of Chemistry; 2014. p. 1–36.

McMahon S, Bertollo N, O'Cearbhaill E, Salber J, Pierucci L, Duffy P, Dürig T, Bi V, Wang W. Bio-resorbable polymer stents: a review of material progress and prospects. Prog Polym Sci. 2018;83:79–96. https://doi.org/10.1016/j.progpolymsci.2018.05.002.

Middleton JC, Tipton AJ. Synthetic biodegradable polymers as orthopedic devices. Biomaterials. 2000;21(23):23352346.

Milosevic M, Stojanovic D, Simic V, Milicevic B, Radisavljevic A, Uskokovic P, Kojic M. A computational model for drug release from PLGA implant. Materials (Basel). 2018;11(12):2416. https://doi.org/10.3390/ma11122416.

Mochizuki M, Hirami M. Structural effects on the biodegradation of aliphatic polyesters. Polym Adv Technol. 1997;8(4):203–9.

Mochizuki A, Niikawa T, Omura I, Yamashita S. Controlled release of argatroban from PLA film—effect of hydroxylesters as additives on enhancement of drug release. J Appl Polym Sci. 2008;108:3353–60. https://doi.org/10.1002/app.27970.

Muhamad II, Sabbagh F, Abdul KN. Polyhydroxyalkanoates: a valuable secondary metabolite produced in microorganisms and plants. In: Plant secondary metabolites, vol. 3. 1st ed. Oakville, MO: Apple Academic Press; 2017. p. 205–34.

Nair LS, Laurencin CT. Biodegradable polymers as biomaterials. Prog Polym Sci. 2007;32(8):762–98.

National Center for Biotechnology Information (2024-S). PubChem Compound Summary for, Exenatide from https://pubchem.ncbi.nlm.nih.gov/compound/Exenatide. Accessed 25 Mar 2024.

Nicolas J, Couvreur P. Synthesis of poly(alkyl cyanoacrylate)-based colloidal nanomedicines. Wiley Interdiscip Rev Nanomed Nanobiotechnol. 2009;1:111–27.

O'Brien MN, Jiang W, Wang Y, Loffredo DM. Challenges and opportunities in the development of complex generic long-acting injectable drug products. J Control Release. 2021;336:144–58., ISSN 0168-3659. https://doi.org/10.1016/j.jconrel.2021.06.017.

Okada M. Chemical syntheses of biodegradable polymers. Prog Polym Sci. 2002;27(1):87–133. https://doi.org/10.1016/S0079-6700(01)00039-9.

Parent M, Nouvel C, Koerber M, Sapin A, Maincent P, Boudier A. PLGA in situ implants formed by phase inversion: critical physicochemical parameters to modulate drug release. J Control Release. 2013;172(1):292–304. https://doi.org/10.1016/j.jconrel.2013.08.024. Epub 2013 Sep 1. PMID: 24001947

Park TG. Degradation of poly(d,l-lactic acid) microspheres: effect of molecular weight. J Control Release. 1994;30:161–73.
Park JH, Ye M, Park K. Biodegradable polymers for microencapsulation of drugs. Molecules. 2005;10(1):146–61. https://doi.org/10.3390/10010146. PMID: 18007283; PMCID: PMC6147704
Pillai G. Nanomedicines for cancer therapy: an update of FDA approved and those under various stages of development. 2014.
Pitt CG, Andrady AL, Bao YT, Sarnuei NKP. Estimation of the rate of drug diffusion in polymers. In: Lee PI, Good WR, editors. Controlled-release technology, pharmaceutical applications. Washington, DC: American Chemical Society; 1987. p. 49–77.
Prabaharan M, Mano JF. Chitosan-based particles as controlled drug delivery systems. Drug Deliv. 2005;12:41–57. https://doi.org/10.1080/10717540590889781.
Ramchandani M, Robinson D. In vitro and in vivo release of ciprofloxacin from PLGA 50:50 implants. J Control Release. 1998;54:167–75.
Ranger GO. Microencapsulation—a brief history and introduction. In: Vandegaer JE, editor. Microencapsulation—processes and applications. New York: Plenum Press; 1974. p. 1–20.
Revalor XS. Revalor®-XS. Merck Animal Health USA. 2024. https://www.merck-animal-health--usa.com/species/cattle/products/revalor-xs. Accessed 25 Mar 2024.
Risperdal Consta. Risperdal Consta®. 2024. https://www.janssenlabels.com/package-insert/product-monograph/prescribing-information/RISPERDAL+CONSTA-pi.pdf. Accessed 18 Mar 2024.
Roberge C, Cros JM, Serindoux J, Cagnon ME, Samuel R, Vrlinic T, Berto P, Rech A, Richard J, Lopez-Noriega A. BEPO®: bioresorbable diblock mPEG-PDLLA and triblock PDLLA-PEG-PDLLA based in situ forming depots with flexible drug delivery kinetics modulation. J Control Release. 2020;319:416–27. https://doi.org/10.1016/j.jconrel.2020.01.022. Epub 2020 Jan 10. PMID: 31931049
Rodríguez-Galán A, Franco L, Puiggalí J. Biodegradable poly(ester amide)s: synthesis and applications. In: Felton GP, editor. Biodegradable polymers: processing, degradation, and applications. Nova Science; 2011.
Rodriguez-Galan A, Franco L, Puiggali J. Handbook of biodegradable polymers: isolation, synthesis, characterization and applications, vol. 26. Wiley; 2011. p. 133–54.
Sagonowsky E. Thanks to complex manufacturing, Novartis expects blockbuster Sandostatin Lar to steer clear of generics in 2020. Fierce Pharma; 2020. https://www.fiercepharma.com/pharma/thanks-to-complex-manufacturing-novartis-expects-blockbuster-sandostatin-lar-to-steer-clear#:~:text=2020%20%7C%20Fierce%20Pharma-,Thanks%20to%20complex%20manufacturing%2C%20Novartis%20expects%20blockbuster%20Sandostatin%20LAR%20to,clear%20of%20generics%20in%202020&text=First%20approved%20in%201998%2C%20Novartis,so%20the%20Swiss%20drugmaker%20figures. Accessed 25 Mar 2024.
Saltzman WM. Drug delivery: engineering principles for drug therapy. Oxford: Oxford University Press; 2001.
Samir A, Ashour FH, Hakim AAA, et al. Recent advances in biodegradable polymers for sustainable applications. npj Mater Degrad. 2022;6:68. https://doi.org/10.1038/s41529-022-00277-7.
Sandostatin LAR. Sandostatin Lar Depot. 2024. https://www.novartis.com/us en/sites/novartis_us/files/sandostatin_lar.pdf. Accessed 18 Mar 2024.
Sanjanwala D, Londhe V, Trivedi R, Bonde S, Sawarkar S, Kale V, Patravale V. Polysaccharide-based hydrogels for drug delivery and wound management: a review. Expert Opin Drug Deliv. 2022;19(12):1664–95. https://doi.org/10.1080/17425247.2022.2152791.
Sartor O. Eligard: leuprolide acetate in a novel sustained-release delivery system. Urology. 2003;61(2 Suppl 1):25–31. https://doi.org/10.1016/s0090-4295(02)02396-8. PMID: 12667884
Schmitt EE, Polistina AR. Surgical sutures. Google Patents. 1964. https://patents.google.com/patent/US3297033A/en. Accessed 25 Mar 2024.

Siegwart DJ, Whitehead KA, Nuhn L, Sahay G, Cheng H, Jiang S, Ma M, Lytton-Jean A, Vegas A, Fenton P, Levins CG, Love KT, Lee H, Cortez C, Collins SP, Li YF, Jang J, Querbes W, Zurenko C, Novobrantseva T, Langer R, Anderson DG. Combinatorial synthesis of chemically diverse core-shell nanoparticles for intracellular delivery. Proc Natl Acad Sci U S A. 2011;108:12996–3001.

Signifor® LAR. Signifor-lar-pi.pdf. 2024. https://signiforlar.com/wp-content/themes/signifor-lar-theme/dist/pdf/signifor-lar-pi.pdf. Accessed 18 Mar 2024.

Sinha VR, Bansal K, Kaushik K, Kumria R, Trehan A. Poly-ε-caprolactone microspheres and nanospheres: an overview. Int J Pharm. 2004;278(1):23.

Sinuva. Sinuva (mometasone furoate) sinus implant. 2024. https://www.accessdata.fda.gov/drugsatfda_docs/label/2017/209310lbl.pdf. Accessed 25 Mar 2024.

Skidmore S, Hadar J, Garner J, Park H, Park K, Wang Y, Jiang X. Complex sameness: separation of mixed poly(lactide-co-glycolide)s based on the lactide:glycolide ratio. J Control Release. 2019;300:174–84., ISSN 0168-3659. https://doi.org/10.1016/j.jconrel.2019.03.002.

Stjerndahl A, Finne-Wistrand A, Albertsson A-C. Industrial utilization of tin-initiated resorbable polymers: synthesis on a large scale with a low amount of initiator residue. Biomacromolecules. 2007;8:937–40. https://doi.org/10.1021/bm0611331.

Storey RF, Sherman J. Kinetics and mechanism of the stannous octoate-catalyzed bulk polymerization of ϵ-caprolactone. Macromolecules. 2002;35:1504–12.

Sun J, Walker J, Beck-Broichsitter M, et al. Characterization of commercial PLGAs by NMR spectroscopy. Drug Deliv Transl Res. 2022;12:720–9. https://doi.org/10.1007/s13346-021-01023-3.

Sze JH, Brownlie JC, Love CA. Biotechnological production of hyaluronic acid: a mini review. 3 Biotech. 2016;6(1):67. https://doi.org/10.1007/s13205-016-0379-9.

Teasdale I, Brüggemann O. Polyphosphazenes: multifunctional, biodegradable vehicles for drug and gene delivery. Polymers. 2013;5(1):161–87. https://doi.org/10.3390/polym5010161.

Tice T. Inspirational chemistry of bioabsorbable long acting injectables. Drug Deliv Transl Res. 2023;13(2):371–7. https://doi.org/10.1007/s13346-022-01199-2.

Tiera MJ, Shi Q, Winnik FM, Fernandes JC. Polycation-based gene therapy: current knowledge and new perspectives. Curr Gene Ther. 2011;11(4):288–306. https://doi.org/10.2174/156652311796150408.

Tønnesen HH, Karlsen J. Alginate in drug delivery systems. Drug Dev Ind Pharm. 2002;28(6):621–30. https://doi.org/10.1081/ddc-120003853.

Trelstar. Full prescribing information - accessdata.fda.gov. 2024. https://www.accessdata.fda.gov/drugsatfda_docs/label/2018/020715s040,021288s035,022427s015lbl.pdf. Accessed 18 Mar 2024.

U.S. Food and Drug Administration. Center for drug evaluation and research inactive ingredients search FAQ. 2024a. https://www.fda.gov/drugs/drug-approvals-and-databases/inactive-ingredients-approved-drug-products-search-frequently-asked-questions. Accessed 25 Mar 2024.

U.S. Food and Drug Administration. Inactive ingredient search for approved drug products. Inactive ingredient search for approved drug products. 2024b. https://www.accessdata.fda.gov/scripts/cder/iig/index.cfm. Accessed 25 Mar 2024.

U.S. Food and Drug Administration. Center for drug evaluation and research inactive ingredients database download. 2024c. https://www.fda.gov/drugs/drug-approvals-and-databases/inactive-ingredients-database-download. Accessed 25 Mar 2024.

U.S. Food and Drug Administration. Commissioner O of the UNIIs, preferred substance names, and their identified synonyms. U.S. Food and Drug Administration. 2024d. https://www.fda.gov/industry/structured-product-labeling-resources/uniis-preferred-substance-names-and-their-identified-synonyms. Accessed 25 Mar 2024.

Ulery BD, Nair LS, Laurencin CT. Biomedical applications of biodegradable polymers. J Polym Sci B Polym Phys. 2011;49:832–64.

US3773919A. Boswell et al. Polylactide-drug mixtures. Google Patents; 1970. https://patents.google.com/patent/US3773919A/en. Accessed 25 Mar 2024.

USP. 2020–2025 complex excipients expert committee work plan. 2023. https://www.usp.org/expert-committees/complex-excipient-expert-committee-work-plan. Accessed 25 Mar 2024.
USP. 2020–2025 complex excipients expert committee work plan. 2024. https://www.usp.org/expert-committees/complex-excipient-expert-committee-work-plan. Accessed 25 Mar 2024
UZEDY. Prescribing-information.pdf. 2024. https://www.uzedy.com/globalassets/uzedy/prescribing-information.pdf. Accessed 25 Mar 2024.
Van de Velde K, Kiekens P. Biopolymers: overview of several properties and consequences on their applications. Polym Test. 2002;21(4):433–42.
Vert M. Polymères de fermentation: Les poly (acide lactique) set leurs précurseurs, les acides lactiques. 2002.
Vicryl® synthetic absorbable sutures. VICRYL® Synthetic absorbable Sutures | Johnson & Johnson Our Story. 1974. https://ourstory.jnj.com/vicryl-synthetic-absorbable-sutures. Accessed 18 Mar 2024.
Vivitrol. Prescribing information. 2024. https://www.vivitrol.com/content/pdfs/prescribing-information.pdf. Accessed 18 Mar 2024.
Vouyiouka S, Papaspyrides C. Mechanistic aspects of solid-state polycondensation. 2012.
Vroman I, Tighzert L. Biodegradable polymers. Materials. 2009;2(2):307–44.
Walker J, Albert J, Liang D, Sun J, Schutzman R, Kumar R, White C, Beck-Broichsitter M, Schwendeman SP. In vitro degradation and erosion behavior of commercial PLGAs used for controlled drug delivery. Drug Deliv Transl Res. 2023;13(1):237–51. https://doi.org/10.1007/s13346-022-01177-8. Epub 2022 Jun 7. PMID: 35672653
Wang T, Lew J, Premkumar J, Poh CL, Win Naing M. Production of recombinant collagen: state of the art and challenges. Eng Biol. 2017;1:18–23. https://doi.org/10.1049/enb.2017.0003.
Webber WL, Lago F, Thanos C, Mathiowitz E. Characterization of soluble, salt loaded, degradable PLGA films and their release of tetracycline. J Biomed Mater Res. 1998;41(1):18–29. https://doi.org/10.1002/(sici)1097-4636(199807)41:1<18::aid-jbm3>3.0.co;2-t.
Wright JC, Burgess DJ. Long-acting injections and implants. New York: Springer; 2012.
Zhang P. USP's updates on developing NF LG polymer monograph specifications: highlights and key challenges. May 2022.
Zhang Z, et al. Biodegradable PGA/PBAT blends for 3D printing: material performance and periodic minimal surface structures. Polymers. 2021;13(21):3757.
Zhao Z, Wang J, Mao HQ, Leong KW. Polyphosphoesters in drug and gene delivery. Adv Drug Deliv Rev. 2003;55(4):483–99. https://doi.org/10.1016/s0169-409x(03)00040-1.
Zilretta. Zilretta (triamcinolone acetonide … accessdata.fda.gov. 2024. https://www.accessdata.fda.gov/drugsatfda_docs/label/2017/208845s000lbl.pdf. Accessed 18 Mar 2024.

Chapter 3
A Roadmap for Production of Drug-Loaded Poly(Lactic-Co-Glycolic Acid) Microparticles: Challenges and Opportunities

Hamed Asadi and Farshad Ramazani

List of Abbreviations

AUC	Area under curve
cGMP	Current good manufacturing practice
*C*max	Max concentration
CMAs	Critical material attributes
CPPs	Critical process parameters
CQA	Critical quality attributes
DCM	Dichloromethane
DP	Drug product
DS	Drug substance
EE	Encapsulation efficiency
FDA	Food and Drug Administration
FMEA	Failure mode and effects analysis
GLP	Good laboratory practice
ICH	International Conference on Harmonization
IJP	Inkjet printing
IND	Investigational new drug
LAI	Long-acting injectable
ME	Membrane emulsification
MPs	Microparticles/microspheres
O/W	Single emulsion solvent evaporation

H. Asadi
Research and Development Department, Nanodaru Pharmaceutical Company, Tehran, Iran

F. Ramazani (✉)
Technical Research and Development, Novartis Pharma AG, Basel, Switzerland
e-mail: farshad.ramazani@novartis.com

F. Ramazani (ed.), *Biodegradable Long Acting Injectables and Implants*,
https://doi.org/10.1007/978-3-031-72302-5_3

PAI	Pre-approval inspection
PLGA	Poly(lactide-co-glycolide)
POC	Proof of concept
PPQ	Process performance qualification
PVA	Polyvinyl alcohol
QbD	Quality by design
QTPP	Quality target product profile
SEM	Scanning electron microscopy
Tox	Toxicology
USP	United State Pharmacopeia
$W/O_1/O_2$	Water-in-oil-in-oil
$W_1/O/W_2$	Water-in-oil-in-water

3.1 Introduction

Poly(lactide-co-glycolide) (PLGA) has been the preferred polymer for long-acting injectable (LAI) formulations over the last three decades. Several microparticles (MPs) made of PLGA formulations that deliver either small molecules or peptides have been marketed (Klose et al. 2008; Lagreca et al. 2020; Nkanga et al. 2020; Park et al. 2019). Formulating protein drugs into PLGA MPs is challenging due to protein instability during the formulation process or within the PLGA acidic environment (Van De Weert et al. 2000). However, one protein drug, somatotropin, was loaded into PLGA MPs and marketed with the trade name of Nutropin Depot® by Genentech, although it discontinued due to high manufacturing costs and complexity.

Several manufacturing methods have been developed so far enabling both hydrophilic and lipophilic drug loading into PLGA MPs. The established methods of PLGA MP manufacturing include emulsification (single and double emulsion), coacervation, and spray drying. Novel methods of PLGA MPs manufacturing include, but not limited to, membrane emulsification (ME), microfluidics, and three-dimensional (3D) printing technology (Ramazani et al. 2016).

The variations in manufacturing methods (such as extraction volume, composition, speed, and temperature) can have a substantial impact on MPs properties (Park et al. 2021). Controlling these critical quality attributes is important to enable successful MPs production. Regardless of the preparation method, variations in formulation composition (including the drug loading, PLGA type and molecular weight, solvents, and their ratios) can lead to MPs with different properties. The composition and manufacturing process determine the final configuration of these MPs. Therefore, it is important to understand both aspects in more detail. In this chapter, we have summarized both established and innovative methods of PLGA MPs production. Different aspects of formulations such as drug loading, drug-release duration, polymer types, and consideration for successful scale-up are discussed. Finally, a comprehensive roadmap for PLGA MPs manufacture from the lab through various stages of clinical development until commercialization is illustrated.

3.2 Manufacture Processes of PLGA Microparticles

Microencapsulation involves creating particles made of polymers, ranging in size from 1 to 1000 µm but ideally 10–125 µm for parenteral use, in which active pharmaceutical ingredients are trapped within a polymer matrix. These particles are commonly referred to as microparticles or microspheres. If the drug is evenly distributed or dissolved within the polymer matrix, it is known as microspheres. On the other hand, if the drug is enclosed within the core of the particle and surrounded by a polymer shell, it is referred to as microcapsules. Approximately all PLGA-based marketed products are microspheres not microcapsules. In the subsequent sections, we have outlined the commonly used manufacturing methods of PLGA MPs in the pharmaceutical industry. In addition, we have summarized few procedures that have high potential for industrial manufacturing of PLGA MPs.

Established Methods of PLGA Microparticles Manufacturing

Emulsion Solvent Evaporation Methods

The emulsion solvent evaporation/extraction method is the commonly used method for the preparation of PLGA MPs (Fig. 3.1). This method can be used for both hydrophilic and lipophilic drugs (Ramazani et al. 2015). For lipophilic drugs, a

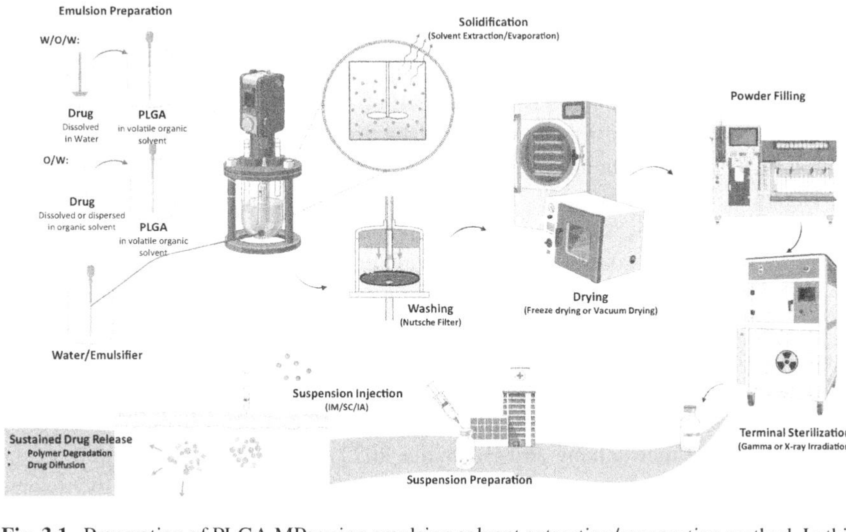

Fig. 3.1 Preparation of PLGA MPs using emulsion solvent extraction/evaporation method. In this approach, the organic phase and water phase are mixed to form an emulsion that will be further transformed to a suspension of solidified MPs by solvent extraction and evaporation. The solidified MPs are then washed, dried, filled into a vial, and terminally sterilized. The final drug product (DP) will be suspended for injection using a proper vehicle and immediately administered in the patients

single emulsion solvent evaporation (O/W) method is used. In this approach, the drug is dissolved in a volatile organic solvent like dichloromethane (DCM) along with PLGA. This organic phase is then mixed with a water phase containing an emulsifier like polyvinyl alcohol (PVA) under shear mixing. Once a stable emulsion is formed, the organic solvent is extracted from the oily droplets into the water phase and then evaporated using elevated temperature and N_2 flow or it can be extracted for example by tangential flow filtration. This process solidifies the microdroplets into MPs. The solidified MPs are collected, washed with water for injection, and vacuum dried or the suspension can be filled into vials and dried using lyophilization method. If aseptic manufacturing is not used, the MPs are filled into glass vials and terminally sterilized. For water-soluble drugs, a water-in-oil-in-water ($W_1/O/W_2$) method can be used. In this case, the drug is dissolved in the W_1 phase and entrapped in the organic phase (PLGA and organic solvent). This inner emulsion is then emulsified with a W_2 phase containing water and an emulsifier to form $W_1/O/W_2$. The rest of the manufacturing process is similar to the O/W emulsion method. However, emulsification of water-soluble drugs in PLGA MPs can be challenging due to low encapsulation efficiency (EE) caused by drug flux in the external water phase during emulsification and hardening of MPs (Park et al. 2019).

There are several ways to improve encapsulation efficiency (EE) of drug in the MPs and minimize the drug waste into water phase. For example, by increasing polymer concentration in organic phase, decreasing the ratio of organic to W_2 phase, addition of salt to W_2 phase, or saturating W_2 phase with drug or organic solvent etc. However, those changes may impact particle size distribution and drug release. Therefore, in design of experiment for early development the impact of those changes needs to be fully investigated (Ramazani et al. 2016).

If above changes did not result in high EE with desired particle size and drug release, an alternative method is the water-in-oil-in-oil ($W/O_1/O_2$) emulsion solvent extraction or coacervation method, where the drug is dissolved in water and mixed with a PLGA solution in an organic solvent. This primary emulsion is then emulsified again using a non-mixable organic oil containing an emulsifier (e.g., cotton oil/lecithin). The organic solvent from the O_1 phase is extracted into the O_2 phase, resulting in polymer coacervation (Ramazani et al. 2016). In cases where the drug is neither soluble in water nor in the organic solvent, the drug can be loaded into PLGA MPs using the solid-in-oil-in-water (S/O/W) method. In this method, the micronized drug is uniformly distributed in the organic phase by a proper homogenizer before emulsification. This method allows for high drug loading compared to the $W_1/O/W_2$ and O/W methods and the solid drug has less chemical interaction with PLGA polymer (Zhang and Bodmeier 2022). However, micronization of the drug is required, which has limitations such as changes in drug crystallinity of some drugs due to jet milling (a common method of micronization) and the need to control foreign visible particles and endotoxin as micronized drugs cannot be sterile filtered anymore (Nkanga et al. 2020) (Fig. 3.1).

Spray Drying

Spray drying can also be used to manufacture PLGA MPs. In this method, the drug can be dissolved or dispersed in an organic solution of PLGA. Water-soluble drugs can be dissolved in water and added to the PLGA solution before spray drying using a suitable nozzle with different diameters and inlet/outlet temperatures. Spray drying often achieves high EE as it does not involve the use of a water phase where most of the non-encapsulated drug is washed away. This method is simple and scalable for microparticle manufacturing. However, it has limitations such as the high manufacturing temperature, which may not be suitable for sensitive drugs or drugs that can interact with PLGA through acylation. In addition, in the initial phase of development, the batch size cannot be scaled down to milligram amounts, and the yield is often low in small-scale production, limiting its use in early-stage drug development (Wan and Yang 2016).

Spinning Disk

The spinning disk method is used for manufacturing triamcinolone acetonide MPs (Zilretta®) for intra-articular injection in osteoarthritis. It involves an annular spinning disk with a flat inclined surface connected to a stirrer motor. The disk has a reservoir in the middle, where the active ingredient is either dispersed as micronized drug crystals or dissolved in an organic solvent containing a biodegradable polymer. Centrifugal force is applied to move the suspension from the reservoir toward the peripheral part of the disk, resulting in increased viscosity due to solvent evaporation. A temperature control system is used to control the evaporation of the solvent from both the reservoir and the disk. Once the solvent is evaporated, the drug-loaded MPs are solidified and can be collected for further processing the bulk (Duvnjak et al. 2023). The bulk MPs are then filled into vial and terminally sterilized. However, if the MPs are degrading by terminal sterilization, aseptic manufacturing needs to be applied.

Novel Microparticles Manufacturing Methods

Controlling the size and distribution of particles is crucial for tailoring the release kinetics of an encapsulated drug as well as injectability and syringeability of the suspension. The size distribution of particles also plays a significant role in determining the fate of the particles in vivo by influencing the foreign body reaction. Smaller MPs (5 μm) are phagocytosed, whereas larger MPs (30 μm) are not. Therefore, achieving control over the average particle size and size distribution can lead to improved batch-to-batch consistency and optimized therapeutic effectiveness. Recent advancements in microsphere manufacturing technology now allow for the creation of monodisperse MPs with precisely controlled particle size distributions. In the following sections, we describe several novel technologies that have potential to be used for large-scale clinical manufacturing PLGA MPs.

Membrane Emulsification

One innovative approach for fabricating monodisperse MPs is membrane emulsification (Kazazi-Hyseni et al. 2015). The underlying principle of microencapsulation using membrane emulsification (ME) is similar to other emulsification methods described previously. In ME, small droplets of the organic phase are formed at the openings of a membrane's pores and are detached from the membrane surface due to the shear stress of the water (continuous) phase. Depending on the specific process, MPs can be formed using either the O/W (oil-in-water) or the $W_1/O/W_2$ methods. In the O/W membrane emulsification approach, a continuous phase (containing water and an emulsifier) flows over the membrane, through which the dispersed phase (a drug dissolved in a polymer solution) is pushed. This process leads to the formation of emulsified droplets with uniform sizes. In the double $W_1/O/W_2$ membrane emulsification method, a primary emulsion (W_1/O) is created by mixing an aqueous drug solution into an organic solution (e.g., PLGA in DCM) using a homogenizer. The W_1/O emulsion is then processed over the membrane to generate monodisperse droplets (Falke et al. 2015; Kazazi-Hyseni et al. 2015; Ramazani et al. 2016). The following steps of solvent evaporation, washing, drying, fill and finish can be performed as mentioned earlier.

The most used membrane for producing monodisperse MPs is Shirasu porous glass (SPG). The surface chemistry of the membranes can be modified to adjust the hydrophilicity/hydrophobicity balance, allowing compatibility with different emulsions (O/W, $W_1/O/W_2$, S/O/W, and $S/O_1/O_2$). The membrane emulsification scale-up is possible, e.g., by increasing the size of membranes, or using multiple devices in parallel. Figure 3.2 shows that the MPs generated by O/W membrane emulsification are narrower in size as compared to conventional O/W emulsifications.

Microfluidics

Another method for manufacturing monodispersed PLGA MPs is microfluidics (Rezvantalab and Keshavarz Moraveji 2019; Xu et al. 2017). As shown in (Fig. 3.3) these microfluidic devices can be categorized into co-flow capillary devices, flow-focusing capillary devices, and a combination of both (Ramazani et al. 2016). Co-flow capillary devices contain three channels; the water phase is introduced into the side channels, while the organic phase containing the drug and polymer is directed into the central channel of the device using syringe pumps with constant and controlled flow rates. At the junction points of the combined microfluidic channels, monodisperse emulsion droplets are continuously formed.

In flow-focusing capillary devices, the organic phase containing the drug and polymer, along with the continuous water phase, are introduced from opposite

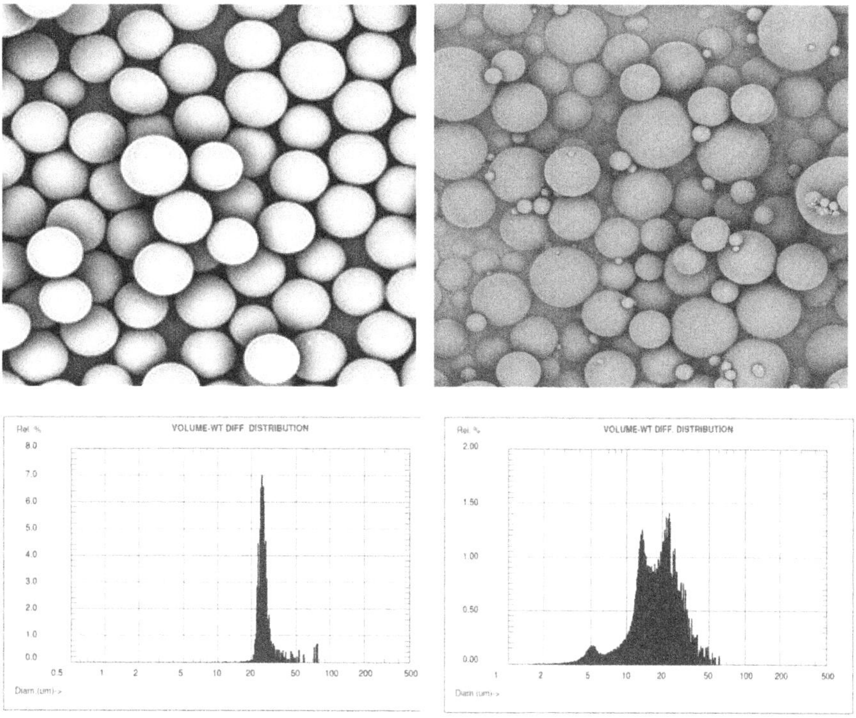

Fig. 3.2 Scanning electron microscopy (SEM) image of PLGA MPs. Left image represents mono-sized PLGA MPs prepared by membrane emulsification (Nanomi) and its particle size distribution, and right image represents PLGA MPs prepared by conventional emulsification method and its particle size distribution. Images were taken from nonpublished data of Farshad Ramazani PhD thesis (Ramazani 2015)

directions into the microfluidic mixing cell. The external aqueous fluid hydrodynamically focuses the internal organic phase through the orifice, causing it to break up and generate monodisperse emulsion droplets.

Microfluidic flow focusing techniques have also been used to prepare double emulsion-based MPs, combining co-flow and flow-focusing apparatus. In this approach, the inner water phase, containing the drug dissolved in water, is pushed through a narrowing circular tube to deliver small water droplets into the organic phase processed in a flow-focusing device. This results in the formation of water-in-oil-in-water ($W_1/O/W_2$) emulsion droplets (Fig. 3.3).

Before scaling up, it is recommended to optimize the microfluidic parameters at the laboratory scale to ensure reproducibility and efficiency. This includes factors such as flow rates, polymer concentration, solvent ratios, and device geometry. The scale-up is possible, e.g., by increasing the number of parallel channels, or using multiple devices in parallel. Recently COVID-19 vaccines manufactured by Moderna were produced by a microfluidics device on a large scale.

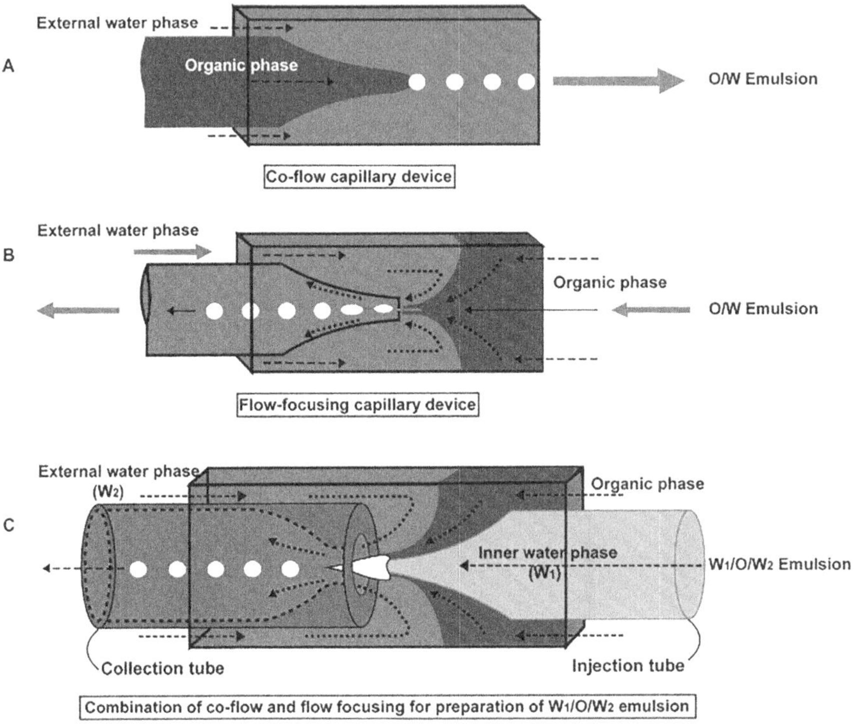

Fig. 3.3 Microfluidic devices for manufacturing PLGA MPs. (**a**) Co-flow capillary, (**b**) flow-focusing and (**c**) combination of flow focusing and co-flow for preparing PLGA MPs with double emulsion method. With permission from Ramazani et al. (2016)

Printing

Another method for manufacturing MPs with desired shapes and sizes is inkjet printing (IJP) (Lee et al. 2012). IJP can be classified into continuous and drop-on-demand approaches. In the continuous approach, a liquid ink (drug and polymer solution dissolved in an organic solvent) is released through an orifice with a defined diameter, creating a constant stream of ink using a high-pressure pump. The ink stream breaks down into micro-droplets through piezoelectric crystal or heating methods. In drop-on-demand IJP, ink droplets are ejected only when required, while in continuous mode printers, droplets are propelled continuously. Ink viscosity and surface tension play a crucial role in the geometry and size of the final MPs (Ramazani et al. 2016).

This method has the potential for scale-up through an increase in droplet ejection frequency or the utilization of multiple parallel piezo nozzle arrays. Consequently, it can achieve the delivery of millions of consistent MPs per second. This process has been applied to produce MPs at a 250 g pilot scale for the clinical trial of a distinctive formulation of octreotide acetate (Palmer et al. 2017).

3.3 Critical Formulation Challenging Regarding PLGA Microparticles Development

Drug Molecule Properties

Drug molecule properties such lipophilicity, potency, and stability are important criteria for MPs development. These properties define if a molecule is suitable to be loaded in PLGA MPs or not. In the following sections, the role of lipophilicity, potency, and stability is discussed.

Lipophilicity

Drug hydrophilicity/lipophilicity not only affects the release profile but also determines the MP preparation method. Most of marketed products are highly lipophilic (low water soluble) that can be encapsulated into PLGA MPs using oil-in-water solvent evaporation method with high efficiency. Lipophilic drugs stay in PLGA droplets, but hydrophilic or amphiphilic drugs diffuse partially into water phase during emulsification and prior to hardening of MPs (Ramazani et al. 2016).

In contrast to lipophilic drugs, hydrophilic drugs are better to be encapsulate using the double emulsion technique (Lagreca et al. 2020). However, a fraction of the starting material will be lost due to partitioning into external water phase using any kinds of water-based emulsification method. Oil-based emulsifications known as coacervation or spray drying method may improve encapsulation of water-soluble drugs.

There are plenty of opportunities to tailor small drug molecule physicochemical properties. For example, in the early stage of drug design, drug lipophilicity could be manipulated by introducing the aromatic ring or coupling fatty acid chain into the drug structure while keeping therapeutic efficacy. For further information about drug substance (DS) properties, refer to Chap. 1 of this book.

Potency

One of the important criteria of a DS that makes it an obvious candidate for the development of LAI is drug potency. It is especially important for PLGA microparticle development as most of marketed PLGA MPs have drug loading <30% meaning 70% of the DP composition is PLGA polymer and 30% drug substance (Nkanga et al. 2020). It worth mentioning that this material must be injected as suspension using a proper vehicle. Therefore, the final drug content of the formulation is even less than 30% considering the volume of the vehicle for injection. As the injection volume is limited depending on the tissue and local tolerability, drug potency can help to minimize the volume of injection.

The ideal drug candidates need to be administered chronically for long periods in which the designing weekly or monthly formulation brings advantages for administration. Up to now, a few number of PLGA MP formulations of potent drugs, mostly peptides, have been approved by the Food and Drug Administration (FDA), which indicates that potency is one of the bottlenecks in developing PLGA MPs (Gonella et al. 2022).

Physicochemical Stability and Drug Polymer Interaction

The drug stability during encapsulation, manufacturing processes, storage, and subsequent release is another critical parameter that must be analyzed. Drug stability will be more important in the case of using intrinsically fragile therapeutic agents such as peptides and proteins (Gonella et al. 2022). The preparation methods involve harsh conditions that may impact the stability of the encapsulated drug, or it may degrade the polymer. For example, temperature for solvent evaporation must be minimized to diminish the risk of degradation for temperature sensitive drugs or drug polymer interaction. The impact of terminal sterilization using different irradiation methods on both polymer and encapsulated drug must be investigated (Duvnjak et al. 2023; Nkanga et al. 2020). PLGA-drug interaction has a great role in controlling release characteristics. The drug acidity–alkalinity could induce degradation by the catalytic effect on hydrolysis and accelerate the release rate. Thus, drug-induced polymer degradation must be evaluated during MP development. It seems logical that basic drugs can catalyze the hydrolysis of ester bonds and accelerate polymer degradation and drug release. However, this group of drugs by shielding the terminal carboxyl residues can slow down polymer degradation and water penetration into the matrix. This phenomenon becomes more pronounced as the quantity of incorporated drug into the MPs increases (Hua et al. 2021).

Since several FDA-approved PLGA-based MPs are loaded with peptide drugs, the chemical peptide-PLGA interaction is an important parameter that must be considered for the successful development of these kinds of formulations. The primary amine groups of peptides and carboxylic acid end groups of PLGA could interact to produce acylated products causing potentially loss of peptide activity, immunogenicity, and toxicity (Hua et al. 2021). To address the problem, many studies have focused on inhibiting the acylation of polypeptide drugs in PLGA by using divalent cations in formulation or neutralizing the inner pH of PLGA microsphere. The peptide-PLGA reactions, mechanisms, and how to stabilize peptide drugs into the PLGA MPs have been elucidated (Houchin and Topp 2008).

Quality Attributes of PLGA Microparticles

Drug Loading Capacity

The drug loading and loading capacity is one of the important critical quality attributes of MPs that can affect MP performance. Generally, the drug loading capacity of the PLGA MP is limited and could be a challenge in development of drugs with relatively low potency and high daily doses. Drug loading capacity could influence the clinical applicability of PLGA-based MPs because of limited injection volume and consequently not reaching the therapeutic dose (Nkanga et al. 2020). To use the maximum loading capacity of the PLGA MPs and optimize encapsulation efficiency, the physicochemical properties of the incorporated drug must be considered in addition to PLGA polymer materials and preparation methods (Su et al. 2021). In general, the encapsulation of hydrophilic drugs into the PLGA MP is more difficult than lipophilic drugs. To address this, the manipulation of the physicochemical properties of a drug such as molecular size, hydrophilicity, and charge is often required. For example, some drugs that are hydrophilic in salt forms can be converted to corresponding free acid, free base, or other salt forms that are more hydrophobic, consequently leading to improved loading (Han et al. 2016). Hence, to design MPs with high drug loading, the effects of encapsulated drug properties and the drug loading capacity of PLGA MPs should be taken into consideration. Drug crystals that are lipophilic and not soluble in common organic solvent for MP preparation, such as dichloromethane and ethyl acetate, are ideal cases for reaching high drug loading. They can be micronized and mixed with PLGA solution for emulsion preparation (Ramazani et al. 2016).

Particles Size Distribution

In the design and development of the polymeric MPs, the particle size and size distribution are other critical quality attributes that must be controlled during manufacturing as they directly correlate to the safety and efficacy of the product. The particle size of MPs has a critical role in drug release kinetic, degradation rate, phagocytosis, and administration. Studies have shown that heterogeneous PLGA MPs with various sizes that have been produced with conventional emulsification methods often show batch-to-batch variation. This nonuniformity in size causes inconsistent drug release that affects drug product's efficacy (Lim et al. 2022). Moreover, there is a linear relationship between the particle size and degradation rate (Hua et al. 2021).

Regardless of the administration route, MPs are cleared from the body by phagocytosis, and it is the body's inherent defense mode against invading pathogens and other nonindigenous particulate matter. Hence, several studies have been conducted to investigate the role of particle size in phagocytosis (Champion et al. 2008). While certain applications such as subcutaneous depot injections necessitate reduced phagocytosis for sustained release, others, such as vaccination, require enhanced phagocytosis. Since most marketed PLGA MPs are aimed for sustained drug release, they are formulated in size of 10–125 micrometers to reduce macrophage phagocytosis and local tissue inflammation (Nkanga et al. 2020). Finally, a broad particle size distribution could cause needle clogging during administration. Therefore, to facilitate administration and improve injectability, control over particle size distribution is a critical aspect. Eventually, each PLGA-based MP should specifically be tailored to fit certain particle size characteristics based on desired purposes and applications.

Drug Release from PLGA Microparticles

Control and customization of drug release from PLGA MPs can be achieved by manipulating several factors, including the size of the MPs, drug loading percentage (drug/polymer ratio), surface porosity, and the composition and properties of the polymer, such as molecular weight, end group, and concentration in the organic solvent (Park et al. 2021). In the following section, we further discuss the role of microstructure of MPs in controlling drug release in addition to above-mentioned physicochemical properties. The microstructure of MPs includes the drug distribution and dispersion forms, PLGA density, and pore size and distribution (Garner et al. 2022). It has been demonstrated that the microstructure of MPs and consequently the drug release from MPs is influenced by formulation composition (drug, PLGA type, and solvent) and preparation processes (emulsification, solidification, and drying) (Wang et al. 2024). Drug release could be controlled by manipulating through multiple steps in the preparation process, which alter the microstructure of MPs, such as homogenization speed during emulsion formation, choosing suitable solvents or co-solvents, selecting appropriate physical forms of drug, and accelerating the hardening rate (Cai et al. 2009; Al-Maaieh and Flanagan 2001). The sum of those parameters (i.e., formulation composition and preparation method) results in MPs with molecularly dispersed or physical mixture of drug crystal with polymer or combination of both. The MPs with molecularly dissolved drug are releasing their payload mainly by diffusion through polymer chains or surface pores. While MPs contain physical mixture of drug often needs polymer degradation to release their payload. In any case, drug release from PLGA MPs is a combination of drug diffusion and polymer degradation. Recent studies pointed that micronized drug crystal loaded PLGA MPs have longer duration of drug release as compared to nanosized and or dissolved drug loaded PLGA MPs (Zhang and Bodmeier 2022).

Drug Distribution and Dispersion within PLGA Matrix

The location of drug molecules within the PLGA microsphere has an important effect on release rate. If the loaded drug were near the microsphere surface, it would be released faster, while internal drugs need more time to pass through PLGA mass transfer channels and release slower than surface-located drugs. As mentioned above, the size and dispersion forms of the drug have a noticeable effect on drug release. The release rate of big-sized or aggregated dispersed drugs is slower than the small-sized loaded drugs due to creating a small surface area inside the PLGA microsphere after release (Xie and Taylor 2016).

Pore Size and Distribution

The pore size and distribution have the main role in the formation of mass transfer channels inside the PLGA microsphere. The pore size of the microsphere is an important factor influencing the water uptake capacity of MPs, which consequently impact drug release and polymer degradation. Various process parameters such as polymer molecular weight, organic to water phase ratio, water phase composition, organic solvent type, and evaporation temperature have influence on MPs surface or matrix structure and porosity. It is known that the temperature of solvent extraction has a significant impact on the pore formation (Vay et al. 2010). The MPs pore size ranges from few nanometers to few micrometers. Although the pores and microsphere porosity are significantly related to the burst release, the pore distribution and their connectivity correspond to the transportation of the encapsulated drug even in the later release phase.

PLGA Density

The PLGA density is another key factor impacting microsphere microstructure. The PLGA density is related to the entanglement of chain segment that defines the difficulty of diffusion and formation of mass transfer channels inside MPs. The rate of drug diffusion through the PLGA is related to the polymer chain mobility, which is mostly affected by formulation composition and manufacturing process. The polymer chain mobility is inversely related to the controlled release capability of PLGA MPs on drugs. The parameters such as PLGA molecular weight, glass transition temperature, terminal group, and block ratio influence polymer chain mobility and density (Hua et al. 2021; Park et al. 2019; Wan et al. 2021). The successful development of a PLGA MP product with desired release kinetic depends on controlling the factors contributing to the microstructure of PLGA MPs.

Residual Solvent

The residual solvent is the other quality attribute of PLGA MPs that should be taken into consideration during the development of DP. Residual solvents are volatile chemicals used during the manufacturing of PLGA MPs and should be removed at the end of the manufacturing process due to the possible harmful effects on patients. However, it may not completely be removed during manufacturing and hence, specific allowable limits are defined in the general chapters of United States Pharmacopeia (USP). Solvents or cosolvents can be classified as Class 1, 2, or 3 depending on their risk to human health. The list of generally used solvents besides their classified class and allowable limits are provided in Table 3.1. Among solvents, dichloromethane (DCM) has been extensively used because of its fast removal from the system owing to its low boiling point (39.8 °C).

The elimination of residual solvents from the formulation presents one of the complexities of the manufacturing process of PLGA MPs. If traces of organic solvent, beyond the acceptable limit, are present in the product, the whole batch fails or may need to be reprocessed to eliminate the residual solvent. Reprocessing of the batch is an additional cost and increases the risk of open handling and contaminations; hence, the solvent should be removed during the manufacturing. Moreover,

Table 3.1 Solvents used during the manufacturing of MPs

Solvents	Solvent class	Limit (PPM)
Dichloromethane	Class 2	600
Chloroform	Class 2	60
Methanol	Class 2	3000
Ethyl acetate	Class 3	5000
Ethanol	Class 3	5000
Acetone	Class 3	5000

the solvent can influence the other quality attributes of MPs such as stability of the encapsulated drug and release profiles (Kamali et al. 2020). Therefore, the manufacturing process for PLGA MPs must be designed in the way to effectively eliminate the organic solvents after MP formation.

Sterility

As requirement for parenteral formulations, PLGA MPs need to be prepared by either aseptic procedures or terminal sterilization. Terminal sterilization is favored by health authorities to ensure the final DP sterility (Duvnjak et al. 2023). Common methods for terminal sterilization include autoclaving, dry heat, ethylene oxide gas, and irradiation, e.g., x-ray, electron beam, and γ-irradiation (Nkanga et al. 2020). However, dry heat and steam sterilization are conducted at high temperatures (e.g., 121 °C), which can melt PLGA MPs as the glass transition temperature of PLGA polymers is often being below 50 °C. Ethylene oxide can leave toxic residues, making it an unsuitable option for terminal sterilization of DPs. Consequently, γ-irradiation, x-ray, and e-beam irradiation are preferred for sterilizing PLGA MPs due to their high efficiency and low thermal impact. γ-irradiation is the most used method that can effectively sterilizes PLGA MPs. X-ray irradiation is equally efficient, with the added benefits of shorter processing times and potentially less damage to the products. Electron beam sterilization, with its low penetration effect, may not be as effective as other irradiation method but can induce less damage to sensitive DS and it hold promise for the future of DP sterilization (Bai et al. 2020). The impact of the standard irradiation dose (i.e., 25 kGy) on PLGA molecular weight reduction and subsequent drug release is generally minimal but needs to be investigated for each project and it is recommended to use higher molecular weight (Mw) polymer to compensate for the Mw loss during manufacturing and terminal sterilization. Some drug molecules may degrade during irradiation; hence, it is crucial to evaluate the effect of irradiation type and dose on the formulation during the early stages of development. In cases where irradiation, with even lower dose, is not an option, aseptic manufacturing needs to be pursued.

Syringeability and Injectability

Efficient injection of PLGA MPs formulations through conventional needles (e.g., 19-30G) is essential for its clinical success. Key factors affecting syringeability and injectability include particle size, shape, density, viscosity, and suspension concentration in addition to needle size shape and geometry. Large PLGA MPs or aggregates in the formulation often lead to needle clogging, which can also result from the bridging effect of small and large MPs in the needle due to high polydispersity (Nkanga et al. 2020). As mentioned above, innovative technologies such as membrane emulsification and microfluidics facilitate the production of uniform MPs that are more easily injectable through smaller needles compared to polydisperse particles although many pharmaceutical companies prefer to use manufacturing methods that produce polydisperse MPs as they are hard to copy by generic companies.

The vehicle selection and its composition have huge impact on injectability. As shown in Table 3.2, vehicle often contains wetting agent, viscosity and osmolarity enhancer, and buffering agent to control the pH in physiological range.

A good vehicle must keep the MPs in suspension to be able to inject them within reasonable time frame. If the MPs sediment inside the needle, they clog the needle. Although larger needle size can solve the problem, it may cause local tissue damage and pain, thereby reducing patient compliance. Most of marketed product are supplied using a vehicle in prefilled syringe plus a vial contain the MPs. Bydureon Bcise©—MPs loaded with Exenatide and supplied with an autoinjector—contains both MPs and the organic-based vehicle for suspension preparation. Upon shaking for few second both products are mixed to form a homogenous single-dose suspension that can be immediately injected into subcutaneous space (Yadav et al. 2024).

3.4 Process Development and Scale-Up Challenges

The complexity and poor reproducibility in large-scale manufacturing are the major obstacles in the development of PLGA MPs. Although the scale-up is not considered a quality attribute of MPs, it can be an obstacle during DP development. For example, Nutropin Depot® developed by Genentech was approved by the FDA in 1999 but was discontinued in 2004 due to the challenges faced during its production and not the feasibility of commercialization. According to the commonly used MPs manufacturing method, the requirement of great resources such as large quantities of quenching liquid, large manufacturing vessels, larger areas, and the manufacturing cost increment make the scale-up of MPs a tedious task for commercialization. As mentioned in Sect. 2, there are three conventional commercial PLGA MPs manufacturing techniques: (1) emulsion-solvent extraction/evaporation, (2) phase separation (coacervation), and (3) spray drying. These methods provided successful production of DPs at a large scale and with adequate control of particle properties as confirmed by approved MPs in the market. However, the batch-to-batch variation and lack of reproducibility are common disadvantages of the emulsion-solvent

Table 3.2 Marketed PLGA MPs: their vehicle composition and the amount of vehicle needed to make a microparticle suspension, needle size, and the route of administration

Drug name	Brand name	MPs injected (mg)	Vehicle injected (mL)	Vehicle composition	Needle size/route of administration
Leuprolide acetate	Lupron Depot®	215.0	1.5	Mannitol (39.7 mg) Stearic acid (10.1 mg) Sodium carboxymethyl cellules (7.5 mg) Polysorbate 80 (1.5 mg) Water for injection	22G/IM
Triptorelin pamoate	Trelstar LA	129.2	2.0	Polysorbate 80 (1.8 mg) Sodium carboxymethyl cellules (27.0 mg) Mannitol (76.0 mg) Water for injection (2.0 mL)	21G/IM
Naltrexone	Vivitrol®	1127.0	3.4	Sodium carboxymethyl cellules (30.0 mg) Polysorbate 20 (1.0 mg) Sodium chloride (9.0 mg) Sodium hydroxide and hydrochloric acid for pH adjustment (pH 5.8–7.2) Water for injection (1.0 mL)	20G 1.5-inch length/IM
Octreotide acetate	Sandostatin® LAR	598.0	2.5	Solvent (prefilled syringe): Sodium carboxymethyl cellules (12.5 mg) Mannitol (15.0 mg) Water for injections (2.5 mL)	19G/IM
Pasireotide pamoate	Signifor® LAR	240.0	1.0	Mannitol (90.0 mg) Sodium carboxymethyl cellules (14.0 mg) Poloxamer 188 (4.0 mg) Water for injections (2.0 mL)	20G 1.5-inch length/IM

(continued)

Table 3.2 (continued)

Drug name	Brand name	MPs injected (mg)	Vehicle injected (mL)	Vehicle composition	Needle size/route of administration
Risperidone	Risperdal Consta®	192.0	2.0	Sodium carboxymethyl cellules (45.0 mg) Sodium chloride (12.0 mg) Water for injections (2.0 mL)	21G 1-inch safety null/safety syringes (for deltoid injection) 20G 2-inch safety null/safety syringes (for gluteal injection)
Exenatide	Bydureon® BCISE	2.0	0.85	Autoinjector contain oil-based vehicle of medium chain triglycerides (MCT) (774.4 mg per dose) Sucrose (0.8 mg per dose)	23G 0.3-inch length/SC
Triamcinolone acetonide	Zilretta	128.0	5.0	Saline 0.9% w/w (5.0 mL) Sodium carboxymethyl cellules (0.5% w/w) Poloxamer 188 (0.1% w/w)	21G 1.5-inch length/ intraarticular injection
Minocycline Hydrochloride	Arestin®	1.0 mg, unit-dose cartridge	NA	Prefilled syringes for implanting in the root of teeth. Packaged in a unit-dose cartridge with a deformable tip.	NA

evaporation method during scale-up. These batch-to-batch variations comes from the alteration of production conditions and inadequate control of mixing, heat, and/or mass transport during the preparation process (Lim et al. 2022). To address this, the control of the mixing process such as shearing rate and stirring rate are necessary; otherwise, they will produce PLGA particles with varying quality attributes (Zheng et al. 2020). Spray drying is a single-step, scalable, and fast process with low manufacturing cost that has made it a great option for MP manufacturing. However, it needs high temperatures during the particle formation process, which limits the applications to heat-stable drugs (Wan et al. 2023). Furthermore, the successful scale-up using the spray-drying method depends on various processing parameters such as degree of atomization, feed solution/material properties and rate, inlet air temperature, and flow rate. Improper control over process variables during scale-up would influence particle size, morphology, EE, residual solvent, and overall yield of product (Bhujel et al. 2021).

Emerging novel MP preparation methods, such as membrane emulsification and microfluidics, have covered the limitations and challenges encountered by conventional methods during scale-up. Although these methods can consistently reproduce PLGA particles with a low polydispersity, they suffer from low production rates, which limit their application for the scale-up production of drug-loaded PLGA MPs (Keohane et al. 2014). Despite advances in PLGA MPs manufacturing methods, there is still no "one-size-fits-all" manufacturing process and each of the manufacturing processes faces different challenges during scale-up.

3.5 Establishing the Roadmap for Development of Drug Product

The objective of PLGA MPs development is to create a high-quality product and a manufacturing process that consistently delivers the right DP to meet the needs of the patients. Addressing challenges associated with formulation and DP development can be achieved through well-structured studies outlined in the roadmap, as delineated in Table 3.3. This roadmap is crafted based on a quality-by-design (QbD) approach, drawing from experience in developing several PLGA MP DPs at both clinical and commercialization stages and the regulatory guidance provided by worldwide regulatory agencies. In addition, recent insights into manufacturing and DP development review are included (Sampathkumar and Kerwin 2023).

Early Phase Development

The initial phase of developing a PLGA MPs DP involves several tasks or gates aimed at preparing and submitting an investigational new drug (IND) application to regulatory authorities for approval to commence phase I clinical trials. These tasks include establishing a manufacturing process for the DS, formulating the DP, conducting analytical development, performing pre-clinical toxicology and safety studies in animals, adhering to current good manufacturing practice (cGMP) standards for manufacturing the DS and product, documenting the technical studies, and submitting the IND application for regulatory review and approval to initiate clinical trials. The DP development roadmap, which focuses on a subset of these activities, is shown in Table 3.3.

Table 3.3 Roadmap for PLGA microparticles development

Preclinical
• Quality target product profile (QTPP) with all necessary information
• Formulation screening (drug loading, particle size, duration of drug release, etc.)
• Stability upon terminal sterilization
• Pharmacokinetics/Pharmacodynamics (PK/PD) study and selection of top formulation
• Animal toxicology studies (non-good laboratory practice/toxicology (GLP-Tox)) to get initial idea about the safety of the compound and local tolerability
• Technical stability study on non-GLP-Tox batch to get an idea about the product stability both DS and DP
• Initial process development and scale up
• Animal toxicology studies (GLP-Tox)
Phase I
• Process development to understand the critical process parameters
• Development of vehicle and matching placebo MPs
• Stability study in representative containers using GLP-Tox/technical batch
• Further scale up if needed using the same manufacturing equipment
• Confirm final formulation for clinical trials
• Container/closure selection and compatibility for drug substance (DS) and drug product (DP)
• DS manufacturing and excipients availability
• Cleaning validation of manufacturing equipment
• Extractable and leachable studies if not covered by risk assessment
• Microbiological and physicochemical hold time especially for sterile manufacture
• Terminal sterilization validation
• If aseptic manufacturing, filter validation (bacterial retention and material adsorption to filter) and aseptic process validation (APV) study
• In-use stability and compatibility studies
• Instruction for use and pharmacy manual preparation
• Defining in process controls
• GMP batch manufacturing, release, and stability study
• Submission to regulatory agencies for clinical trial approval
Phase II
• Refine QTPP if needed
• Develop further analytical methods if needed
• Define vehicle and microparticle characteristics for phase III and commercial use
• Develop commercial container/closure system or prefilled syringe if applicable
• Stability studies on Good Manufacturing Practice (GMP) samples
• Formulation robustness studies
• Further process development and scale up production
• Update in-use stability to support phase III
• Device compatibility and definition of combination product pathway if applicable
• Resupply batch manufacture if needed

(continued)

Table 3.3 (continued)

Phase III
• Refine QTPP if needed (for commercialization)
• Establish pivotal study and commercial manufacturing site
• Refine in-process control and Process analytical technology (PAT) tools
• DP technical run including mixing homogeneity study using placebo/active
• Process risk assessment (hazards or failure mode and effects analysis (FMEA) approach)
• Process characterization studies
• Extractable and leachable risk assessment and studies
• Transportation studies for product impact
• Resupply batch manufacture if needed
Commercial
• Finalize critical quality attributes (CQAs)
• Ensure alignment of DP characteristics with commercialization target
• Manufacturing site risk assessment (i.e., failure mode and effects analysis (FMEA)) for commercial phase
• Risk mitigation studies
• Validation plan finalization
• Finalize control strategy for DP process
• Commercial manufacturing site readiness
• Supply chain and raw material risk assessment and risk mitigation
• Facilities and equipment qualification
• DP process validation cGMP studies
• DP process performance qualification (PPQ) runs at manufacturing site
• DP Stability—Three batches with at least one batch in secondary packaging
• Continuous process verification
• PPQ batch stability monitoring
• Transport validation using actual shipping lanes
• Microbial challenge studies
• Preparation of pre-approval inspection

Quality-by-Design Approach

As mentioned, the roadmap establishment of DP should be based on the QbD approach, and according to the International Conference on Harmonization (ICH) quality guidelines Q8 (R2), the QbD is "a systematic approach for development that begins with predefined objectives and emphasizes product and process understanding and process control, based on sound science and quality risk management" (Yu et al. 2014). The QbD approach offers a perfect opportunity for the pharmaceutical companies to ensure the quality of pharmaceutical products. The QbD approach in the drug development process starts by defining the quality target product profile (QTPP), subsequently identifying the critical quality attributes (CQAs), and finally, understanding the effect of the critical material attributes (CMAs) and critical process parameters (CPPs) linking to CQA in a systematic way. The QTPP is a prospective summary of DP characteristics that are defined to elaborate various aspects of the safety and quality of the developed DP. After defining QTPP, CQA is the second step that must be identified. Based on ICH Q8, the CQAs represent the physical, chemical, biological, and microbiological properties or characteristics that lie within an appropriate limit, range, or distribution to ensure the desired product quality. Although there can be many quality attributes required to be considered during DP development, the CQAs are those that directly affect patient safety or efficacy. The typical examples of QTTP with CQAs for the parenteral polymeric MPs are provided in Table 3.4.

Table 3.4 A typical example of QTPP along with CQA for parenteral microparticle formulations based on the QbD approach

QTPP elements	CQAs	Target	Justification
Dosage form	No	Injectable microparticles	Must be injected (e.g., IM) after reconstitution
Rout of administration	No	Parenteral (e.g., IM)	Parenteral to have rapid action, reduce dosing frequency, maintain plasma concentration, and improve bioavailability
Dosage strength	No	Based on label claim	Depends on pharmacokinetic of encapsulated drug and therapeutic concentration
Morphology/ appearance	Yes	Free-flowing dry powder without agglomeration and surface porosity, without foreign visible particles	To make a homogeneous suspension for injection without significant burst release. Control foreign visible particles based on pharmacopoeia
Particle size and distribution	Yes	Prefer to be in the size range 10–125 μm, narrow size distribution (PDI < 0.7)	To reduce macrophage phagocytosis and local tissue inflammation Narrow PDI for better syringeability or injectability
Encapsulation efficiency and drug loading	Yes	>70% (w/w) for encapsulation efficiency >25% (w/w) for drug loading	Economic feasibility Flexibility in dosing Influence on the efficacy. Influence on the clinical applicability (limited injection volume and not reaching the therapeutic dose)
Identification	No	Chromatogram	Shows safety of drug product
Assay/potency	Yes	As per ICH guidelines	Assay/potency of the product is critical factor and ensure it is within the acceptable limits throughout stability period
Impurity	Yes	As per ICH guidelines	To confirm that the total impurities will not change significantly as it may affect the safety and efficacy of the product
Residual solvent	Yes	Acceptable limit of solvent residue, USP <467>	To ensure safety and prevent risk to human health Influence on encapsulated drug stability and release profiles

(continued)

Table 3.4 (continued)

QTPP elements	CQAs	Target	Justification
Drug release	Yes	Sustained/controlled release	To ensure safety and efficacy of drug and to reduces dosing frequency
Water content	Yes	<1% w/w	Can affect drug stability
Content uniformity	Yes	Acceptable limit of uniformity of dosage units, USP <905>	Variability in content uniformity affects safety and efficacy
Injectability	Yes	Narrow size distribution Desired composition of diluent	Lead to needle clogging Keep the MPs in suspension to be able to inject them within reasonable time frame
Sterility	Yes	Must be sterile	Not compliant with sterility limit directly impact patient safety
Bacterial endotoxin	Yes	Meets relevant quality standard based on pharmacopeia	To ensure safety of the product
Stability/shelf life	Yes	As per ICH guidelines	To ensure safety and efficacy
In vivo PK study	No	Max concentration (*C*max), area under curve (AUC)	To ensure safe, effective, and prolonged release of encapsulated drug
Container closure system	Yes	Compatible with product No leakage Keep sterility	To keep product quality throughout shelf life
Yield (%)	No	>90%	To confirm economic feasibility

It is very important to analyze and control all the CQAs of PLGA MPs during drug development with respect to its costs and patient compliance; however, among them, the drug-release patterns and impurities (chemical degradation of encapsulated drug) are considered high-risk CQAs. In general, the CQAs of the DPs are influenced by the variation of the material attributes and process-related parameters. The impact of CPPs on CQAs calls for monitoring and controlling them to ensure that the desired quality can be achieved. In addition to CPPs, the physical or chemical characteristics and the quality attributes of raw materials (CMAs) are other key elements that need to be controlled to ensure achieving the desired quality. For example, Table 3.5 summarizes the CPPs and CMAs that influence a key CQA of PLGA MP, i.e., drug release with respect to methods exploited for the development of PLGA MPs.

Late Phase Development

Most of the process development activities for PLGA microparticles are performed in preclinical and phase I. This is because any changes in the process can have a huge impact on the quality of DP and may call for equivalent study to ensure that

Table 3.5 The CPPs and CMAs of various preparation methods that influence drug release duration: the main CQA of PLGA MPs

Preparation method	CQA	CPPs on drug release	CMAs on the drug release
Emulsion solvent evaporation	Drug release	• Organic/water ratio • Surfactant concentration in water phase • pH and salt concentration of water phase • Polymer concentration • Emulsification power • Mixing rate during solvent evaporation • Temperature and pressure (solidification rate)	• Polymer type (molecular weight, end group, L:G ratio) • Organic solvent types • Drug water solubility • Presence of additives
Spray drying	Drug release	• Polymer concentration • Inlet/outlet temperature • Feed rate • Atomization energy input • Drying flow rate • Nozzle type	• Polymer type (molecular weight, end group, L:G ratio) • Drug water solubility • Drug temperature stability
Membrane emulsification	Drug release	• Transmembrane pressure • Flow velocity • Polymer concentration • Mixing rate during solvent evaporation • Temperature and pressure (solidification rate)	• Polymer type (molecular weight, end group, L:G ratio) • Organic solvent types • Drug water solubility • Presence of additives
Microfluidic	Drug release	• Total flow rates • Flow rate ratio • Polymer concentration • Temperature and pressure (solidification rate) • Mixing rate while solvent evaporation	• Polymer type (molecular weight, end group, L:G ratio) • Organic solvent types • Drug water solubility • Presence of additives

the quality of those MPs is the same. To avoid any further in vivo or clinical study, it is recommended to define the MP factors such as drug loading, particle size, polymer type, sterilization, and in general all the critical process parameter at the beginning and do not change them significantly through the course of development especially after proof of concept (POC) study.

Other parameters that need to be further looked at in the late phase toward commercialization include process performance qualification (PPQ) runs to demonstrate that the process can consistently produce a quality product by incorporating as much variation as expected during production operations. Some variations such as different PLGA polymer batch, speed, and temperature of MP formation are important to be evaluated. Therefore, a formulation is considered "robust" if all CQAs of the DS and DP remain within their respective end-of-shelf-life acceptance criteria throughout the entire claimed formulation composition.

Other critical steps, in mid- to late-stage development, are defining commercial container/closure system, prefilled syringe, or applicator for administration of PLGA MPs and performing all the necessary stability and compatibility test with the new setting. For commercialization, pre-approval inspection (PAI), where FDA inspects the manufacturing facility to ensure compliance with cGMP, is needed. For FDA inspectors, it is important to ensure that methods, facilities, and controls used in the entire process of PLGA MPs manufacturing are sufficient to ensure and maintain its identity, strength, quality, and purity. It is always important to align with clinical team, understand their needs, and plan in advance for any resupply manufacture if needed in each stage of development.

3.6 Conclusion

PLGA MPs can extend the release of several different drug types from small molecules to peptides with different water solubility. Few manufacturing methods are already established in pharmaceutical companies to produce PLGA MPs for clinical applications. Nevertheless, the development of PLGA MPs is difficult and involves intensive process development and scale-up. Crafting well-structured roadmap based on quality-by-design approach can help scientists to better manage such complex projects. It starts with predefined objectives and emphasizes product and process understanding and process control, based on rational science and quality risk management approaches. Most of the process development activities for PLGA MP manufacture are performed during the early phase development. Therefore, it is recommended to define the MPs' characteristics and all the critical process parameters at the early phase and not change them significantly throughout the course of development to avoid any potential equivalent study.

References

Al-Maaieh A, Flanagan DR. Salt and cosolvent effects on ionic drug loading into microspheres using an O/W method. J Control Release. 2001;70(1–2):169–81. https://doi.org/10.1016/S0168-3659(00)00347-3.

Bai Y, Moeinzadeh S, Kim S, Park Y, Lui E, Tan H, Zhao W, Zhou X, Yang YP. Development of PLGA-PEG-COOH and gelatin-based microparticles dual delivery system and E-beam sterilization effects for controlled release of BMP-2 and IGF-1. Part Part Syst Charact. 2020;37(10):2000180. https://doi.org/10.1002/ppsc.202000180.

Bhujel R, Maharjan R, Kim NA, Jeong SH. Practical quality attributes of polymeric microparticles with current understanding and future perspectives. J Drug Deliv Sci Technol. 2021;64:102608. https://doi.org/10.1016/j.jddst.2021.102608.

Cai C, Mao S, Germershaus O, Schaper A, Rytting E, Chen D, Kissel T. Influence of morphology and drug distribution on the release process of FITC-dextran-loaded microspheres prepared with different types of PLGA. J Microencapsul. 2009;26(4):334–45. https://doi.org/10.1080/02652040802354707.

Champion JA, Walker A, Mitragotri S. Role of particle size in phagocytosis of polymeric microspheres. Pharm Res. 2008;25(8):1815–21. https://doi.org/10.1007/s11095-008-9562-y.

Duvnjak M, Villois A, Ramazani F. Biodegradable long-acting injectables: platform technology and industrial challenges. Handb Exp Pharmacol. 2023;284:133–50. https://doi.org/10.1007/164_2023_651.

Falke LL, van Vuuren SH, Kazazi-Hyseni F, Ramazani F, Nguyen TQ, Veldhuis GJ, Maarseveen EM, Zandstra J, Zuidema J, Duque LF, Steendam R, Popa ER, Kok RJ, Goldschmeding R. Local therapeutic efficacy with reduced systemic side effects by rapamycin-loaded subcapsular microspheres. Biomaterials. 2015;42 https://doi.org/10.1016/j.biomaterials.2014.11.042.

Garner J, Skidmore S, Hadar J, Park H, Park K, Qin B, Wang Y. Surface analysis of sequential semi-solvent vapor impact (SAVI) for studying microstructural arrangements of poly(lactide-co-glycolide) microparticles. J Control Release. 2022;350 https://doi.org/10.1016/j.jconrel.2022.08.052.

Gonella A, Grizot S, Liu F, López Noriega A, Richard J. Long-acting injectable formulation technologies: challenges and opportunities for the delivery of fragile molecules. Expert Opin Drug Deliv. 2022;19(8):927–44. https://doi.org/10.1080/17425247.2022.2105318.

Han FY, Thurecht KJ, Whittaker AK, Smith MT. Bioerodable PLGA-based microparticles for producing sustained-release drug formulations and strategies for improving drug loading. Front Pharmacol. 2016;7:185. https://doi.org/10.3389/fphar.2016.00185.

Houchin ML, Topp EM. Chemical degradation of peptides and proteins in PLGA: a review of reactions and mechanisms. J Pharm Sci. 2008;97(7) https://doi.org/10.1002/jps.21176.

Hua Y, Su Y, Zhang H, Liu N, Wang Z, Gao X, Gao J, Zheng A. Poly(lactic-co-glycolic acid) microsphere production based on quality by design: a review. Drug Deliv. 2021;28(1):1342–55. https://doi.org/10.1080/10717544.2021.1943056.

Kamali H, Atamanesh M, Kaffash E, Mohammadpour F, Khodaverdi E, Hadizadeh F. Elimination of residual solvent from PLGA microspheres containing risperidone using supercritical carbon dioxide. J Drug Deliv Sci Technol. 2020;57 https://doi.org/10.1016/j.jddst.2020.101702.

Kazazi-Hyseni F, Van Vuuren SH, Van Der Giezen DM, Pieters EH, Ramazani F, Rodriguez S, Veldhuis GJ, Goldschmeding R, Van Nostrum CF, Hennink WE, Kok RJ. Release and pharmacokinetics of near-infrared labeled albumin from monodisperse poly(D,L-lactic-co-hydroxymethyl glycolic acid) microspheres after subcapsular renal injection. Acta Biomater. 2015;22:141–54. https://doi.org/10.1016/j.actbio.2015.04.030.

Keohane K, Brennan D, Galvin P, Griffin BT. Silicon microfluidic flow focusing devices for the production of size-controlled PLGA based drug loaded microparticles. Int J Pharm. 2014;467(1–2) https://doi.org/10.1016/j.ijpharm.2014.03.051.

Klose D, Siepmann F, Elkharraz K, Siepmann J. PLGA-based drug delivery systems: importance of the type of drug and device geometry. Int J Pharm. 2008;354(1–2):95–103. https://doi.org/10.1016/J.IJPHARM.2007.10.030.

Lagreca E, Onesto V, Di Natale C, La Manna S, Netti PA, Vecchione R. Recent advances in the formulation of PLGA microparticles for controlled drug delivery. Prog Biomater. 2020;9(4):153–74. https://doi.org/10.1007/s40204-020-00139-y.

Lee BK, Yun YH, Choi JS, Choi YC, Kim JD, Cho YW. Fabrication of drug-loaded polymer microparticles with arbitrary geometries using a piezoelectric inkjet printing system. Int J Pharm. 2012;427(2) https://doi.org/10.1016/j.ijpharm.2012.02.011.

Lim YW, Tan WS, Ho KL, Mariatulqabtiah AR, Kasim NHA, Abd Rahman N, Wong TW, Chee CF. Challenges and complications of poly(lactic-co-glycolic acid)-based long-acting drug product development. Pharmaceutics. 2022;14(3) https://doi.org/10.3390/pharmaceutics14030614.

Nkanga CI, Fisch A, Rad-Malekshahi M, Romic MD, Kittel B, Ullrich T, Wang J, Krause RWM, Adler S, Lammers T, Hennink WE, Ramazani F. Clinically established biodegradable long acting injectables: an industry perspective. Adv Drug Deliv Rev. 2020;167:19–46. https://doi.org/10.1016/j.addr.2020.11.008.

Palmer D, Bamsey K, Groves R, Patil P, Jones H, McAleer L, Seaman P. Printing particles: a high-throughput technique for the production of uniform, bioresorbable polymer microparticles and encapsulation of therapeutic peptides. Chem Eng Sci. 2017;166 https://doi.org/10.1016/j.ces.2017.03.038.

Park K, Skidmore S, Hadar J, Garner J, Park H, Otte A, Soh BK, Yoon G, Yu D, Yun Y, Lee BK, Jiang X, Wang Y. Injectable, long-acting PLGA formulations: analyzing PLGA and understanding microparticle formation. J Control Release. 2019;304:125–34. https://doi.org/10.1016/j.jconrel.2019.05.003.

Park K, Otte A, Sharifi F, Garner J, Skidmore S, Park H, Jhon YK, Qin B, Wang Y. Formulation composition, manufacturing process, and characterization of poly(lactide-co-glycolide) microparticles. J Control Release. 2021;329:1150–61. https://doi.org/10.1016/j.jconrel.2020.10.044.

Ramazani F. Polymeric particles for sustained and local drug delivery. Utrecht Institute for Pharmaceutical Sciences; 2015. https://www.researchgate.net/profile/Farshad-Ramazani/publication/290542201_Polymeric_particles_for_sustained_and_local_drug_delivery/links/5699dcdf08aea14769437ff1/Polymeric-particles-for-sustained-and-local-drug-delivery.pdf

Ramazani F, Chen W, Van Nostrum CF, Storm G, Kiessling F, Lammers T, Hennink WE, Kok RJ. Formulation and characterization of microspheres loaded with imatinib for sustained delivery. Int J Pharm. 2015;482(1–2) https://doi.org/10.1016/j.ijpharm.2015.01.043.

Ramazani F, Chen W, Van Nostrum CF, Storm G, Kiessling F, Lammers T, Hennink WE, Kok RJ. Strategies for encapsulation of small hydrophilic and amphiphilic drugs in PLGA microspheres: state-of-the-art and challenges. Int J Pharm. 2016;499(1–2):358–67. https://doi.org/10.1016/j.ijpharm.2016.01.020.

Rezvantalab S, Keshavarz Moraveji M. Microfluidic assisted synthesis of PLGA drug delivery systems. RSC Adv. 2019;9(4):2055–72. https://doi.org/10.1039/C8RA08972H.

Sampathkumar K, Kerwin BA. Roadmap for drug product development and manufacturing of biologics. J Pharm Sci. 2023; https://doi.org/10.1016/j.xphs.2023.11.004.

Su Y, Zhang B, Sun R, Liu W, Zhu Q, Zhang X, Wang R, Chen C. PLGA-based biodegradable microspheres in drug delivery: recent advances in research and application. Drug Deliv. 2021;28(1):1397–418. https://doi.org/10.1080/10717544.2021.1938756.

Van De Weert M, Hennink WE, Jiskoot W. Protein instability in poly(lactic-co-glycolic acid) microparticles. Pharm Res. 2000;17(10):1159–67. https://doi.org/10.1023/A:1026498209874.

Vay K, Scheler S, Frieß W. New insights into the pore structure of poly(d,l-lactide-co-glycolide) microspheres. Int J Pharm. 2010;402(1–2):20–6. https://doi.org/10.1016/j.ijpharm.2010.09.014.

Wan F, Yang M. Design of PLGA-based depot delivery systems for biopharmaceuticals prepared by spray drying. Int J Pharm. 2016;498(1–2):82–95. https://doi.org/10.1016/j.ijpharm.2015.12.025.

Wan B, Andhariya JV, Bao Q, Wang Y, Zou Y, Burgess DJ. Effect of polymer source on in vitro drug release from PLGA microspheres. Int J Pharm. 2021;607 https://doi.org/10.1016/j.ijpharm.2021.120907.

Wan B, Bao Q, Burgess D. Long-acting PLGA microspheres: advances in excipient and product analysis toward improved product understanding. Adv Drug Deliv Rev. 2023;198:114857. https://doi.org/10.1016/j.addr.2023.114857.

Wang M, Wang S, Zhang C, Ma M, Yan B, Hu X, Shao T, Piao Y, Jin L, Gao J. Microstructure formation and characterization of long-acting injectable microspheres: the gateway to fully controlled drug release pattern. Int J Nanomed. 2024;19:1571–95. https://doi.org/10.2147/IJN.S445269.

Xie T, Taylor LS. Dissolution performance of high drug loading celecoxib amorphous solid dispersions formulated with polymer combinations. Pharm Res. 2016;33(3) https://doi.org/10.1007/s11095-015-1823-y.

Xu J, Zhang S, MacHado A, Lecommandoux S, Sandre O, Gu F, Colin A. Controllable microfluidic production of drug-loaded PLGA nanoparticles using partially water-miscible mixed solvent microdroplets as a precursor. Sci Rep. 2017;7(1) https://doi.org/10.1038/s41598-017-05184-5.

Yadav P, Singh Y, Chauhan D, Yadav PK, Kedar AS, Tiwari AK, Shah AA, Gayen JR, Chourasia MK. Development and approval of novel injectables: enhancing therapeutic innovations. Expert Opin Drug Deliv. 2024;21(4) https://doi.org/10.1080/17425247.2024.2351987.

Yu LX, Amidon G, Khan MA, Hoag SW, Polli J, Raju GK, Woodcock J. Understanding pharmaceutical quality by design. AAPS J. 2014;16(4) https://doi.org/10.1208/s12248-014-9598-3.

Zhang C, Bodmeier R. A comparative study of PLGA microparticle properties loaded with micronized, nanosized or dissolved drug. Int J Pharm. 2022;628:122313. https://doi.org/10.1016/j.ijpharm.2022.122313.

Zheng Y, Sheng F, Wang Z, Yang G, Li C, Wang H, Song Z. Shear speed-regulated properties of long-acting docetaxel control release poly (lactic-co-glycolic acid) microspheres. Front Pharmacol. 2020;11:1286. https://doi.org/10.3389/fphar.2020.01286.

Chapter 4
Biodegradable In Situ Forming and Preformed Implants

Mazda Rad-Malekshahi and Zahra Kheyri

List of Abbreviations

2P	2-pyrrolidone
12-HOA	12-Hydroxyoctadecanoic acid
AOT	Bis-(2-ethylhexyl) sodium sulfosuccinate
APF DDM	ARBURG plastic free-forming droplet deposition modeling
BA	Benzyl alcohol
BB	Benzyl benzoate
BMP	Bone morphogenetic protein
CAB	Cholesterol 4-(2-anthryloxy)butanoate
D-DBS	1,3:2,4-di-O-benzylidene-D-sorbitol
DDOA	2,3-bis-n-Decyloxyanthracene
DMSO	Dimethyl sulfoxide
EMA	European Medicines Agency
EOP	Elementary osmotic pump
FDA	Food and Drug Administration
FDM	Fused deposition modeling
HFMB	Hydrogel-forming microneedle bases
HME	Hot melt extrusion
HPC	Hydroxypropylcellulose
IDDS	Implantable drug delivery systems
IMD	Implantable medical device
ISFD	In situ forming depot
ISFIs	Injectable in situ forming implants
LAIs	Long-acting injections
LCST	Lower critical solution temperature
LNG	Levonorgestrel

M. Rad-Malekshahi (✉) · Z. Kheyri
Department of Pharmaceutical Biomaterials and Medical Biomaterials Research Center, Faculty of Pharmacy, Tehran University of Medical Sciences, Tehran, Iran
e-mail: m-radmalekshahi@sina.tums.ac.ir

F. Ramazani (ed.), *Biodegradable Long Acting Injectables and Implants*,
https://doi.org/10.1007/978-3-031-72302-5_4

MNs	Microneedles
NMP	N-methyl-2-pyrrolidone
PCL	Poly(caprolactone)
PDE	Permissible daily exposure
PEG	Polyethylene glycol
PEO	Poly(ethylene oxide)
PGA	Poly(glycolic acid)
PDLLA	Poly(dl-lactide)
PLA	Poly(lactic acid)
PLA-PCL	Poly(lactide)/(caprolactone) copolymers
PLGA	Poly(lactic-co-glycolic acid)
PNIPAAM	Poly(N-isopropyl acrylamide)
POE	Poly(ortho ester)
PPO	Poly(propylene oxide)
PPOP	Push-pull osmotic pump
SC	Stratum corneum
TA	Triacetin

4.1 Introduction

Although orally administered medication is preferred by patients, it remains a challenge for many modern pharmaceuticals such as peptides, proteins, or highly lipophilic small molecules that lack oral bioavailability. In addition, alternative delivery methods like pulmonary, nasal, oral, transdermal, ocular, and rectal routes often face issues related to enzymatic degradation or poor absorption (Fogueri and Singh 2009). Parenteral drug delivery has garnered significant attention in the realm of medical advancements. However, numerous drugs possess high potency but short half-lives, necessitating delivery methods that eliminate the need for frequent injections. These methods offer several advantages, including enhanced patient compliance, minimized plasma fluctuations, and reduced side effects through lower overall dosages. Consequently, there has been substantial research into developing new parenteral dosage forms that offer convenient administration, as well as streamlined production processes (Parent et al. 2013). Parenteral preparations are sterile medications specifically designed for administration via injection, implantation, or infusion to humans or animals (Ibrahim et al. 2021). Among these approaches, long-acting injections (LAIs) have been developed to extend effective drug concentration for longer periods. Various marketed LAI technology platforms, including implants and in situ forming depots (gels/implants), have been reviewed (Nkanga et al. 2020). In addition to solid or preformed implants, in situ forming implants have been developed as "intelligent" systems, which must then be injected into the body as a liquid that solidifies at the injection site (Parent et al. 2013). Many advances have been made in the injection of biodegradable in situ molding systems. Before injection, these systems have low viscosity. They can be easily injected into patients as they are liquid materials containing drugs dissolved/suspended in organic solvents and water-insoluble biodegradable polymers. After injection, the solvent diffuses into the subcutaneous fluid and the polymer solidifies to form a drug or semisolid reservoir at the injection site. From the patient's perspective, in situ forming implants (ISFIs) are less invasive and therefore less painful than solid implants; hence, this can improve patient compliance and comfort as it eliminates the need for invasive procedures when placing and removing implants (Kempe and Mäder 2012).

In situ forming systems can be divided into three groups based on their formation process: (1) in situ cross-linking systems; (2) in situ solidifying organogels; (3) in situ precipitation systems (Parent et al. 2013; Kempe and Mäder 2012; Packhaeuser et al. 2004; Hatefi and Amsden 2002; Agarwal and Rupenthal 2013).

Atrigel® delivery system is at the core of in situ forming implant drug products available for clinical use. These systems are thick solutions containing water-insoluble biodegradable polymers and biocompatible solvents. To produce medicinal products such as Atridox®, Eligard®, Sublocade®, and Perseris®, drugs are added to, dissolved, or suspended in Atrigel® systems. In addition, poly(DL-lactide) (PDLLA), poly(lactic-co-glycolic acid) (PLGA), and poly (lactide)/(caprolactone) copolymers (PLA-PCL) are polymers commonly used in Atrigel® system (Wang and Burgess 2021; Juvekar and Kathpalia 2017; Malik et al. 2010).

In this chapter, we will introduce design considerations such as solvents, polymers, chemicals, and additives that can affect phase separation, depot structure, and drug release with the aim of supporting the development of many manufacturing facilities. Furthermore, we will examine the available products, both preformed and in situ forming implants, discussing their relevant advantages and disadvantages.

4.2 Preformed Implants

Preformed (or classical) implants are solid cylindrical rods 10–35 mm long and 1–3 mm in diameter. These implants are produced by using biodegradable or non-biodegradable polymers for subcutaneous or local administration (e.g., OZURDEX®). Depending on the size of the implant, it is inserted surgically or injected using special equipment or large needles. Nonbiodegradable implants often require surgical removal after treatment, unlike biodegradable implants, which slowly break down into metabolizable monomers (e.g., PLGA decomposes into lactic and glycolic acids) (Parent et al. 2013; Nkanga et al. 2020; Kempe and Mäder 2012; Stewart et al. 2018; Saraf et al. 2013).

There is no appropriate classification system for implantable drug delivery systems (IDDS); this is mainly because there are complex implants that do not neatly fit into specific categories. However, IDDS can generally be divided into two main categories: passive and active implants, which are further classified into biodegradable and nonbiodegradable depending on the materials used in their construction. Active implants are characterized by their use of active and energy-dependent mechanisms to generate the force required for drug release. Dynamic drug delivery implants are predominantly of the pump-type variety. The primary type of polymeric active implants are osmotic pumps (Stewart et al. 2018). In 1955, Australian pharmacologists developed the Rose–Nelson osmotic pump, the first implantable osmotic pump designed to deliver drugs to the guts of animals such as sheep and cattle (Almoshari 2022). The Rose–Nelson osmotic pump evolved into a simplified version known as the elementary osmotic pump (EOP) as a single chamber osmotic pump, introduced by Theeuwes in 1974, which is a basic device comprised of a compressed tablet. It consists of an osmotic core containing the drug, enclosed by a semi-permeable membrane featuring a critical-size orifice for drug delivery. Upon contact with aqueous fluids, the dosage form absorbs water at a rate dictated by the fluid permeability of the membrane and the osmotic pressure of the core formulation. The imbibition of water is influenced by the fluid permeability of the membrane and the osmotic pressure of the compressed tablet. This osmotic absorption of water results in the creation of a saturated drug solution within the core, which is released at a controlled rate through the delivery orifice in the membrane. Typically, EOP delivers 60–80% of its content at a constant rate, exhibiting a short lag time of 30–60 min, and is suitable for moderately soluble drugs (Shahi et al. 2015; Prabakaran et al. 2003). Another kind of system is the push-pull osmotic pump (PPOP), a multiple-chamber osmotic pump that comprises two compartments: the upper compartment (drug compartment) housing the drug with osmotically active agents and the lower compartment (push compartment) containing polymeric osmotic agents. Upon exposure to an aqueous environment, both compartments absorb water simultaneously. As the lower compartment lacks any orifice, it expands and displaces the diaphragm into the upper drug chamber, facilitating drug delivery through the delivery orifice. This system is suitable for delivering both highly water-soluble and practically water-insoluble drugs, such as oxybutynin hydrochloride, nifedipine, and glipizide (Shahi et al. 2015; Patel and Mehta 2013; Malaterre et al. 2009).

Passive implantable drug delivery systems (IDDS) have no moving parts; instead, they rely on drug diffusion to achieve a consistent drug release over time. Such IDDS can be crafted from either biodegradable or nonbiodegradable materials. Generally, these implants are created by mixing drug molecules with biocompatible polymers with no mechanical parts (Stewart et al. 2018; Kleiner et al. 2014). Within passive IDDS, two common types of implants can be distinguished based on the location of the drug in the device: reservoir-type implants and monolithic implants (Zheng and Pokorski 2021). Monolithic implants (or matrix implants) disperse the drug within a matrix, typically composed of a biocompatible polymer. In contrast, reservoir-type implants consist of a drug-loaded core enclosed by a permeable membrane that controls drug release. Typically, this membrane is constructed from nonbiodegradable materials like silicones, poly(urethanes), poly(acrylates), or copolymers such as poly(ethylene vinyl acetate). However, there are also examples of reservoir-type implants using biodegradable rate-controlling membranes (Stewart et al. 2018). Nonbiodegradable implantable drug delivery systems (DDSs) are widely used in contraceptive delivery. These DDSs are designed to be durable and efficient throughout their operational lifetime. However, an important limitation of nonbiodegradable implants is the necessity for removal once they have depleted their drug payloads. While the materials used in their construction are generally recommended for long-term biocompatibility, they can sometimes lead to infections, tissue damage, or disfigurement. Therefore, once all the drugs have been released, they are typically extracted to mitigate potential adverse effects (Stewart et al. 2018). Biodegradable polymeric implants were developed to address the limitations of nonbiodegradable implants. These implants are composed of biodegradable polymers such as poly(lactic acid) (PLA), poly(lactic-co-glycolic acid) (PLGA), and poly(caprolactone) (PCL). These polymers can break down into smaller components that are either excreted or absorbed by the body without requiring additional surgery (Stewart et al. 2018; Kumar and Pillai 2018).

4.3 Methods of Implant Manufacture

When selecting a manufacturing process to produce implantable drug delivery systems, several factors must be considered, including cost, efficiency, and differences in the characteristics of the resulting implants. Implants can be manufactured by several techniques, such as compression, solvent casting, hot melt extrusion, injection molding, 3D printing, electrospinning, and other manufacturing techniques (Stewart et al. 2018; Magill et al. 2023).

Thermoplastic polymers like PLA or PLGA can be used to manufacture implants by processes such as hot molding, injection molding, compression, or extrusion. Implants manufactured by different techniques are unlikely to have identical microporous structures and exhibit different degradation rates. Consequently, they will have different in vitro and in vivo release profiles (Stewart et al. 2018). Fialho et al. performed a comparison of hot-forming and compression methods for manufacturing intraocular implants and found that the manufacturing method significantly affected polymer degradation and, consequently, drug release from the resulting implants. Compressed implants exhibited a higher rate of drug release compared to their molded counterparts (Fialho and Da Silva Cunha 2005). In the following sections, we discuss different manufacturing methods of solid implants.

Compression

Instead of utilizing heat and/or solvents, this technique exerts force on the material (drug and polymer together) until it undergoes flow and conforms to the shape of the die being imposed upon it. Solid implants produced through this method generally exhibit higher densities (Larrañeta et al. 2021; Ganesh et al. 2022). One advantage of compression as a manufacturing method is that it does not require heat or solvents, making it well suited to produce implants containing heat- or solvent-sensitive substances such as proteins or peptides. However, implants produced by compression typically have a faster release rate than those produced by other manufacturing methods. In such cases, additional measures, such as coating the implant, may be necessary to prolong the duration of drug release. In addition, as shown by Fialho et al., compressed implants tend to have an uneven surface with numerous pores and channels, which could lead to irregular drug release from implants manufactured by this process (Stewart et al. 2018).

Solvent Casting

In this approach, the polymer and/or drug are dissolved in an appropriate organic solvent, and the resulting mixture is cast into or around a mold with the desired shape and size for the implant (Larrañeta et al. 2021; Stewart et al. 2018). The solvent is permitted to evaporate, leaving the polymer mixture solidified in the form of the implant. The capacity to produce implants in diverse shapes and sizes with a uniform distribution of drugs/particles represents a primary benefit of this technique (Utomo et al. 2022). However, a drawback is associated with the necessity of using an organic solvent, which may pose toxicity concerns. Although some studies have indicated no significant cytotoxicity issues in the final product (Picco et al. 2022, 2023; Korelidou et al. 2022), this may not hold true on an industrial scale where the solvent volume would be much larger, raising potential environmental impacts (Larrañeta et al. 2021; Stewart et al. 2018). An alternative to solvent casting is particulate leaching, where solid particles (such as salt) are mixed with the polymer solution. As the solvent evaporates, the salt particles become embedded in the polymer matrix. When soaked in water, the salt particles leach out, leaving a porous structure (Holmberg 2020). Özalp et al. employed solvent casting to produce discs with vancomycin incorporated in PLA and PLGA (Ozalp et al. 2002). In a comparative study of different methods, Prata et al. dissolved PCL and dexamethasone in a mixture of chloroform and acetone. This mixture was poured into metallic molds, and after the solvents evaporated, small discs were cut out (Prata et al. 2018).

Injection Molding

This method involves using heat and pressure to inject a molten mixture of drug and polymer into a mold of the desired shape and then allowed to cool and solidify. This technique is highly versatile in terms of mold shape and can be easily scaled up using larger machines. Furthermore, the pressure and heat applied during the process can help in sterilizing the product and enhance the interactions between the drug and polymer, which is beneficial in some cases for the release of the drug (Magill et al. 2023).

Hot Melt Extrusion

Hot melt extrusion (HME) is a widely employed technique in pharmaceutical manufacturing. Through the regulation of melting temperature and mixing, materials are uniformly dispersed and subsequently extruded through a die (Ganesh et al. 2022). The method enhances the dissolution of poorly soluble drugs to improve bioavailability and controls the drug release during implant formation (Maniruzzaman et al. 2012). Furthermore, this method avoids the use of solvents during production, thereby enhancing the biocompatibility of the implants (Larrañeta et al. 2021). The HME process is easily scalable and can be seamlessly adapted to industrial settings without altering the final properties of the product (Larrañeta et al. 2021; Maniruzzaman et al. 2012). However, there are certain limitations associated with this technique. The utilization of high temperatures renders thermally labile drugs unsuitable, and the stability of polymers must be carefully evaluated (Maniruzzaman et al. 2012). Similarly, the selection of polymers is restricted to those possessing specific physical properties required by the process (Maniruzzaman et al. 2012).

In order to use this method, the polymers used must necessarily be thermoplastic in nature. Aliphatic polyesters, such as PLA, PGA, and PLGA, fall into this category and are therefore suitable for processing by hot melt extrusion (Breitenbach 2002). A major advantage of this technique is that it does not require solvents. However, it can potentially lead to the degradation of heat-sensitive drugs. Nevertheless, this does not preclude its use in the manufacture of implants containing thermally labile drugs (Stewart et al. 2018).

Research by Repka et al. has shown that hydrocortisone, a thermally labile active ingredient, has been successfully incorporated into hydroxypropylcellulose (HPC) films produced by melt extrusion. In particular, products such as Zoladex®, Depot-Profact®, and Implanon® are manufactured using this method, which offers the advantage of continuous processing and high throughput rates (Stewart et al. 2018; Breitenbach 2002; Fialho and Da Silva Cunha 2005).

3D Printing

Three-dimensional (3D) printing technology is used in the production of dental implants, prostheses, and orthopedic implants. This method is cost-effective, offers reproducibility, and is highly adaptable, making it a promising candidate for the production of implantable drug delivery systems (Stewart et al. 2018).

Before the printing process begins, the product is designed using computer-aided design software, which provides great flexibility in the design of the implant. The implant may be hollow to contain an active ingredient, or it may consist of a uniform drug mixture that can be printed in the desired configuration. It is important to choose these aspects carefully, considering factors such as the properties of the drug and the polymer, including stability and crystallinity, as the choice of 3D printing method depends on these characteristics (Magill et al. 2023). Various 3D printing methods include stereolithography, selective laser sintering, fused deposition modeling (FDM), laminated object fabrication, inkjet-based techniques, and bioprinting (Larrañeta et al. 2021). 3D printing holds significant promise, particularly for rapidly producing prototypes for research purposes. However, their suitability for large-scale mass production remains unclear. Notably, a pivotal step toward establishing 3D printing as a viable manufacturing technique for commercial products occurred in 2015 with the approval of the Food and Drug Administration (FDA) for a 3D-printed drug product (Stewart et al. 2018).

Both 3D printing techniques—ARBURG plastic free-forming droplet deposition Modeling (APF DDM) and fused deposition modeling (FDM)—can be used to prepare controlled-release PLGA implants. However, because APF DDM deposits individual droplets rather than continuous filaments, the resulting implant geometry can differ even when using the same composition and theoretical design. These structural differences can affect drug-release conditions, such as mesh porosity, leading to variations in drug-release kinetics. The significance of these effects depends on the specific implant design and processing conditions. Therefore, the properties of the implants should be carefully fine-tuned to meet the requirements of each application, particularly regarding drug-release kinetics and mechanical properties. It is crucial to note that the actual filling density and porosity of 3D-printed PLGA implants play a key role in determining the release patterns and the mobility of the drug within the polymeric system (Bassand et al. 2022).

Electrospinning

To make polymer and drug flow through the electrospinning apparatus, they must be in liquid form, and there are two approaches to achieve this: melting and dissolving. The former involves exposing the polymers to high temperatures, whereas the latter employs organic solvents to liquefy them. It is important to note that high temperatures can potentially degrade certain drugs if the polymer blend includes drugs, which can limit drug options or necessitate different methods of drug incorporation. The choice between these methods also depends on the desired drug-release profile, with melting typically resulting in a longer, more linear release pattern with less initial burst compared to the in-solution method. Furthermore, various techniques can be employed to introduce drugs into the fibers, each offering distinct kinetic profiles and material properties, as elaborated by Luraghi et al. (2021) and Magill et al. (2023). Electrospinning has been applied in a wide array of fields including antibiotics, anticancer therapy, ocular treatment, cardiovascular disease, and wound healing (Stewart et al. 2018).

Other Manufacturing Techniques

Most implantable devices detailed in the literature rely on polymer-based materials, making the previously mentioned methods suitable for their preparation. However, when implants are constructed from metals or ceramics, alternative manufacturing methods are required. Techniques such as computer numerical control and sintering methods are typically employed in such instances (Magill et al. 2023; Baracu and Dinu Gugoasa 2021).

4.4 Mechanism of Drug Release from Implantable Polymeric Drug Delivery System

Drug delivery systems utilizing swelling, osmotic pressure, or passive diffusion depend on essential factors such as the molecule's solubility and diffusion coefficient within the polymer, the drug load, and the polymer in vivo degradation rate, all of which influence the release kinetics of the drugs (Gulati et al. 2016). In the following sections, the mechanism of drug release has been explained.

Mechanism of Drug Release from Nonbiodegradable Implants

Nondegradable polymers are widely used in the production of transdermal films and implant devices, among other biomedical uses (Pillai and Panchagnula 2001). Well-known polymers such as poly(urethanes), silicone, and poly (ethylene vinyl acetate) are often employed for constructing these devices. Although various forms of nondegradable implants are accessible in the market, those utilizing membrane-enclosed reservoirs and matrix-controlled systems dominate the field. In such nondegradable implants, passive diffusion is the main mechanism driving solute transport (Magill et al. 2023; Hussain et al. 2021; Kleiner et al. 2014).

Reservoir-type systems contain a drug compartment that is separated by a polymer membrane that acts as a barrier, which controls the flow of the drug. These systems maintain a constant release rate that is not affected by concentration gradients, and this release rate is generally determined by the thickness and permeability of the rate-controlling polymeric membrane, resulting in a zero-order release pattern. Matrix-type systems, on the other hand, involve a rate-controlling medium, typically a polymer, in which the drug is evenly dissolved or dispersed. Drug release from the preloaded matrix usually follows a first-order pattern. In these systems, drug release is predominantly governed by Fickian diffusion, influenced by factors such as concentration gradients, diffusion distance, and the degree of swelling. Matrix-type systems allow for gradual drug diffusion through the polymeric material, leading to prolonged drug release. However, the release kinetics of matrix-type systems depends on the volume fraction of the drug within the matrix, meaning that the release rate is directly related to the amount of drug encapsulated in the matrix (Hussain et al. 2021).

Mechanism of Drug Release from Biodegradable Implants

Implantable drug delivery devices generally consist of a drug depot encased in a polymer or a mixture of drug and polymer. After insertion at the desired site in the body, the drug is gradually released at a predetermined rate as the polymer decomposes. In a reservoir system, the release of the drug is determined by the rate of polymer degradation or by dissolution of the drug followed by diffusion through the polymer wall, or by a combination of both mechanisms. In a drug-polymer blend, on the other hand, drug release is influenced by diffusion, swelling, or erosion. The rate of drug release from the system depends on factors such as the solubility and permeability of the drug in the polymer, the amount of drug loaded, and the in vivo degradation rate of the polymer (Hussain et al. 2021; Dash and Cudworth 1998).

Degradation of the polymer and subsequent release of the drug may occur by one or more mechanisms. These mechanisms include hydrolysis, in which bonds such as ester bonds break down in the polymer backbone; enzymatic degradation, in which hydrolysis-sensitive bonds such as amide bonds are broken down in the presence of a catalyst; oxidation; and physical degradation, in which bonds break down due to physical effects such as swelling or mechanical stress. The duration of polymer degradation can vary significantly depending on factors such as the molecular weight of the polymer and surface properties which affect the drug release from a given formulation. In addition, degradation is affected by in vivo factors such as pH and temperature. Therefore, a comprehensive evaluation of the in vivo degradation time of polymers is required (Hussain et al. 2021; Dash and Cudworth 1998; Ulery et al. 2011).

4.5 Microneedles (MNs)

Microneedles (MNs) represent minimally invasive devices that offer a painless and bloodless means of bypassing the skin's stratum corneum (SC) barrier (Paredes et al. 2021; Hutton et al. 2018). These MNs, ranging in height from 10 to 900 μm and manufactured using microfabrication techniques in various geometries and materials, have been extensively explored for their potential to enhance transdermal drug and vaccine delivery. When an array of MNs facilitates the diffusion of drug molecules through the SC and other skin layers, enabling localized or systemic drug delivery. Consequently, compared to conventional oral and injectable methods, this approach presents several advantages, including protection against gastrointestinal degradation, avoidance of first-pass hepatic metabolism, improved bioavailability, painless application, reduced risk of infection, and the convenience of self-administration by patients (Magill et al. 2023; Vora et al. 2022).

Moreover, by utilizing biodegradable polymers with adjustable characteristics, polymeric MNs enable controlled drug delivery (Vora et al. 2021). There are various types of implantable MNs with different delivery mechanisms including dissolving MNs loaded with nano/microparticles, fast-separable solid implantable MNs, and hydrogel-forming MNs (Magill et al. 2023).

Implantable Microtip-Based MNs

Biodegradable tip-based microneedles (MNs) for implantation are commonly constructed using various biodegradable polymers such as PLA, chitosan, PGA, PCL, or PLGA (Peng et al. 2021; McAlister et al. 2021). When these slow-dissolving MN tips are inserted into the skin, the polymer biodegrades after a short period of MN application, facilitating drug release into the intradermal space (Magill et al. 2023).

PLGA, PLA, and silk protein have previously been employed to craft the tips of these MN arrays due to their excellent biocompatibility, robust mechanical properties, and safety profiles. These materials have been utilized for delivering various substances, including levonorgestrel, finasteride, and growth hormone. Within implanted PLGA MN tips, the cargo can be released in a sustained manner to provide a long-lasting effect. According to reports, the release duration can span from days to 3 months. Several factors, such as drug loading, the length of the polymeric chains, and the presence of porogens (e.g., trehalose), may influence the rate of drug release from implanted PLGA tips (Magill et al. 2023; Zhao et al. 2020). The drug-release kinetics from PLGA MN tips can be readily adjusted, as suggested by Aung Than et al., by blending PLGA polymers with varying molecular weights and copolymer ratios (Than et al. 2018).

One of the pioneering studies demonstrating the capability of this system for in vivo drug release investigations was conducted by Li et al. (2019b). This research involved the utilization of swiftly detachable PLGA tips loaded with levonorgestrel, achieving a remarkable 30 days of drug release in vivo. This approach has since been further explored with optimizations in various parameters within the MN array formulation (Lee et al. 2021; Li et al. 2019a). Figure 4.1 shows the design and functionality of the engineered MN patch with a porous interface, facilitating the swift

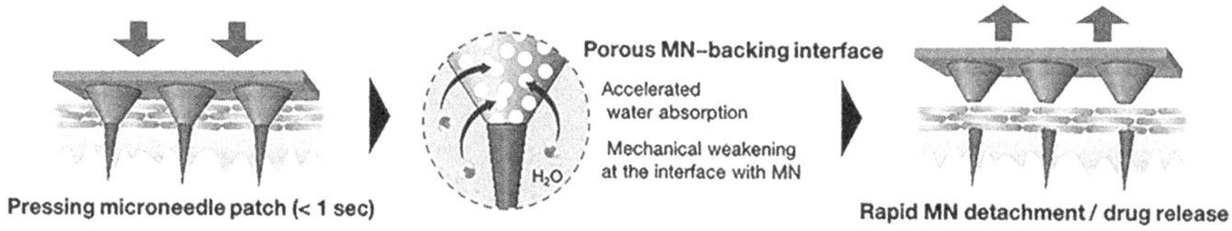

Fig. 4.1 Design of a microneedle patch featuring a porous backing. (Used with permission from Yunki Lee et al. (2021))

detachment of microneedles (MNs). During compression, the MNs are designed to penetrate the skin, and upon removal from the skin while under tension, they separate from the patch backing. This mechanism allows for the immediate detachment of MNs during the application of the MN patch (Lee et al. 2021).

In a recent study conducted by Yunki Lee et al. (2021), a porous microneedle (MN) patch was developed to enable the sustained delivery of contraceptive hormones through the skin. This innovative approach involved the incorporation of a porous structure into the back of the patch, allowing it to maintain sufficient mechanical strength for MN insertion into the skin while ensuring easy detachment of the MNs under tension. Upon removal of the patch, the MNs promptly separated from the patch-backing layer, a result achieved through the mechanical weakening at the porous interface and the accelerated dissolution of the water-soluble backing material. These detached MNs, constructed from biodegradable polymers with varying concentrations of PLGA and diverse molecular weights, exhibited sustained release of the contraceptive hormone levonorgestrel (LNG). This sustained release ranged from a few days to well over a month.

The MN patch system offered several noteworthy advantages for contraceptive application: (1) the capability for self-administering drug-loaded MNs into the skin, (2) an exceptionally brief patch wear time (less than 1 s), (3) the elimination of biohazardous sharp waste, and (4) the extended release of contraceptive hormone spanning from several days to more than a month. To summarize, the MN patch with a porous backing, combined with sustained drug delivery via MNs embedded in the skin, presents a straightforward and convenient solution for long-acting contraception. This innovation holds the potential to enhance accessibility to and compliance with contraceptive methods (Lee et al. 2021).

A recent advancement in PLGA-tip MN arrays introduced core-shell micro-tips to achieve even more sustained drug release compared to conventional PLGA tips, with in vitro testing demonstrating a sustained drug release of 6 months for levonorgestrel delivery (Li et al. 2022).

In more recent developments, researchers have experimented with two-layer and three-layer MN designs to reduce application times and expedite the implantation of PLGA/PLA tips. Notably, Peng et al. introduced an innovative microneedle patch that combines implantable PLGA tips with hydrogel-forming microneedle bases (HFMB) utilizing a dissolvable material. The use of preformed (HFMB) not only improved insertion capability but also greatly enhanced ex vivo drug delivery efficiency. This approach achieved delivery of up to 80% of the loaded drug and enabled a faster implantation process within a minute compared to the traditional dissolving baseplate PLGA-tipped microneedle design. Consequently, these novel implantable MN patches hold the potential for long-acting drug delivery applications (Magill et al. 2023).

4.6 Preformed Implants Advantages and Disadvantages

The advantages and disadvantages of implantable drug delivery systems are indicated in Table 4.1 (Hussain et al. 2021; Mohammad Zaki et al. 2012).

Table 4.1 Advantages and disadvantages of implantable drug delivery systems

	Explanation
Advantages	
Local drug delivery	Medications are released directly at the implant site, primarily affecting the disease area close to implantation
Enhanced patient compliance	Patients no longer must remember the timing and frequency of their medication
Minimized systemic side effects	Controlled and sustained release enables localized dosing at the site of action, reducing adverse effects in other body parts; this approach avoids fluctuations in drug concentration in the bloodstream associated with repeated intermediate doses
Lower dosing requirements	Local implantation of drugs for specific sites can bypass the initial liver metabolism, leading to reduced doses needed to achieve systemic effectiveness
Improved drug stability	Implants protect drugs that are prone to rapid degradation in the gastrointestinal tract and liver
Suitable for chronic conditions	Implants are ideal for treating chronic illnesses, eliminating the need for extended hospital stays or constant medical supervision
Intermittent release	Highly programmable pumps (nonbiodegradable implants) can provide intermittent drug release based on factors like heart rhythm or metabolic demand; implant technology offers flexibility in materials and manufacturing processes, allowing controlled delivery of both hydrophilic and lipophilic drugs
Disadvantages	
Surgical procedure required	The insertion of these nonbiodegradable implants necessitates a surgical procedure, either major or minor, which can be invasive for the patient Biodegradable implants often need large needle size to inject
Removal process for nonbiodegradable implants	Nonbiodegradable polymeric implants can be removed from the body at the end of the treatment, but this requires another surgical procedure
Risk of device malfunction for nonbiodegradable implants	If, for any reason, the device malfunctions during treatment, surgical intervention may be required to remove the device from the patient's body
Limited to potent drugs	The small size of these devices is designed to minimize patient discomfort, limiting their use to potent drugs that require only small amounts
Potential for adverse reactions	Because these devices deliver a high concentration of the drug to the implantation site, there is a risk of adverse reactions due to this localized high drug concentration or large size and the nature of implant

4.7 Translation of Implantable Drug Delivery Systems (IDDS) into Clinical Practice

Numerous considerations play a pivotal role in the clinical adoption of IDDS, encompassing factors such as the foreign body response, scalability of manufacturing, and ensuring sterility. Regulatory agencies like the FDA and European Medicines Agency (EMA), among others, meticulously assess these elements before granting approval for the commercial application and clinical use of an IDDS (Magill et al. 2023).

Following the design, optimization, and in vitro evaluation of an IDDS, potential challenges are encountered during in vivo experiments, most notably the foreign body responses. The implantation of these devices often triggers a host response, resulting in what is known as a foreign body reaction. This reaction occurs when the body identifies the implanted device as foreign, prompting the activation of innate immune system cells and initiating an inflammatory and fibrotic process (Magill et al. 2023).

The process begins with the adsorption of plasma proteins onto the implant's surface, followed by the deposition of a protein layer (e.g., fibrinogen, fibronectin, vitronectin), ultimately leading to the formation of a fibrous capsule around the implant. This capsule prevents the implant from functioning as intended. The failure rate of implantable devices varies based on factors such as surface characteristics, design, and features, with some types of implantable devices estimated to have a failure rate of 10%. Such failures can pose life-threatening risks to patients (Magill et al. 2023).

The surface characteristics of IDDS, including attributes like porosity, roughness, and charge, exert a substantial impact on the occurrence of foreign body reactions. Furthermore, the size and configuration of the implant are pivotal determinants in the body's response. Hence, when choosing materials, manufacturing methods, size, and shape for an IDDS, these factors necessitate meticulous consideration due to their potential influence on the body's reaction to foreign bodies (Magill et al. 2023). This local tolerability of the LAIs is discussed extensively in Chap. 9 of this book.

Scale-up manufacturing presents another crucial consideration for IDDS. Some IDDSs were developed using manufacturing techniques that are challenging to adapt for industrial production, including methods like 3D printing or electrospinning (Magill et al. 2023; Omer et al. 2021; Kjar and Huang 2019).

In addition, some of the works in this context have utilized sophisticated methods to create implants, which involve numerous steps that may not be easily applicable in an industrial setup for mass production. It is worth noting that even if the manufacturing techniques can be adapted, large-scale production may alter the characteristics of the final products, resulting in performance-related problems. Sterility represents another vital aspect demanding consideration during the translation of IDDS for clinical use (Stewart et al. 2023; Tipnis and Burgess 2018). Various established methods are available for sterilizing implantable devices, including dry heat, steam, ethylene oxide, hydrogen peroxide, ozone, or radiation (Tipnis and Burgess 2018). It is worth noting that not all materials are suitable for dry heat or steam sterilization, as moisture or high temperatures could potentially damage the IDDS or its drug payload. Ethylene oxide and gamma/electron beam sterilization are widely employed for sterilizing medical devices since they do not necessitate high temperatures (Tipnis and Burgess 2018). However, they have certain limitations. Ethylene oxide, for instance, may leave behind toxic residues post-sterilization (Tipnis and Burgess 2018). On the other hand, gamma or electron beam radiation can induce alterations in material properties such as chemical composition, crystallinity, molecular weight, or density (Tipnis and Burgess 2018). The impact of radiation largely hinges on the dose and the specific materials present in the IDDS. The FDA establishes endotoxin and sterility limits for implants based on their intended use and tissue interface. The endotoxin limits vary depending on the anatomical locations that the implantable medical device (IMD) may potentially contact. For devices in contact with the cardiovascular system, applicable to most anatomical sites, the limit is set at 20 endotoxin unit (EU) per patient. Devices with potential contact with cerebrospinal fluid (including brain tumors) have a lower endotoxin limit of 2.15 EU. In cases where multiple implants are inserted in a single patient, the EU limit per implant is determined by dividing the total EU limit per patient by the number of devices implanted. There is no publicly available data identifying the most effective approach for consistently meeting the FDA requirement of endotoxin limits below 20 EU per patient, or 4 EU per device (assuming a total of up to five devices implanted per patient) (Dominas et al. 2021).

4.8 Injectable In Situ Forming Implants (ISFIs)

The challenges associated with preformed implants have motivated researchers to explore the development of in situ forming implants/gels, which offer a cost-effective, quick, and user-friendly approach to controlled drug release through injection. In situ forming implants (ISFIs) are formulations typically composed of biodegradable polymers in liquid or semisolid form, which solidify or jellify spontaneously upon injection at the target site (Nkanga et al. 2020). The use of biodegradable polymers to create in situ forming implants for drug delivery was first reported in the early 1990s (Agarwal and Rupenthal 2013).

In general, injectables are liquid or semisolid that can be administered via a syringe and fixed to form a paste. Gelling mechanisms fall into three main groups (Ibrahim et al. 2021; Nkanga et al. 2020; Kempe and Mäder 2012; Agarwal and Rupenthal 2013):

(i) **In situ cross-linking systems,** in which cross-linking of polymer chains is initiated by light irradiation or by physical and chemical cross-linking agents.

(ii) **In situ solidifying organogel systems**, which are hot melts that work well when cooled to body temperature, or lyotropic liquid crystals that self-assemble in an aqueous solution.

(iii) **In situ phase separation systems,** where polymers become insoluble due to a combination of physical forces such as hydrogen bonding, hydrophobic interactions, and ionic bonds between polymer chains.

In Situ Cross-Linking Systems

In situ cross-linking systems can be achieved through various mechanisms, including temperature-induced cross-linking (thermo-responsive), light absorption (photo-responsive), ionic interactions between anionic polymers and small cations (ion-responsive), or the presence of enzymes (enzymatic-responsive). These cross-linked polymer networks provide advantages by allowing control over the diffusion of hydrophilic macromolecules and facilitating the sustained release of peptides and proteins over extended periods (Ibrahim et al. 2021).

In Situ Solidifying Organogels

In situ solidifying organogel systems consist of amphiphilic lipids that are insoluble in water. These lipids are heated and dissolved in an organic solvent like ethanol. Upon contact with water, they swell and form various types of lyotropic crystals. The properties of these crystals depend on several factors such as the type of drug loaded, the structural properties of the lipid, the amount of water present, and the temperature (Ibrahim et al. 2021; Yaghmur et al. 2013).

Organogels are typically created through a process involving the heating of a mixture containing the gelator and the liquid component to produce an organic solution or dispersion, followed by the cooling of this mixture, which causes it to solidify into a gel. The initial heating step facilitates the dissolution of the gelator in the liquid phase. Upon cooling, the solubility of the gelator in the liquid decreases, leading to reduced interactions between the gelator and the solvent. This reduction in interactions causes the gelator molecules to separate from the solution. These gelator molecules then self-assemble into distinct structures such as tubules, rods, and fibers due to interactions between them. These assembled structures entangle and connect with each other, forming a three-dimensional network that immobilizes the fluid phase, resulting in the formation of a gel. This three-dimensional network acts as the structural framework of the gel, providing it with strength and resilience. The connections between the gelator aggregates play a crucial role in gel formation; without these connections, the gel state may not be achieved even if there are numerous gelator aggregates present. A few examples of organogelators include: cholesterol 4-(2-anthryloxy)butanoate (CAB); 1,3:2,4-di-O-benzylidene-D-sorbitol

(D-DBS); 2,3-bis-n-Decyloxyanthracene (DDOA); 12-hydroxyoctadecanoic acid (12-HOA) (Murdan 2005).

In addition, some organogels are created by introducing a third component into an organic solution. For instance, an organic solution of bis-(2-ethylhexyl) sodium sulfosuccinate (AOT) can be gelled by adding phenolic species (Xu et al. 1993; Tata et al. 1994). Similarly, solutions of lecithin in an organic solvent like iso-octane can be gelled by adding small amounts of a polar substance such as water, glycerol, ethylene glycol, or formamide (Scartazzini and Luisi 1988; Shchipunov and Shumilina 1996). Complex organogels have also been developed through the gelation of a water-in-oil (w/o) microemulsion using gelatin as a hydrogelator (Haering and Luisi 1986). The use of organogels in drug delivery is limited due to the toxicity of the ingredients and the lack of information about the purity and stability of oils and waxes. Another disadvantage is the high initial viscosity of the cubic phase, which must be heated to mix, and the possibility of different melting points of waxes and oils, which can cause phase separation (Kempe and Mäder 2012).

In Situ Phase Separation Systems

Another way to create an in situ forming gel is to take advantage of the precipitation that occurs when the solubility of polymers decreases with pH, temperature, or removal of solvents (Kempe and Mäder 2012).

pH-Induced Gelling Systems

Variations in the pH of the solution can induce significant changes in the polymeric network's conformation. This transformation from a liquid state to a gel due to pH fluctuations is commonly observed in polymers containing ionizable functional groups (Ibrahim et al. 2021). An example of this phenomenon is seen in chitosan, a biocompatible and biodegradable cationic polymer. It dissolves in acidic solutions and undergoes hydrogel formation when the pH exceeds 6 due to the deprotonation of its primary amino group. However, these gels can be unstable without additives, which may lead to the rapid release of low molecular weight drugs (Kempe and Mäder 2012).

Thermoplastic Pastes

Thermoplastic pastes are composed of polymers with a melting point or glass transition temperature ranging from 25 °C to 65 °C. Prior to injection, these polymers are heated above their glass transition or melting point, rendering them in a liquid state. Inside the body, they solidify into a semisolid depot as they cool down to body temperature. When injected at a temperature between 37 °C and 65 °C, these polymers behave as viscous fluids and can be easily administered. Drugs are incorporated into the molten polymer through simple mixing, eliminating the need for organic solvents. Initially, biodegradable thermoplastic pastes were created using polymers like polyanhydrides, PCL, PLA, and their blends with PEG, as well as PLA–PEG–PLA triblock copolymers. However, some of these polymers have high transition points, resulting in injection temperatures that exceed 60 °C. These high-temperature injections caused significant pain and led to necrosis at the injection site (Alonso et al. 1993). In addition, scar tissue formation around the depot hindered drug diffusion, and these implants exhibited slow drug release. To tackle these challenges, Heller and colleagues devised a biodegradable semisolid polymer known as poly(ortho ester) (POE) (Heller et al. 2000; Einmahl et al. 2001b). POEs with low molecular weights and softening temperatures ranging from 35 °C to 45 °C have exhibited excellent biocompatibility in both animal and human studies (Einmahl et al. 2001a). These polymers maintain a semisolid state at room temperature, enabling the straightforward inclusion of therapeutic agents through simple mixing, eliminating the need for heating or solvents. To enable injection through a needle, their intrinsic viscosity should not exceed 0.8 dl/g. POEs with intrinsic viscosities below 0.5 dl/g were found to be less effective in extending drug release (Kempe and Mäder 2012).

Thermally Induced Gelling Systems

Thermally triggered gelling systems experience a transition from a solution to a gel state when the temperature rises in an aqueous environment. Thermosensitive polymers are particularly appealing in this context because they can undergo this gelation process under physiological conditions without the need for organic solvents, co-polymerization agents, or external triggers. However, these temperature-induced phase transitions are influenced by the temperature-dependent nature of specific molecular interactions, such as hydrogen bonds or hydrophobic effects. When the temperature reaches the lower critical solution temperature (LCST), the hydrogen bonding between water and a polymer becomes less favorable compared to interactions between the polymer molecules themselves. Consequently, the solvated macromolecules lose their water content, and interactions between the polymers become more prominent, forming a network-like structure. This transition results in a significant increase in the viscosity of the system. Ideally, aqueous polymer solutions should flow freely at room temperature and transform into a gel state at body temperature. The shape of the gel depot and the drug release pattern, including the initial release of the incorporated drug, are directly linked to the kinetics of the transition from solution to gel. This transition should occur rapidly. It is worth noting that both synthetic and naturally occurring materials can exhibit temperature-controlled gelation behavior (Kempe and Mäder 2012; Ruel-Gariépy and Leroux 2004; Joo et al. 2009). Synthetic polymers including copolymers of poly(ethylene oxide) (PEO) and poly(propylene oxide) (PPO), commonly referred to as poloxamers, as well as copolymers of poly(N-isopropyl acrylamide) (PNIPAAM), are extensively researched for drug delivery applications. However, their practical use is limited due to their nonbiodegradable nature. In aqueous solutions, PNIPAAM undergoes a reversible transition from a liquid state to a gel state at 32 °C. By introducing salts and surfactants, it is possible to shift the lower critical solution temperature (LCST) of PNIPAAM to 37 °C (Kempe and Mäder 2012; Ruel-Gariépy and Leroux 2004).

Among the commercially accessible PEO-PPO-PEO copolymers, poloxamer 407 (sold as Pluronic® F127) has been identified as having lower toxicity. Aqueous solutions within the concentration range of 16–30 wt.% develop semisolid, transparent gels and have been investigated for a variety of biomedical purposes (Kempe and Mäder 2012; Jeong et al. 2002).

Phase Separation Systems by Solvent Exchange

Dunn et al. introduced the Atrigel® technology, which involves dissolving polymers (the most commonly used polymers are poly(dl-lactide) (PLA), poly(lactic-co-glycolic acid) (PLGA), and lactide/caprolactone copolymers (PLA-PCL)) in organic solvents with high water miscibility such as N-methyl-2-pyrrolidone (NMP), dimethyl sulfoxide (DMSO), propylene glycol, acetone, tetrahydrofuran (THF), 2-pyrrolidone, ethyl acetate, glycofurol, or low molecular weight PEG. NMP is commonly used as it is a good solvent for PLGA type polymers and also show low toxicity. The viscosity of the implant forming solutions is influenced by both the polymer concentration (polymer concentrations in formulations typically range from 10 wt.% to 80 wt.%) and its molecular weight. Phase inversion dynamics, influenced by solvent properties, lead to rapid solidification of polymer solutions upon contact with aqueous media, usually occurring within seconds to minutes. This fast phase inversion and high solvent affinity to the precipitant result in implants with large pores, leading to high initial drug release (Kempe and Mäder 2012; Wang and Burgess 2021).

4.9 Liquid Crystalline System

Latitude Pharmaceuticals Inc. provides a phospholipid gel depot (PG Depot) that consists of 20% to 80% phospholipids nanodispersed in water and can be easily injected with a 25 G needle. This system can accommodate a wide range of drugs, including small molecules, proteins, and peptides. Initially, a phospholipid nanodispersion with high water content is created, and a gel phase is then achieved by removing the water in a subsequent step. The depot effect can be adjusted to last from 1 to 7 days (Rahnfeld and Luciani 2020). Lyotropic liquid phases are characterized by the self-assembly of amphiphilic lipids into various structures, including lamellar, hexagonal, or cubic phases, as well as reversed micellar cubic phases. While these systems have high viscosity, which is advantageous for topical applications, it can hinder their injectability. To address this, cubic ("cubosomes") or hexagonal nanoparticles ("hexosomes") can be formed by dispersing the phases in aqueous solutions, maintaining the liquid crystalline structure and increasing surface area while reducing viscosity (Singhvi et al. 2018). Another approach to improve injectability involves using a less viscous precursor with minimal water content; upon injection, body fluids are absorbed in vivo, forming a viscous depot system in situ. The only liquid crystalline system that has reached clinical studies is based on the FluidCrystal® technology marketed by Camurus AB. Buvidal® (buprenorphine weekly or monthly) was approved in the European Union and Australia and tentatively approved in the United States (Brixadi™) for the treatment of opioid use disorder. This depot system uses a phosphatidylcholine and glyceroldioleate precursor solution that forms an in situ reversed nonlamellar liquid crystalline structure with a customizable release profile. Other products in various stages of clinical trials, such as octreotide for treating acromegaly and neuroendocrine tumors, leuprolide for treating prostate cancer, and setmelanotide for treating genetic obesity, further demonstrate the significant potential of liquid crystalline phases as depot systems (Xia et al. 2019; Chen et al. 2015; Li et al. 2019c).

4.10 Key Parameters of In Situ Implant Composition

Solvent

The solvent type is an important factor influencing the initial dissolution of the drug/polymer and the phase inversion rate of the matrix. The ideal solvent or solvent mixture used to prepare ISFI systems should have the required properties such as viscosity, hydrophilicity, ability to dissolve polymers, and safety. When choosing an organic solvent for the preparation of the ISFI system, it is important to consider its ability to dissolve biodegradable polymers, its biocompatibility, and its ability to facilitate the conversion of the polymer into ISFI systems. Lastly, it is important to be aware of any harmful effects of metabolic products of solvents (Ibrahim et al. 2021; Kapoor et al. 2012).

The viscosity of the organic solvent used to prepare the ISFI system can affect the ease of injection (injectability or syringability) (Patki et al. 2021). For example, systems containing nonaqueous solvents have high viscosity, making them difficult to inject from a syringe, so the syringe must be warmed up before injection (Ibrahim et al. 2021).

To make ISFI systems injectable, good solvents can be used which have a high affinity to the polymer. This helps to enhance the dissolution profile and reduce the overall viscosity of the polymeric solutions used in the ISFI formulation. The use of good solvents in the preparation of ISFI systems can provide benefits such as lower injected volumes and higher drug loading in the polymeric samples. This is advantageous as it helps to overcome injection volume limitations and decreases the amount of solvent needed for administration (Parent et al. 2013; Bode et al. 2018).

Organic solvents used to dissolve PLGA polymers can be divided into two groups: highly water-miscible solvents (e.g., NMP, DMSO, acetone, propylene glycol, tetrahydrofuran, 2-pyrrolidone, and benzyl alcohol (BA)) and nonmiscible solvents (such as benzyl benzoate (BB), ethyl acetate, triacetin, and triethyl citrate). Highly water-soluble organic solvents, often referred to as strong solvents, play a crucial role in creating uniform gel depots and facilitating easy injection. This is because they promote the rapid formation of a phase-inverting system within seconds or minutes. Such a system creates a highly hydrophilic environment and a low-viscosity solution, making injection administration straightforward. Conversely, hydrophobic solvents are considered weak solvents. Using these solvents can lead to the gradual development of a phase-inverting system over weeks or months. This system reduces the interaction between the polymeric solution and water, thereby slowing down the liquid-liquid phase separation and, subsequently, the gelation rate (Ibrahim et al. 2021).

Ensuring biocompatibility and low toxicity is crucial when selecting and incorporating organic solvents into ISFI systems (Parent et al. 2013). For example, while NMP solvent is considered safe for pharmaceutical use, its high miscibility in water is undesirable because it leads to rapid exchange between NMP and water during the preparation of ISFI systems. Ibrahim et al. (2020) observed that this unfavorable rapid solidification of NMP-ISFI systems and immediate release of NMP into the surrounding medium resulted in the rapid diffusion of dissolved risperidone. Consequently, significant amounts of both the drug and solvent may be released initially within minutes or hours, potentially causing harmful local or systemic irritations of mucous membranes, skin, and muscles upon injection (Camargo et al. 2013; Karp et al. 2021). As a result, water-immiscible solvents are a viable option, but they may hinder ISFI injectability, posing challenges related to biocompatibility. Therefore, a preferable alternative is to use a combination of water-immiscible and water-miscible solvents, which helps achieve a reasonable viscosity, appropriate phase inversion, and a reduced burst release profile (Ibrahim et al. 2021).

Table 4.2 indicates that only two solvents, NMP and DMSO, have been used in the marketed products. NMP and DMSO are used due to excellent solubilizing, inertness, stability, and acceptable safety/toxicology profiles. According to ICH Guidance for Industry Q3C, NMP falls into Class 2 solvents in pharmaceutical products, with a maximum daily intake of 5.3 mg. Atridox®, Eligard®, Sublocade®, and Perseris® contained different amounts of NMP that exceed the permissible daily exposure (PDE) of NMP. Meanwhile, DMSO belongs to class 3 solvents, which may be considered less toxic and of lower risk to human health. In May 2023, the FDA approved a new risperidone formulation, UZEDY™, as an extended-release injectable suspension containing DMSO (45%) based FDA label, which equals 75 mg for 50 mg risperidone/0.14 mL to 375 mg DMSO for 250 mg risperidone/0.7 mL. Furthermore, it would be advisable to limit DMSO intake to 100 mg per day to prevent elevated toxicity levels. However, the half-life, lack of acute toxicity at clinically relevant doses, and the extended time between doses of administration have a significant impact on their safety risk (Wang et al. 2012; Johnson 2020; FDA 1998; Markowicz-piasecka et al. 2024; Bekkem et al. 2013).

Table 4.2 List of commercially available in situ forming implant drug products with permission (Wang and Burgess 2021; Markowicz-piasecka et al. 2024)

Product/ approval date	Powder syringe	Liquid syringe	Needle (Gauge-inch)	Duration
Atridox® 1988	50 mg doxycycline hyclate	450 mg (36.7% PLA dissolved in 63.3%/284.85 mg NMP)	23 (blunt cannula)	7 days
Eligard® 2002	7.5 mg leuprolide acetate	0.25 mL: 82.5 mg PLGA (L/G 50:50), 160 mg NMP	20–5/8	1 month
	22.5 mg leuprolide acetate	0.375 mL: 158.6 mg PLGA (L/G 75:25), 193.9 mg NMP	20–5/8	3 months
	30 mg leuprolide acetate	0.5 mL: 211.5 mg PLGA (L/G 75:25) 258.5 mg NMP	20–5/8	6 months
	45 mg leuprolide acetate	0.375 mL: 165 mg PLGA (L/G 85:15) 165 mg NMP	18–5/8	9 months
Sublocade® 2017	N/A	0.5 mL: 100 mg buprenorphine, 178 mg PLGA (L/G 50:50), 278 mg NMP	19–5/8	1 month
		1.5 mL: 300 mg buprenorphine, 533 mg PLGA (L/G 50:50), 833 mg NMP		
Perseris® 2018	90 mg risperidone	0.6 mL: 228 mg PLGA (L/G 80:20), 282 mg NMP	18–5/8	1 month
	120 mg risperidone	0.8 mL: 304 mg PLGA (L/G 80:20), 376 mg NMP		
UZEDY™ 2023	50/75/100/125/150/200/250 mg risperidone	Dimethyl sulfoxide (45% w/w, methoxy-poly (ethylene glycol)-co-poly (D,L-lactide) (15% w/w), and poly (D,L-lactide)-co-poly (ethylene glycol)-co-poly (D,L-lactide) (10% w/w)	21–5/8	1 or 2 months

Polymer

Polymer characteristics to control drug release from in situ forming implants are discussed in detail in Chap. 2 of this book.

Drug

The solubility, loading, and stability of the drug must be considered when designing an ISFI system. The drug may be dissolved or dispersed in the system, forming a homogeneous or heterogeneous solution, which affects the release profile of the system (Ibrahim et al. 2021). The solubility of risperidone in organic solvents, such as NMP and DMSO, has an impact on the formation of PLGA-ISFI systems. The homogeneous system in NMP (drug dissolved) shows a higher release of risperidone, whereas the heterogeneous system in DMSO shows a slower release with greater drug retention and lower breakpoint (Wang et al. 2012; Ibrahim et al. 2020). Furthermore, injectable formulations containing a higher ratio of drug to polymer offer the benefit of potentially reducing the volume of the formulation when considering a fixed dose for administration (Bittner et al. 2018). This reduction in volume is economically advantageous and more comfortable for patients as it can minimize the duration of pain experienced during injection. Patki et al. (2021) observed that increasing the loading of remdesivir was possible by lowering the concentration of PLGA, subsequently reducing the overall effort required for the injection process and the associated pain (Patki et al. 2021).

The drug must be compatible with the polymers and solvents and there must be no chemical interaction between the drug and ISFI components. However, if the drug contains water or forms hydrogen bonds with the polymer chain, it can cause polymer degradation (Ibrahim et al. 2021; Tang et al. 2018).

Moreover, the polymer and/or solvent used may influence the degradation or deactivation of the drug. Dong et al. (2006) observed that the degradation of PLGA in leuprolide acetate-loaded ISFI formulations was faster when using the hydrophilic NMP solvent compared to the hydrophobic triacetin solvent. This difference was attributed to the state of leuprolide acetate in the respective solvents; the drug in a dissolved state could release its bound water, participating in PLGA degradation. In contrast, when leuprolide acetate was in a dispersed state, it retained lower water content, resulting in more stable drug-loaded PLGA-ISFI systems. Therefore, it was reported that the drug remained chemically stable when dispersed in the polymeric solution rather than being dissolved in the PLGA solutions (Dong et al. 2006).

However, PLGA-based ISFI systems are not suitable for drugs that degrade upon exposure to water or are highly susceptible to the acidic environment produced following the degradation of the polymeric texture. As a result, modified strategies, such as the introduction of stabilizers, basic excipients, or PEGylation, have shown significant success in preparing stable PLGA-ISFI systems with satisfactory drug release profiles (Ibrahim et al. 2021).

Additives

Additives can be used to control the release of drugs in ISFI systems and are classified as hydrophilic, hydrophobic, and amphiphilic. Examples of hydrophilic additives are mannitol and polyvinylpyrrolidone, which can improve the miscibility of liquids but lead to a significant initial burst release of drugs. These additives can promote phase separation, which, in turn, enhances diffusion capabilities. The rapid depletion of these additives during the initial burst release can affect water penetration and the overall system's morphology (Ibrahim et al. 2021).

However, hydrophobic additives may reduce drug distribution and affect overall morphology. Examples of these additives are stearic acid, glyceryl monostearate, methyl heptanoate, ethyl heptanoate, and ethyl nonanoate. They hold more solvent, create less porosity, and reduce initial burst and sustained release (Ibrahim et al. 2021).

Sterilization

Due to the high viscosity of implants formed in situ, sterilization by filtration is often less recommended. Gamma irradiation is usually chosen as a suitable sterilization method for polymer solutions. However, this method can result in a loss of molecular weight (Mw) of the polymer, potentially leading to an increased initial burst release. This problem can be mitigated by using a polymer with a slightly higher initial Mw than the targeted polymer. It is important to note that gamma irradiation may present stability problems for certain drugs and the impact of terminal sterilization needs to be investigated during the initial phase of development (Nkanga et al. 2020; Wang and Burgess 2021).

4.11 Delivery of Biological Agents by In Situ Forming Implants

For drug candidates to be suitable for in situ forming implant formulation development, they should generally be highly potent with slow plasma clearance, good physicochemical stability, and a wide therapeutic index. This is important to handle the potential high plasma levels during burst release and the low concentrations during the slow release phase (Wang and Burgess 2021).

Despite these considerations, the advantages of mild manufacturing conditions and the wide range of available polymers for creating in situ forming implants have led to the formulation of various biological agents, including proteins, peptides, and genes, resulting in improved delivery and therapeutic outcomes in preclinical studies (Wang and Burgess 2021).

Considering the exposure of biological agents to an organic solvent environment when incorporated into in situ forming implants, potential interactions between the agents and solvents should be taken into account (Wang and Burgess 2021). Studies have indicated that the presence of NMP (N-Methyl-2-pyrrolidone) can lead to protein denaturation and degradation (Wang and Burgess 2021). As alternatives to NMP, other solvents such as DMSO (dimethyl sulfoxide), triacetin, benzyl benzoate, benzyl alcohol, PEG 400 (polyethylene glycol 400), and glycofurol have been suggested. For instance, in the case of bone morphogenetic protein (BMP) in situ forming implants, the use of a stabilizer (acetic acid) has enabled successful formulation with NMP as the solvent (Wang and Burgess 2021). Despite the disruptive effects of NMP on the secondary/tertiary structure of BMP2, it has been retained as a solvent due to its ability to enhance the bioactivity of BMPs within the body (Wang and Burgess 2021; Miguel et al. 2009).

In the case of lysozyme in situ forming implants, successful preparation has been achieved by homogenizing mannitol, a protein stabilizer, together with lysozyme into a solution composed of PLA (poly(lactic acid)), benzyl benzoate, and benzyl alcohol (Singh and Singh 2004). Both conformational and biological activities of lysozyme were notably enhanced. In addition, the presence of mannitol had a significant effect in reducing both burst release and the overall release rate of lysozyme. Polyols play a protective role for proteins in aqueous solutions through mechanisms involving hydration and the formation of hydrogen bonds (Allison et al. 1999).

In case of protein-based in situ forming implants, necessity for stabilizers depends on the choice of solvent. For instance, lysozyme was effectively incorporated into a PLGA/DMSO/ethyl acetate/water-based implant without the use of stabilizers (Singh and Singh 2004). DMSO was chosen due to its capacity to dissolve both lysozyme and PLGA. Ethyl acetate, which does not dissolve the protein, was added to fine-tune lysozyme solubility, and a small amount of water was introduced to control dissolution. Throughout the formulation process, lysozyme maintained its biological activity, regardless of the proportions of the ternary solvent blend. Lysozyme was evenly dispersed within the resulting formulation and exhibited controlled release over a 21-day period (in vitro study) (Wang and Burgess 2021). However, it needs to mention that for successful implementation of such a research work, local toxicology study needs to be performed to show the suitability of ethyl acetate.

Insulin has also been integrated into in situ forming implants. While mannitol was required to preserve lysozyme's bioactivity in the PLA/benzyl benzoate/benzyl alcohol system (Singh and Singh 2004), insulin did not necessitate the use of stabilizers (Kang and Singh 2005). Furthermore, a soluble tumor necrosis factor was successfully incorporated into a PLGA/bovine serum albumin/glycofurol-based in situ forming implant without requiring a stabilizer. Glycofurol, being a water-miscible, pharmaceutically acceptable solvent, is notably less toxic and irritating compared to NMP. It is suitable for use in parenteral products administered via intravenous, intranasal, and intramuscular routes. The in situ forming implant formulation released tumor necrosis factor over a 30-day period in vitro and exhibited enhanced treatment efficacy in vivo (Wang and Burgess 2021).

There have been a few examples of in situ forming implants designed for gene delivery (Jeon et al. 2011; Eliaz and Szoka 2002). In one approach, plasmid DNA was incorporated into a PLGA/glycofurol system (Eliaz and Szoka 2002), initially using a small amount of water to dissolve the DNA. This formulation enabled the controlled release of plasmid DNA over a 60-day period, ensuring continuous gene expression at the injection site. In another formulation, a PLGA/tetraglycol system was utilized, which could only deliver the gene for 7 days in vitro. Compared to other biologics, delivering peptides is a less challenging task due to their relatively higher stability. Successful examples in this category include the commercial product Eligard® and an LR12 peptide-based in situ forming implant (Wang and Burgess 2021; Parent et al. 2017).

4.12 Current Marketed In Situ Forming Implant Products

The few commercially available in situ forming implant drug products (**Atridox®**, **Eligard®**, **Sublocade®**, and **Perseris®**) are based on the Atrigel® technology. **UZEDY™** is a recent FDA-approved product from BEPO® technology as outlined in Table 4.2. These products utilize Atrigel® systems, which consist of viscous polymer solutions that are mixed with the drug prior to injection as double syringe, or the drug will be dissolved in the polymer solution as a single unite syringe and are therefore manufactured as prefilled syringes (Malik et al. 2010; Markowicz-piasecka et al. 2024). Since both the drug and the polymer are in a solution state in these products, there is a potential for degradation and chemical reactions between them compared to formulations in a dry, solid state. Consequently, the drug and polymer solutions are typically kept in separate syringes, with one syringe containing the liquid Atrigel® delivery system (colorless to yellow), and the other syringe containing the drug powder. Perseris® as an example is depicted in Fig. 4.2. For administration, liquid syringe and powder syringe are connected via luer lock fittings and their contents are thoroughly mixed by alternately pushing the materials back and forth until a homogeneous solution or suspension is obtained. The combined contents are then transferred into the liquid syringe. Afterward, the two syringes are separated, and a safety needle is attached for the injection process. This two-syringe mixing approach ensures maximum stability of both the drug and the polymer, and it allows the drug to undergo sterilization by gamma irradiation while in a dry state, further contributing to its stability. Consequently, these products typically have a shelf life of 1.5–2 years when stored at 2–8 °C. Notably, Sublocade® is the only product that comes in a single syringe, with the drug already dissolved in the Atrigel® system (Wang and Burgess 2021).

Atridox® is an in situ forming implant for subgingival use containing the antibiotic doxycycline hyclate using Atrigel® technology. It is a single-dose product supplied in a two-syringe mixing system. Syringe A contains 450 mg of Atrigel®

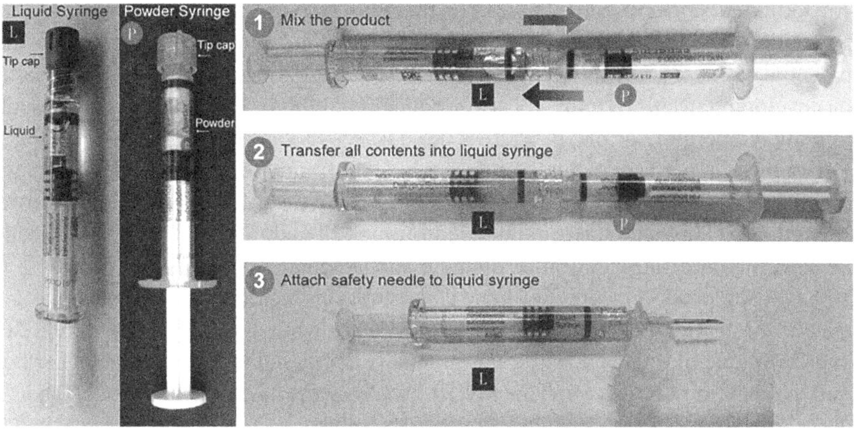

Fig. 4.2 In situ forming implant products with two-syringe mixing system and instruction for use (Perseris®). (Used with permission from Wang and Burgess (2021))

matrix, a depot that converts to a gel on the spot and is 36.7% poly (DL -lactide) (PLA) dissolved in 63.3% N-methyl-2-pyrrolidone. Syringe B contains 50 mg of doxycycline hyclate, which is equal to 42.5 mg doxycycline. After reconstitution, the product for subgingival administration appears as a pale-yellow viscous liquid with a doxycycline hyclate concentration of 10%. It is recommended for periodontal patients every 4 months. Clinical studies have shown that topical administration of Atridox® is effective in the treatment of periodontitis and is a less invasive alternative to procedures such as scaling and root planning (Nkanga et al. 2020). In addition, another clinical study conducted by Zeidner et al. (2008) found that a single subcutaneous application of Atridox® was a more effective prophylactic antibiotic choice compared to oral doxycycline for patients with co-infections of *Anaplasma phagocytophilum* and *Borrelia burgdorferi*.

Eligard® follows a similar design concept to Atridox® in that it utilizes an in situ forming gel based on the Atrigel® formulation. However, Eligard® is intended for the treatment of advanced prostatic cancer and contains leuprolide acetate as the active ingredient. The Atrigel® delivery system, consisting of PLGA dissolved in N-methyl-2-pyrrolidone (NMP), is mixed with leuprolide acetate powder upon use. This mixture forms a suspension that solidifies as a depot within the body, regulating the release of leuprolide. Eligard® is available in various forms with differences in PLGA composition and drug content. Clinical studies have demonstrated that Eligard® can sustain the release of leuprolide over the intended duration, which can range from 4 to 26 weeks depending on the formulation used. The leuprolide serum concentration remains relatively constant, typically ranging from 0.1 to 2 ng/mL for all dosages. However, a notable burst release occurs within 3–5 h after injection, with levels ranging from 25.3 ng/mL for 7.5 mg Eligard® to 150 ng/mL for 30 mg Eligard®. This initial burst release may be attributed to slower and incomplete solid depot formation in vivo, causing leuprolide to be rapidly washed out from the depot.

An additional concern with Eligard® is the use of the organic solvent N-methyl-2-pyrrolidone (NMP), which, although used in many FDA-approved products, has the potential to cause local tissue irritation. Consequently, ongoing research is focused on developing in situ forming gels from aqueous solutions, triggered by factors such as body temperature or pH, to address these issues (Nkanga et al. 2020).

Sublocade® is an extended-release buprenorphine in situ forming implant designed for treating moderate to severe opioid use disorder. It is available in two strengths: 100 mg and 300 mg, intended for monthly administration. Clinical studies have demonstrated that after a single subcutaneous injection of RBP-6000 at 300 mg (Sublocade®), buprenorphine concentrations reached their peak at approximately 17 h, ranging from 4 to 7 ng/mL. Following this initial burst release, buprenorphine plasma concentrations gradually decreased to approximately 1–2 ng/mL, maintaining this level for roughly 50 days. However, with repeated Sublocade® doses, the burst release became less, and the Sublocade® 300 mg dose reached a steady-state concentration after six monthly injections. At steady state, following the six injections, buprenorphine concentrations were maintained at around 8 ng/mL, which is four times higher than the concentration at steady state after the initial dose. No in vitro release data was available for Sublocade® (Wang and Burgess 2021).

Perseris® is another application of the Atrigel® delivery system, designed for the release of risperidone as a treatment for schizophrenia in adults. When mixed, risperidone forms a suspension within the Atrigel® delivery system, which is composed of PLGA (L/G 80:20 molar ratio) and NMP. This mixture subsequently solidifies into a depot within the body. Similar to Eligard®, Perseris® exhibits two distinct peaks in risperidone plasma levels. The first peak occurs approximately 4–6 h after administration, primarily due to the leakage of the drug during the formation of the depot (with a maximum concentration, C_{max}, of 10.9 ng/mL for a 120 mg dose). Subsequently, around day 10–14 post-administration, a second peak of similar magnitude to the initial burst release is observed. This second peak is likely a result of the onset of PLGA matrix degradation and bulk erosion (Nkanga et al. 2020).

UZEDY™ is the first subcutaneous, long-acting formulation of risperidone for treating schizophrenia in adults to reach the market based on BEPO® technology. BEPO® is an in situ forming depot (ISFD) technology that functions through a solvent-exchange mechanism. The drug delivery depot's matrix is made from a combination of diblock (DB) and triblock (TB) polyethylene glycol-polyester copolymers. This mixture allows for extensive customization of the drug release to achieve desired pharmacokinetics. The release rate and profile can be fine-tuned by altering the composition of either the TB or DB copolymers, or by adjusting their relative ratios, among other factors. Furthermore, the polymeric composition of the formulation significantly affects the solvent exchange rate between the organic solvent and the surrounding aqueous medium. This exchange rate, in turn, determines the internal structure of the resulting depot and the release pattern of the therapeutic agent (Roberge et al. 2020).

In May 2023, the FDA approved a new risperidone formulation, UZEDY™ (Teva and MedinCell), as an extended-release injectable suspension. This formulation uses technology to control the steady release of risperidone, achieving therapeutic blood concentrations within 6–24 h of a single dose. UZEDY™ is a sterile, white to off-white opaque viscous suspension available in various strengths: 50 mg/0.14 mL, 75 mg/0.21 mL, 100 mg/0.28 mL, 125 mg/0.35 mL, 150 mg/0.42 mL, 200 mg/0.56 mL, and 250 mg/0.7 mL. Each kit includes one single-dose prefilled syringe and one 21 gauge, 5/8-inch needle. The product allows for flexible 1- and 2-month dosing intervals (Markowicz-piasecka et al. 2024). UZEDY™ contains inactive ingredients such as dimethyl sulfoxide (45% w/w), methoxy-poly (ethylene glycol)-co-poly (D,L-lactide) (15% w/w), and poly (D,L-lactide)-co-poly (ethylene glycol)-co-poly (D,L-lactide) (10% w/w) (Markowicz-piasecka et al. 2024).

4.13 Some Challenges to Be Addressed

In situ forming implants present a promising pharmaceutical approach for controlled and sustained delivery, which offers less frequent administration, avoiding painful surgical procedures, and administration both locally and systemically. Yet despite their potential, few products have come to market. This limited commercialization may be due to several challenges that need to be overcome (Parent et al. 2013):

1. Rapid drug release: An important issue is the rapid initial release of the drug from these implants, which can be problematic.
2. Safety and tolerability: Ensuring the safety and tolerability of in situ forming implants is critical, as adverse effects may hinder their widespread use.
3. Reproducibility: Achieving consistent and reproducible results in the formulation and manufacturing of these implants is another challenge.
4. Sterilization and stability: Ensuring the sterilization and long-term stability of these implants is essential for their safety and effectiveness.

While certain in situ forming implant (ISFI) products are already on the market, the injection of organic solvent(s) continues to raise concerns. Selecting the most suitable solvent is challenging, due to the limited toxicological data available for their parenteral administration. Nonetheless, NMP (N-methyl-2-pyrrolidone), DMSO (dimethyl sulfoxide), BA (benzyl alcohol), BB (benzyl benzoate), and glycofurol have been utilized in injectable products for human use (Strickley 2004; Mottu et al. 2000), while 2P (2-pyrrolidone) and TA (triacetin) have been employed in veterinary products (Food and Drug Administration 2008).

Toxicity evaluations of ISFI appear to exhibit variations based on factors such as the administration route, employed methodology, evaluation timing, and the specific animal model utilized. There is a clear need for comprehensive toxicological studies with standardized methodologies to enhance the prospects of ISFI applications. Despite the anticipated costs, such studies would not only contribute to ISFI development but also benefit other drug delivery systems, particularly for drugs with limited water solubility. Novel solvents like low molecular weight poly(ethylene glycol) (PEG) have been introduced (Schoenhammer et al. 2009a, b, 2010).

Nonetheless, it is noteworthy that conventional PEG can speed up the degradation of PLGA in solutions through a trans-esterification mechanism. To address this, significant improvement was achieved by capping the polymer ends with alkyl groups. Furthermore, transient edema was observed with PEG (Parent et al. 2013). An alternative strategy to enhance the safety of in situ systems involves minimizing or eliminating the requirement for organic solvents by creating new polymers. For instance, researchers have synthesized alkyl-substituted polylactides using the monomer 2-hydroxyoctanoic acid (Asmus et al. 2011). These resulting biodegradable polymers, referred to as "hexyl-substituted polylactides" or "hex-PLA," yield viscous solutions that can be injected with minimal or no intervention of NMP (less than 5%) and have demonstrated favorable biocompatibility (Asmus et al. 2013). However, the use of potentially harmful organic solvents remains a significant concern for these formulations. Hence, conducting toxicological studies and exploring new solvents could contribute to the further advancement of ISFI systems.

Moreover, an essential aspect of drug delivery systems is the consistency of drug-release properties, which depends on the uniformity of shape, size, and structure. The structure of ISFI is significantly influenced by the composition and the rate of phase inversion. In contrast, factors related to the formulation, the administration method, and the environmental conditions impact the shape and size of these implants (Kempe and Mäder 2012; Parent et al. 2013). Given that in situ systems solidify at the injection site, they are understandably sensitive to environmental variations. Consequently, strong correlations between polymer precipitation and drug release have been achieved in both in vitro and in vivo. However, establishing connections between in vitro and in vivo drug releases remains a challenging endeavor (Parent et al. 2013).

Besides potential differences in the composition of the surrounding external tissue (including proteins, salts, and acids), one of the primary influencing factors in vivo is the stiffness of the tissue. Originally, in situ forming implants (ISFI) were designed for subcutaneous or intramuscular injection. However, emerging applications in preclinical studies are expanding to more unconventional injection sites, such as directly into specific areas like the eye, the brain, physiological pockets, tumors, or bone defects. This is possible because ISFI can readily adapt to the surrounding tissue, ensuring a high level of contact. Nevertheless, the ultimate shape of the injected implant can impact diffusion conditions and, consequently, drug release. Ultimately, the development of injection devices designed to regulate the depth and rate of injection could potentially expand the clinical applications of ISFI in the future (Parent et al. 2013).

A critical component of in situ systems is the polymer, with PLA and PLGA being highly popular due to their established safety records and wide availability. However, these synthetic polymers come with a higher cost, and variations in suppliers or batch-to-batch differences can result in distinct properties. For instance, the copolymer microstructure, characterized by the alternation of GA/LA units, is significantly influenced by the polymerization conditions (Okada 2002).

Moreover, despite the non-toxicity and favorable in vivo degradation characteristics of PLGA, potential interactions between the polymer and drugs or the creation of an acidic microenvironment during degradation may pose additional challenges. Shifting toward other polymers with surface erosion or homogeneous bulk erosion properties could be a viable approach (Parent et al. 2013).

4.14 Conclusion

Implantable drug delivery systems (IDDSs) offer an exceptionally appealing therapeutic approach that overcomes the constraints associated with traditional drug administration methods. Biodegradable in situ forming implants have gained considerable popularity due to their remarkable advantages, which include ease of manufacture, convenient routes of administration, reduced dosing frequency, and improved patient compliance. In further studies, it is important to investigate novel solvent or polymer combinations that show excellent drug release properties with high biocompatibility. In addition to delivering small molecules, these systems hold a great potential for delivering large molecules such as therapeutic proteins and oligonucleotides.

References

Agarwal P, Rupenthal ID. Injectable implants for the sustained release of protein and peptide drugs. Drug Discov Today. 2013;00(00):1–13. https://doi.org/10.1016/j.drudis.2013.01.013.

Allison SD, Chang B, Randolph TW, Carpenter JF. Hydrogen bonding between sugar and protein is responsible for inhibition of dehydration-induced protein unfolding. Arch Biochem Biophys. 1999;365(2):289–98.

Almoshari Y. Osmotic pump drug delivery systems—a comprehensive review. Pharmaceuticals. 2022;15(11):1430.

Alonso PE, Perula LA, Rioja LF. Pain-temperature relation in the application of local anaesthesia. Br J Plast Surg. 1993;46(1):76–8.

Asmus LR, Gurny R, Möller M. Solutions as solutions--synthesis and use of a liquid polyester excipient to dissolve lipophilic drugs and formulate sustained-release parenterals. Eur J Pharm Biopharm. 2011;79(3):584–91.

Asmus LR, et al. In vivo biocompatibility, sustained-release and stability of triptorelin formulations based on a liquid, degradable polymer. J Control Release. 2013;165(3):199–206.

Baracu AM, Dinu Gugoasa LA. Review—recent advances in microfabrication, design and applications of amperometric sensors and biosensors. J Electrochem Soc. 2021;168(3):037503.

Bassand C, Benabed L, Charlon S, et al. 3D printed PLGA implants: APF DDM vs. FDM. JCR. 2022;353:864–74.

Bekkem A, Selby G, Chakrabarty JH. Retrospective analysis of intravenous DMSO toxicity in transplant patients. Biol Blood Marrow Transplant. 2013;19(2):S313. https://doi.org/10.1016/j.bbmt.2012.11.466.

Bittner B, Richter W, Schmidt J. Subcutaneous administration of biotherapeutics: an overview of current challenges and opportunities. BioDrugs. 2018;32(5):425–40. https://doi.org/10.1007/s40259-018-0295-0.

Bode C, Kranz H, Siepmann F, Siepmann J. In-situ forming PLGA implants for intraocular dexamethasone delivery. Int J Pharm. 2018;548(1):337–48.

Breitenbach J. Melt extrusion: from process to drug delivery technology. Eur J Pharm Biopharm. 2002;54(2):107–17.
Camargo JA, et al. Injectable PLA-based in situ forming implants for controlled release of Ivermectin a BCS class II drug: solvent selection based on physico-chemical characterization. Drug Dev Ind Pharm. 2013;39(1):146–55.
Chen Y, et al. Phytantriol-based in situ liquid crystals with long-term release for intra-articular administration. AAPS PharmSciTech. 2015;16(4):846–54.
Dash AK, Cudworth GC. Therapeutic applications of implantable drug delivery systems. J Pharmacol Toxicol Methods. 1998;40(1):1–12.
Dominas C, Deans K, Packard R, Jonas O. Preparation and sterilization of an implantable drug-delivery microdevice for clinical use. MethodsX. 2021;8(May):101382. https://doi.org/10.1016/j.mex.2021.101382.
Dong WY, Körber M, López Esguerra V, Bodmeier R. Stability of poly(D,L-Lactide-co-glycolide) and leuprolide acetate in in-situ forming drug delivery systems. J Control Release. 2006;115(2):158–67.
Einmahl S, et al. Therapeutic applications of viscous and injectable poly(ortho esters). Adv Drug Deliv Rev. 2001a;53(1):45–73.
Einmahl S, Schwach-Abdellaoui K, Heller J, Gurny R. Poly(ortho esters): recent developments for biomedical applications. Chimia. 2001b;55(3):218–22.
Eliaz RE, Jr Szoka FC. Robust and prolonged gene expression from injectable polymeric implants. Gene Ther. 2002;9(18):1230–7.
FDA. Appendix 6. Toxicological data for class 3 solvents. Guidance compliance regulatory information. 1998. p. 57. https://www.fda.gov/downloads/Drugs/GuidanceComplianceRegulatoryInformation/Guidances/UCM073403.pdf.
Fialho SL, Da Silva Cunha A. Manufacturing techniques of biodegradable implants intended for intraocular application. Drug Deliv. 2005;12(2):109–16.
Fogueri LR, Singh S. Smart polymers for controlled delivery of proteins and peptides: a review of patents. Recent Pat Drug Deliv Formul. 2009;3(1):40–8.
Food and Drug Administration. New animal drug application. NADA; 2008. p. 141–265.
Ganesh K, et al. A review of manufacturing techniques for subcutaneous drug delivery implants. Procedia CIRP. 2022;110(C):331–6.
Gulati K, et al. Drug-releasing nano-engineered titanium implants: therapeutic efficacy in 3D cell culture model, controlled release and stability. Mater Sci Eng C. 2016;69:831–40. https://doi.org/10.1016/j.msec.2016.07.047.
Haering G, Luisi PL. Hydrocarbon gels from water-in-oil microemulsions. J Phys Chem. 1986;90:5892–5.
Hatefi A, Amsden B. Biodegradable injectable in situ forming drug delivery systems. J Control Release. 2002;80:9–28.
Heller J, et al. Poly(ortho esters) - their development and some recent applications. Eur J Pharm Biopharm. 2000;50(1):121–8.
Holmberg A. Developing an implantable drug delivery system for ecological studies on fish. 2020. http://urn.kb.se/resolve?urn=urn:nbn:se:umu:diva-169961.
Hussain S, Solanki D, Yadav R, Khan Y. Implantable drug delivery system: an overview. IJPPR. 2021;20(4):116–32. www.ijppr.humanjournals.com
Hutton ARJ, et al. Transdermal delivery of vitamin K using dissolving microneedles for the prevention of vitamin K deficiency bleeding. Int J Pharm. 2018;541(February):56–63.
Ibrahim TM, El-Megrab NA, El-Nahas HM. Optimization of injectable PLGA in-situ forming implants of anti-psychotic risperidone via Box-Behnken design. J Drug Deliv Sci Technol. 2020;58(March):101803. https://doi.org/10.1016/j.jddst.2020.101803.
Ibrahim TM, El-Megrab NA, El-Nahas HM. An overview of PLGA in-situ forming implants based on solvent exchange technique: effect of formulation components and characterization. Pharm Dev Technol. 2021;26(7):709–28.
Jeon O, Krebs M, Alsberg E. Controlled and sustained gene delivery from injectable, porous PLGA scaffolds. J Biomed Mater Res A. 2011;98(1):72–9.

Jeong B, Kim SW, Bae YH. Thermosensitive sol-gel reversible hydrogels. Adv Drug Deliv Rev. 2002;54(1):37–51.
Johnson J. Pharmacology/toxicology NDA review and evaluation. 2020. 0–14.
Joo MK, Park MH, Choi BG, Jeong B. Reverse thermogelling biodegradable polymer aqueous solutions. J Mater Chem. 2009;19(33):5891–905.
Juvekar S, Kathpalia H. Solvent removal precipitation based in situ forming implant for controlled drug delivery in periodontitis. J Control Release. 2017;251:75–81. https://doi.org/10.1016/j.jconrel.2017.02.022.
Kang F, Singh J. In vitro release of insulin and biocompatibility of in situ forming gel systems. Int J Pharm. 2005;304(1–2):83–90.
Kapoor DN, Katare OP, Dhawan S. In situ forming implant for controlled delivery of an anti-HIV fusion inhibitor. Int J Pharm. 2012;426(1–2):132–43.
Karp F, et al. In situ formed implants, based on PLGA and Eudragit blends, for novel Florfenicol controlled release formulations. J Pharm Sci. 2021;110(3):1270–8.
Kempe S, Mäder K. In situ forming implants—an attractive formulation principle for parenteral depot formulations. J Control Release. 2012;161(2):668–79. https://doi.org/10.1016/j.jconrel.2012.04.016.
Kjar A, Huang Y. Application of micro-scale 3D printing in pharmaceutics. Pharmaceutics. 2019;11(8):390.
Kleiner LW, Wright JC, Wang Y. Evolution of implantable and insertable drug delivery systems. J Control Release. 2014;181(1):1–10. https://doi.org/10.1016/j.jconrel.2014.02.006.
Korelidou A, et al. 3D-printed reservoir-type implants containing poly(lactic acid)/poly(caprolactone) porous membranes for sustained drug delivery. Biomater Adv. 2022;139(July):213024.
Kumar A, Pillai J. "Implantable Drug Delivery Systems. In Nanostructures for the Engineering of Cells, Tissues and Organs." Elsevier: Amsterdam, The Netherlands. 2018;473–511.
Larrañeta E, Raghu Raj Singh T, Donnelly RF. Overview of the clinical current needs and potential applications for long-acting and implantable delivery systems. In: Long-acting drug delivery systems: pharmaceutical, clinical, and regulatory aspects. Woodhead Publishing; 2021. p. 1–16.
Lee Y, et al. Immediate detachment of microneedles by interfacial fracture for sustained delivery of a contraceptive hormone in the skin. J Control Release. 2021;337:676–85.
Li W, Tang J, et al. Long-acting reversible contraception by effervescent microneedle patch. Sci Adv. 2019a;5(11):eaaw8145.
Li W, Terry RN, et al. Rapidly separable microneedle patch for the sustained release of a contraceptive. Nat Biomed Eng. 2019b;3(3):220–9. https://doi.org/10.1038/s41551-018-0337-4.
Li Y, et al. In situ phase transition of microemulsions for parenteral injection yielding lyotropic liquid crystalline carriers of the antitumor drug bufalin. Colloids Surf B Biointerfaces. 2019c;173:217–25. https://www.sciencedirect.com/science/article/pii/S0927776518306349
Li W, et al. Core-shell microneedle patch for six-month controlled-release contraceptive delivery. J Control Release. 2022;347:489–99.
Luraghi A, Peri F, Moroni L. "Electrospinning for Drug Delivery Applications: A Review." J Control Release. 2021;(334):463–84.
Magill E, et al. Solid implantable devices for sustained drug delivery. Adv Drug Deliv Rev. 2023;199:114950.
Malaterre V, Ogorka J, Loggia N, Gurny R. Approach to design push-pull osmotic pumps. Int J Pharm. 2009;376(1–2):56–62.
Malik K, Singh I, Nagpal M, Arora S. Atrigel: a potential parenteral controlled drug delivery system. Pharm Sin. 2010;1(1):74–81. http://www.pelagiaresearchlibrary.com/der-pharmacia-sinica/vol1-iss1/DPS-2010-1-1-74-81.pdf
Maniruzzaman M, Boateng JS, Snowden MJ, Douroumis D. A review of hot-melt extrusion: process technology to pharmaceutical products. ISRN Pharm. 2012;2012:1–9.
Markowicz-piasecka M, et al. Long-acting injectable antipsychotics—a review on formulation and in vitro dissolution. Pharmaceutics. 2024;16(1):28.

McAlister E, et al. The role of microneedle arrays in drug delivery and patient monitoring to prevent diabetes induced fibrosis. Adv Drug Deliv Rev. 2021;175:113825. https://doi.org/10.1016/j.addr.2021.06.002.
Miguel BS, et al. N-methyl pyrrolidone as a potent bone morphogenetic protein enhancer for bone tissue regeneration. Tissue Eng Part A. 2009;15(10):2955–63.
Mohammad Zaki AJ, Patil SK, Baviskar DT, Jain DK. Implantable drug delivery system: a review. Int J PharmTech Res. 2012;4(1):280–92.
Mottu F, Laurent A, Rufenacht DA, Doelker E. Organic solvents for pharmaceutical parenterals and embolic liquids: a review of toxicity data. PDA J Pharm Sci Technol. 2000;54(6):456–69.
Murdan S. Organogels in drug delivery. Expert Opin Drug Deliv. 2005;2(3):489–505.
Nkanga CI, Fisch A, Rad-Malekshahi M, et al. Clinically established biodegradable long acting injectables: an industry perspective. Adv Drug Deliv Rev. 2020;167:19–46. https://doi.org/10.1016/j.addr.2020.11.008.
Okada M. Chemical syntheses of biodegradable polymers. Prog Polym Sci. 2002;27(1):87–133. https://www.sciencedirect.com/science/article/pii/S0079670001000399
Omer S, Forgách L, Zelkó R, Sebe I. Scale-up of electrospinning: market overview of products and devices for pharmaceutical and biomedical purposes. Pharmaceutics. 2021;13(2):1–21.
Ozalp Y, Ozdemir N, Hasirci V. Vancomycin release from poly(D,L-lactide) and poly(lactide-co-glycolide) disks. J Microencapsul. 2002;19(1):83–94.
Packhaeuser CB, Schnieders J, Oster CG, Kissel T. In situ forming parenteral drug delivery systems: an overview. Eur J Pharm Biopharm. 2004;58(2):445–55.
Paredes AJ, et al. Novel tip-loaded dissolving and implantable microneedle array patches for sustained release of finasteride. Int J Pharm. 2021;606:120885.
Parent M, et al. PLGA in situ implants formed by phase inversion: critical physicochemical parameters to modulate drug release. J Control Release. 2013;172(1):292–304. https://doi.org/10.1016/j.jconrel.2013.08.024.
Parent M, et al. One-week in vivo sustained release of a peptide formulated into in situ forming implants. Int J Pharm. 2017;521(1–2):357–60.
Patel KN, Mehta TA. A review on oral osmotically driven systems. Int J Pharm Pharm Sci. 2013;5(3):996–1004.
Patki M, Palekar S, Reznik S, Patel K. Self-injectable extended release formulation of Remdesivir (SelfExRem): a potential formulation alternative for COVID-19 treatment. Int J Pharm. 2021;597(January):120329. https://doi.org/10.1016/j.ijpharm.2021.120329.
Peng K, et al. Dissolving microneedle patches loaded with amphotericin B microparticles for localised and sustained intradermal delivery: potential for enhanced treatment of cutaneous fungal infections. J Control Release. 2021;339:361–80.
Picco CJ, et al. 3D-printed implantable devices with biodegradable rate-controlling membrane for sustained delivery of hydrophobic drugs. Drug Deliv. 2022;29(1):1038–48.
Picco CJ, et al. Development of 3D-printed subcutaneous implants using concentrated polymer/drug solutions. Int J Pharm. 2023;631(September 2022):122477.
Pillai O, Panchagnula R. Polymers in drug delivery Omathanu Pillai and Ramesh Panchagnula. Curr Opin Chem Biol. 2001;5:447–51.
Prabakaran D, Singh P, Kanaujia P, Vyas SP. Effect of hydrophilic polymers on the release of diltiazem hydrochloride from elementary osmotic pumps. Int J Pharm. 2003;259(1–2):173–9.
Prata AI, Coimbra P, Eugénia Pina M. Preparation of dexamethasone ophthalmic implants: a comparative study of in vitro release profiles. Pharm Dev Technol. 2018;23(3):218–24.
Rahnfeld L, Luciani P. Injectable lipid-based depot formulations: where do we stand? Pharmaceutics. 2020;12(6):1–28.
Roberge C, et al. BEPO®: bioresorbable diblock MPEG-PDLLA and triblock PDLLA-PEG-PDLLA based in situ forming depots with flexible drug delivery kinetics modulation. J Control Release. 2020;319:416–27. https://doi.org/10.1016/j.jconrel.2020.01.022.
Ruel-Gariépy E, Leroux J-C. In situ-forming hydrogels--review of temperature-sensitive systems. Eur J Pharm Biopharm. 2004;58(2):409–26.

Saraf S, et al. Poly(ethylene glycol)-poly(lactic-co-glycolic acid) based thermosensitive injectable hydrogels for biomedical applications. J Control Release. 2013;172(3):715–29. https://doi.org/10.1016/j.jconrel.2013.10.006.

Scartazzini R, Luisi PL. Organogels from lecithins. J Phys Chem. 1988;92(3):829–33.

Schoenhammer K, Petersen H, Guethlein F, Goepferich A. Injectable in situ forming depot systems: PEG-DAE as novel solvent for improved PLGA storage stability. Int J Pharm. 2009a;371(1–2):33–9.

Schoenhammer K, Petersen H, Guethlein F, Goepferich A. Poly(ethyleneglycol) 500 dimethylether as novel solvent for injectable in situ forming depots. Pharm Res. 2009b;26(12):2568–77.

Schoenhammer K, et al. Biocompatibility of an injectable in situ forming depot for peptide delivery. J Pharm Sci. 2010;99(10):4390–9.

Shahi S, Zadbuke N, Jadhav A, Borde S. Osmotic controlled drug delivery systems: an overview. Pharm Technol Innov. 2015;03(15):32–49.

Shchipunov YA, Shumilina EV. Lecithin organogels: role of polar solvent and nature of intermolecular interactions. Colloid J. 1996;58(1):117–25.

Singh S, Singh J. Controlled release of a model protein lysozyme from phase sensitive smart polymer systems. Int J Pharm. 2004;271(1–2):189–96.

Singhvi G, Banerjee S, Khosa A. Lyotropic liquid crystal nanoparticles: a novel improved lipidic drug delivery system. In: Organic materials as smart nanocarriers for drug delivery. Elsevier; 2018. https://doi.org/10.1016/B978-0-12-813663-8.00011-7.

Stewart SA, Domínguez-Robles J, Donnelly RF, Larrañeta E. Implantable polymeric drug delivery devices: classification, manufacture, materials, and clinical applications. Polymers. 2018;10(12):1379.

Stewart SA, Domínguez-Robles J, Donnelly RF, Larrañeta E. Evaluation of sterilisation techniques for 3D-printed implantable devices. RPS Pharm Pharmacol Rep. 2023;2(1):1–12.

Strickley RG. Solubilizing excipients in Oral and injectable formulations. Pharm Res. 2004;21(2):201–30.

Tang H, et al. Recent Development of PH-responsive polymers for cancer nanomedicine. Molecules. 2018;24(1):4.

Tata M, John VT, Waguespack YY, Mcpherson GL. Microstructural characterization of novel phenolic organogels through high-resolution NMR-spectroscopy. J Phys Chem. 1994;98:3809–17.

Than A, et al. Self-implantable double-layered micro-drug-reservoirs for efficient and controlled ocular drug delivery. Nat Commun. 2018;9(1):1–12. https://doi.org/10.1038/s41467-018-06981-w.

Tipnis NP, Burgess DJ. Sterilization of implantable polymer-based medical devices: a Review. Int J Pharm. 2018;544(2):455–60.

Ulery BD, Nair LS, Laurencin CT. Biomedical applications of biodegradable polymers. J Polym Sci B Polym Phys. 2011;49(12):832–64.

Utomo E, et al. Development of intranasal implantable devices for schizophrenia treatment. Int J Pharm. 2022;624(July):122061.

Vora LK, et al. Microneedle array systems for long-acting drug delivery. Eur J Pharm Biopharm. 2021;159(December 2020):44–76. https://doi.org/10.1016/j.ejpb.2020.12.006.

Vora L, Moffatt K, Donnelly R. Long-lasting drug delivery systems based on microneedles. In: Long-acting drug delivery systems: pharmaceutical, clinical, and regulatory aspects. Elsevier; 2022. p. 249–87.

Wang X, Burgess DJ. Drug release from in situ forming implants and advances in release testing. Adv Drug Deliv Rev. 2021;178(xxxx):113912. https://doi.org/10.1016/j.addr.2021.113912.

Wang L, et al. Design of a long-term antipsychotic in situ forming implant and its release control method and mechanism. Int J Pharm. 2012;427(2):284–92.

Xia M-q, et al. Pharmacokinetics of sinomenine hydrochloride cubic liquid crystal injection based on microdialysis technology. J Drug Deliv Sci Technol. 2019;52:553–8. https://www.sciencedirect.com/science/article/pii/S1773224719301091

Xu XD, Ayyagari M, Tata M, McPherson GL, John VT. Formation of novel organogels by the addition of phenols to AOT micelles in isooctane. J Phys Chem. 1993;97:11350–3.
Yaghmur A, Rappolt M, Larsen SW. In situ forming drug delivery systems based on lyotropic liquid crystalline phases: structural characterization and release properties. J Drug Deliv Sci Technol. 2013;23(4):325–32. https://doi.org/10.1016/S1773-2247(13)50049-9.
Zeidner NS, et al. A sustained-release formulation of doxycycline hyclate (Atridox) prevents simultaneous infection of Anaplasma Phagocytophilum and Borrelia Burgdorferi transmitted by tick bite. J Med Microbiol. 2008;57(Pt 4):463–8.
Zhao X, et al. Exploring trehalose on the release of levonorgestrel from implantable PLGA microneedles. Polymers. 2020;12(1):59.
Zheng Y, Pokorski JK. Hot melt extrusion: an emerging manufacturing method for slow and sustained protein delivery. Wiley Interdiscip Rev Nanomed Nanobiotechnol. 2021;13(5):1–23.

Chapter 5
Suspension-Based Long-Acting Injectables

Noor Al-Rifai and René Holm

List of Abbreviations

API	Active pharmaceutical ingredient
CFD-DEM	Computational fluid dynamics—discrete element methods
CQA	Critical quality attributes
DLS	Dynamic light scattering
FBRM	Focused beam reflectance measurements
FDA	Food and Drug Administration
HIV	Human immunodeficiency virus
HPH	High-pressure homogenization
IA	Intra-arterial
IB	Intra-bursal
ICH	International Council for Harmonisation of Technical Requirements for Pharmaceuticals for Human Use
IL	Intra-lesional
IM	Intramuscular
IV	Intravenous
IVIVC	In vitro–in vivo correlations
LAIs	Long-acting injectables
PA	Periarticular

N. Al-Rifai (✉)
Janssen Pharmaceutica NV, A Johnson & Johnson Company, Pharmaceutical Product Development & Supply, Beerse, Belgium
e-mail: NAlRifai@its.jnj.com

R. Holm (✉)
Department of Physics, Chemistry, and Pharmacy, University of Southern Denmark, Odense, Denmark
e-mail: reho@sdu.dk

F. Ramazani (ed.), *Biodegradable Long Acting Injectables and Implants*,
https://doi.org/10.1007/978-3-031-72302-5_5

PAT	Process analytical technologies
PEG	Polyethylene glycol
PSD	Particle size distribution
PVP	Polyvinylpyrrolidone
QbD	Quality-by-design
SC	Subcutaneous
SE	Stress energy
SN	Stress number
ST	Soft tissue administration

5.1 Introduction

Aqueous suspensions, also occasionally termed aqueous crystalline suspensions, are formulations in which the active pharmaceutical ingredient (API) is distributed in an aqueous medium at a predetermined particle size, typically between 0.1 μm and 10 μm (Holm et al. 2023; Kipp 2004; Bauer et al. 2023). This type of formulation depends on the API's high lipophilicity and low water solubility as the drug-retaining method. The creation of suspension-based long-acting injectables (LAIs) has been the subject of high research activity in recent years, as evidenced by the availability of a large number of papers, patents and commercial products using the technology.

The interest in LAIs is in general motivated by the convenience and preference of the patient for less frequent dosing (Blackwood et al. 2020; Margolis et al. 2017; Burke et al. 2018; Garris et al. 2022). For an LAI, a 2-week dose or more needs to be included into a volume that preferably can be injected by one administration; hence, the obtainable concentration in an LAI may have an influence on its acceptability among the patients the formulation is intended to help, i.e. the potential dose that can be contained and therefore the frequency of administration. Aqueous suspensions can contain very high API contents, up to 400 mg/mL as in the case of Depo-Provera. This makes the formulation approach stand out among the other LAI formulation strategies that are injected, such as oil solutions and microspheres, where less API is typically incorporated (Alidori et al. 2024).

The main method for adjusting the release rate of LAI suspensions relies on adjusting the crystalline API's particle size distribution (PSD). The Noyes-Whitney/Nernst-Brunner equation (Eq. 5.1) (Dokoumetzidis and Macheras 2006) provides a fundamental explanation to this, through its description of dissolution rate:

$$\frac{dm}{dt} = \frac{D \cdot A}{h}\left(C_s - C_t\right), \tag{5.1}$$

where m is the total dissolved mass, A is the total exposed particle surface area, D is the diffusion coefficient of the solute in the solvent, h is the thickness of the diffusion layer surrounding the particle, C_s is the saturated solubility and C_t is the actual concentration of dissolved drug at any given time (Dokoumetzidis and Macheras 2006; Siepmann and Siepmann 2013). From a conceptual perspective, the formulation technology is therefore well understood, which in combination with the potential for a high drug load and a regulatory path for the formulation makes it a widely applied approach in the pharmaceutical industry when a new LAI is considered.

This chapter aims to provide a holistic overview of the state-of-the-art product and process development, challenges and opportunities (covering the type of products and applicable therapeutic areas), formulation optimization approaches and factors that are critical to process understanding and which can guide process development/scale-up, predictive modelling that can facilitate resource reduction and process analytical technologies—all elements that, when combined, can contribute towards an effective control strategy (Fig. 5.1).

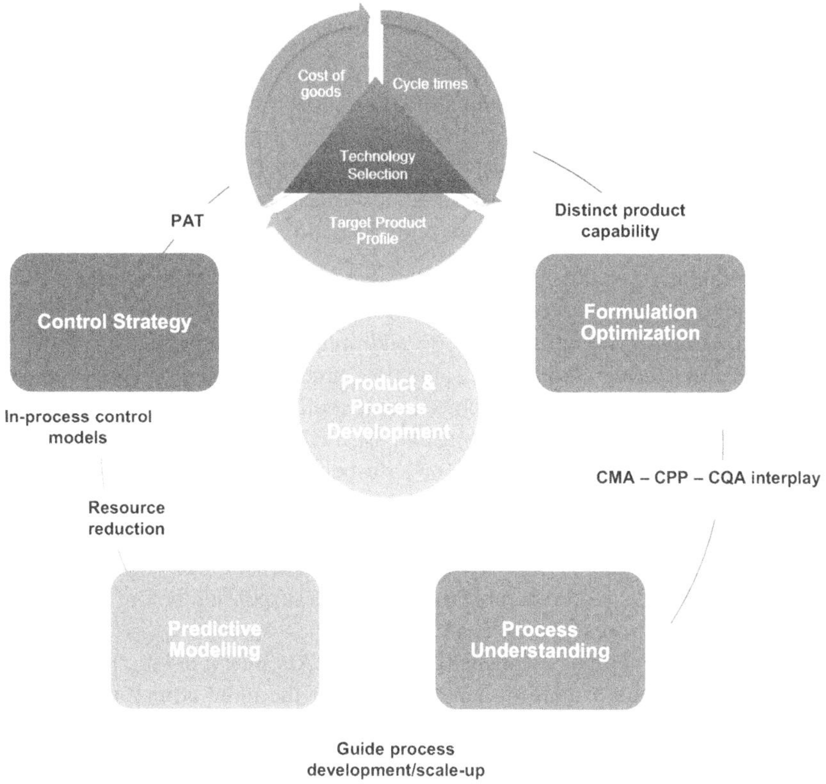

Fig. 5.1 Factors associated with suspension-based LAI product and process development

5.2 Manufacturing Methods of Micro- and Nanosuspensions

There are two primary categories of manufacturing methods for aqueous suspension-based LAIs: bottom-up (building particles from molecules) and top-down (breaking down large particles into smaller ones). Figure 5.2 outlines the various technologies used to produce micro- and nano-crystals through these methods. In the industry today, the production of long-acting suspensions mainly utilizes top-down approaches such as high-pressure homogenization and wet stirred media milling, while bottom-up methods are still in the early stages of development and are primarily used in research settings (Al-Rifai and Holm 2021).

Top-down manufacturing of LAIs is generally considered scalable and robust. However, these processes are energy intensive, and only a fraction of the applied energy goes into effective comminution, which generates:

- A heat transfer challenge due to heat generation, which mandates significant cooling/temperature control
- Potential for chemical product degradation due to mechanical or thermal stress
- Potential wear of both the grinding media and equipment (though avoidable via careful material selection)
- Unwanted changes in physical or chemical properties, e.g. process induced amorphization or creation of hydrophobic surfaces, which can impact the material processability and drug-release properties of the final product

The formulation (stabilizer) can play a central role in a comminution process, where agglomeration or recrystallization of nano/microsized fragments is competing with the top-down comminution. In addition, the formula needs to allow for optimal manufacturing while at the same time producing a milled suspension that would be stable on shelf life—two objectives that often have different requirements. Following the formation of the suspension, the final dosage form can take the form of a suspension that is ready-to-use or a suspension that is lyophilized (freeze-dried). Lyophilization is a technique used to stabilize a suspension that is physically unstable, preventing the growth of particle size during storage. If a lyophilized suspension is chosen, the end goal is to ensure that the reconstituted lyophilized formulation has a particle size comparable to the original suspension before it underwent the lyophilization process (Holm et al. 2023).

Manufacturing methods for suspension-based LAIs shown in Fig. 5.2 are briefly described below with a more in-depth analysis of the more common methods in Sect. 5.4.

Top-Down Comminution

Top-down comminution involves the reduction of particle size through mechanical forces utilizing various aids: rotating equipment parts, grinding media or motion. The following are the most common techniques for LAI manufacturing and the most described processes in the literature are discussed further in Sect. 5.4.3.3.

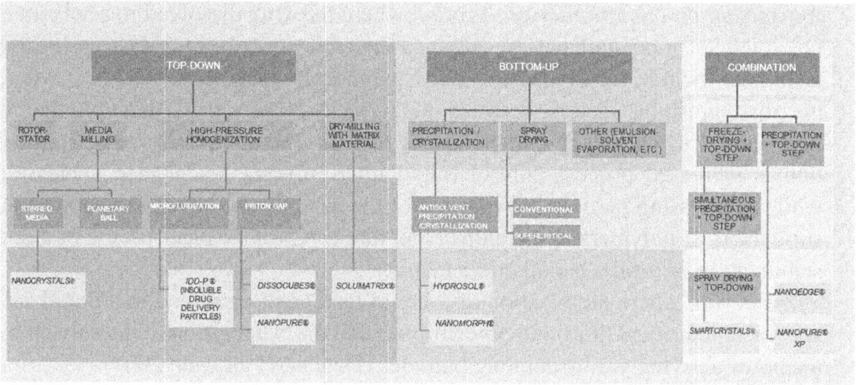

Fig. 5.2 Overview of the different technologies available for the production of micro and nano-crystals. (Adapted from Al-Rifai and Holm 2021)

- *Rotor-stator-type homogenizers* can deaggregate and for very brittle compounds, may be able to micronize. The ability of this manufacturing equipment to comminute is largely dependent on the particle properties.
- *Media milling* utilizes grinding media for comminution and includes stirred media mills and planetary ball mills.
- *High-pressure homogenization* equipment can be classified into two main types:
  - Microfluidization: This relies on a jet stream that is passed through a micro-scale channel thereby generating high pressures of 1000–2000 bar causing particle breakage.
  - Piston-gap homogenization: This operates on a similar high-pressure principle and forcing the suspension through a small gap that generates the high pressures. Essentially the main difference to the microfluidizer is the design of the valve/interaction chamber. Two different types of technology exist which depend on the dispersion medium type: Dissocubes® and Nanopure ® (see Sect. 5.4.3.3.2.1).
- *Dry milling* is a popular choice and can be appropriately deployed as a terminal step in a drug substance manufacturing process. However, it has limitations due to the lowest attainable particle size due to the relatively lower levels of energy achievable in comparison to the wet comminution techniques.

Bottom-Up Synthesis

The following are the most described techniques in the literature for bottom-up synthesis of nano/micro-suspensions for drug delivery applications and are reviewed in more depth in these texts (Chan and Kwok 2011; Sinha et al. 2013) and in Sect. 5.4.3.3.3:

- *Antisolvent precipitation/crystallization* where the drug dissolved in a solvent is precipitated (or crystallized) by adding a miscible antisolvent. Variations of this technique include the Hydrosol® and Nanomorph® (see Sect. 5.4.3.3.3) where the distinguishing factors between the typical coarse crystallization that is applied in drug substance manufacturing and the nanoparticle-forming application is increased process control (rapid micro-mixing to enhance nucleation while suppressing particle growth) and/or the use of stabilizers (ionic/non-ionic surfactants or polymers in the antisolvent that can provide electrostatic or steric stabilization or form a nanoparticle matrix).
- *Spray drying* where atomization of a liquid feed (API dissolved in a liquid solvent or supercritical fluid) into fine droplets and the evaporation of the solvent by means of a drying gas, generating particles (Date and Patravale 2004).
- *Emulsion-solvent evaporation* where an amphiphilic polymer solution is mixed with a drug-containing volatile immiscible organic solvent to form an emulsion, which is then nanosized with an appropriate technique (e.g. sonication) and converted into a polymeric nanoparticle suspension after the evaporation of the solvent from the polymer solution (Wright and Burgess 2012; Lee et al. 2006; Zhang et al. 2016).

Combination Methods

Combination methods also exist and mostly aim to alter the API millability by applying some form of bottom-up pretreatment (e.g. spray drying or freeze drying) that changes the brittleness of the API and renders it more amenable to comminution during a subsequent milling step. The reader is referred to Salazar et al. (2014) for a more in-depth review of combinative particle size reduction technologies.

5.3 Application of Micro- and Nanosuspensions in Long-Acting Injectables

Marketed Products Based on Micro- and Nanosuspensions

The US Food and Drug administration (FDA) gave the first marketing approval for an intramuscular (IM) administered LAI in 1952. It was an antibiotic meant to be injected every second or fourth week. Within that same decade, four products were approved, utilizing either an aqueous suspension or oil solution technology. Since then, numerous other LAI products have made their way onto the market, where aqueous suspensions are the most utilized. A detailed summary of the FDA-approved LAI suspension products, along with their dosage, frequency of administration and formulation composition, can be found in Table 5.1.

Table 5.1 FDA-approved LAI suspensions (IM: intramuscular; IA: intra-articular; SC: subcutaneous; IB: intra-bursal; PA: periarticular; IV: intravenous). (Adapted from Holm 2023)

Trade name	Active ingredient	Product presentation	Route of administration[a]	Year of FDA approval	Maximum dose	Dosing interval	Indication	Excipients in formulation
Abilify Asimtuffi®	Aripiprazole	Solid crystalline	IM	2023	960 mg (syringe)	Two months	Schizophrenia	Carboxymethylcellulose sodium, polyethylene glycol 400, povidone, sodium chloride, sodium phosphate monobasic monohydrate, sodium hydroxide (to adjust pH) and water for injection
Abilify Maintena®	Aripiprazole	Lyophilized powder	IM	2013	400 mg (vial)	Monthly	Schizophrenia	Carboxymethyl cellulose sodium, mannitol, sodium phosphate monobasic monohydrate, sodium hydroxide
Aristada®	Aripiprazole lauroxil	Solid crystalline	IM	2015	1064 mg (syringe)	Two months	Schizophrenia	Sorbitan monolaurate, polysorbate 20, sodium chloride, sodium phosphate dibasic anhydrous, sodium phosphate monobasic
Aristada initio®[b]	Aripiprazole Lauroxil	Solid crystalline	IM	2018	675 mg (syringe)	Initial dosing	Schizophrenia	Polysorbate 20, sodium chloride, sodium citrate dihydrate, sodium phosphate dibasic anhydrous, sodium phosphate monobasic dihydrate
Bicillin® L-A	Penicillin G benzathine	Solid crystalline	IM	1950	1, 2 and 4 mL (0,6, 1,2 and 2,4 M units	Two weeks	Syphilis, prophylaxis	Sodium citrate buffer and, as w/v, approximately 0.5% Lecithin, 0.6% carboxymethylcellulose, 0.6% povidone, 0.1% methylparaben and 0.01% propylparaben
Depo-Medrol®	Methylprednisolone acetate	Solid crystalline	IM, IL, ST	1959	80 mg/mL (1 mL vial)	Single dose	Arthritis	Polyethylene glycol 3350, myristyl-gamma-picolinium chloride, sodium chloride

(continued)

Table 5.1 (continued)

Trade name	Active ingredient	Product presentation	Route of administration[a]	Year of FDA approval	Maximum dose	Dosing interval	Indication	Excipients in formulation
Depo-Provera®	Medroxyprogesterone acetate	Solid crystalline	IM	1960	400 mg/mL (2.5 mL vial)	Weekly	Metastatic endometrial or renal carcinoma	Polyethylene glycol 3350, sodium sulfate anhydrous, myristyl-gamma-picolinium chloride
Depo-subQ Provera 104®	Medroxyprogesterone acetate	Solid crystalline	SC	2004	104 mg (syringe)	Three months	Birth control	Methylparaben, propylparaben, sodium chloride, polyethylene glycol, polysorbate 80, monobasic sodium phosphate dibasic sodium phosphate and methionine povidone
Invega Sustenna®	Paliperidone palmitate	Solid crystalline	IM	2009	234 mg (syringe)	Monthly	Schizophrenia	Polysorbate 20, polyethylene glycol 4000, citric acid monohydrate, sodium hydroxide, sodium dihydrogen phosphate monohydrate, disodium hydrogen phosphate anhydrous
Invega Trinza®	Paliperidone palmitate	Solid crystalline	IM	2015	819 mg (syringe)	Three months	Schizophrenia	Polysorbate 20, polyethylene glycol 4000, citric acid monohydrate, sodium hydroxide, sodium dihydrogen phosphate monohydrate, disodium hydrogen phosphate anhydrous
Invega Hafyera™	Paliperidone palmitate	Solid crystalline	IM	2021	1560 mg (syringe)	Six months	Schizophrenia	Polysorbate 20, polyethylene glycol 4000, citric acid monohydrate, sodium hydroxide, sodium dihydrogen phosphate monohydrate, disodium hydrogen phosphate anhydrous
Ryanodex®[c]	Dantrolene sodium	Lyophilized powder	IV	2014	250 mg (vial)	Prior to surgery	Malignant hyperthermia	Mannitol, polysorbate 80, povidone K12, sodium hydroxide or hydrochloric acid
Kenalog®-40	Triamcinolone acetonide	Solid crystalline	IM or IA	1958	40 mg/mL (10 mL vial)	Several weeks	Anti-inflammatory	Sodium chloride, benzyl alcohol, carboxymethylcellulose sodium, polysorbate 80

Zyprexa Relprevv®	Olanzapine pamoate	Solid crystalline	IM	2009	405 mg (vial)	Monthly	Schizophrenia	Carboxymethylcellulose sodium, mannitol, polysorbate 80, sodium hydroxide and/or hydrochloric acid
Vocabria®	Cabotegravir	Solid crystalline	IM	2021	600 mg (vial)	Monthly or two momths	HIV-1 infection	Mannitol, polysorbate 20, polyethylene glycol (PEG) 3350
Rekambys®	Rilpivirine	Solid crystalline	IM	2021	900 mg	Monthly or two months	HIV-1 infection	Citric acid monohydrate, glucose monohydrate, poloxamer 338, sodium dihydrogen phosphate monohydrate, sodium hydroxide

[a]IM; intra-muscular; IL; intra-lesional; ST; soft tissue administration SC; subcutaneous; IA; intra-arterial; IV; intravenous

[b]The formulation releases the API over an extended period, but it is only intended for one injection at the start of the treatment with Aristada® or if a dose of Aristada® was missed

[c]Ryanodex is an immediate release formulation for intravenous administration and is therefore not an LAI. It was included into the table despite this as it is an aqueous nanosuspension; hence, the formulation composition may still be of interest

The available market for aqueous LAI suspensions is dominated by two indications, namely schizophrenia and HIV; however, there are also antibiotic products and injectables for contraception and corticosteroids. The first FDA-approved drug product based on injectable suspensions came in 1950, which was Bicillin® L-A administered once every 2 weeks and Kenalog-40 in 1958. Following the initial approval of a few LAI suspensions, the LAI landscape was not extensively expanded with additional aqueous suspensions for decades; however, in the early 2000s, the industry got a number of products approved, Depo-subQ Provera 104 in 2004 and Invega Sustenna in 2009. Subsequently, the FDA approved Zyprexa Relprevv (olanzapine pamoate) for the treatment of schizophrenia as a monthly intramuscular injection, Abilify Maintena (a lyophilized formulation for reconstitution, for the treatment of schizophrenia as a monthly intramuscular injection) and Aristada (a micro-suspension formulation of aripiprazole lauroxil for the treatment of schizophrenia as a monthly or bimonthly intramuscular injection). In more recent years, Invega Trinza and Hafyera received FDA approval, extending the duration of paliperidone palmitate to 3 and 6 months, respectively, for the treatment of schizophrenia. It is interesting to note that Abilify Asimtuffi®, which is a third-generation product in the Abilify Maintaina family, was approved recently. The first product brought to the market was a lyophilized cake for reconstitution with sterile water supplied in two individual vials for monthly injection. This was quickly followed up with a dual chamber syringe where the reconstitution of the lyophilized cake was done inside the syringe and in the latest product, a prefilled syringe with 960 mg dispersed in 2.5 mL water for bimonthly injection have been marketed. When comparing the composition of the different Abilify LAI formulations, it is notable that the latest contains polyethylene glycol (PEG) and polyvinylpyrrolidone (PVP), both polymers which form a thixotropic solution, and are therefore characterized by their ability to transform from a semisolid or gel-like state into a more fluid consistency when agitated or sheared. Using this approach, lyophilization no longer seems needed to obtain a stable product.

In the human immunodeficiency virus (HIV) treatment domain, Rekambys (rilpivirine) and Apretude (cabotegravir) mark the first long-acting regimens for HIV treatment with injection every month or every second month. Most recently, Cabenuva, comprising co-packaged cabotegravir and rilpivirine nanosuspensions as separate drug products, has been approved as the first complete long-acting treatment for HIV, with a monthly or bimonthly dosing frequency.

Micro and Nanosuspensions in Clinical Trials

There were 49 ongoing clinical trials using long-acting injectables registered in July 2024 (www.clinicaltrials.gov 2024), and not all were using LAI suspensions, but the majority. Most of the studies using aqueous LAI suspensions were on HIV and in particular prevention of HIV as the end target. A few studies also investigated the application of LAIs for other indications. In the studies, particularly two commercial products were tested, namely, long-acting injectable cabotegravir and rilpivirine. No reports of clinical studies investigating aqueous LAI suspensions that were not already regulatory approved were found.

Novel Applications of Micro- and Nanosuspensions in Preclinical Studies

The aqueous suspension technology is relatively well developed from a clinical and nonclinical perspective, i.e. the release mechanism is well understood and regulatory approved. Novel applications of suspensions for LAI usage therefore seem to go in the direction of applying the technology to new molecules and new indications where longer treatment times are needed. To the best of our knowledge, most of the innovation activity with respect to suspensions occur industrially and not academically; hence, it is less available in the public domain.

5.4 Road Mapping and Industrial Challenges

Like most pharmaceutical processes, there is a high interdependence between the compound material attributes, formulation characteristics, process parameters and the product quality for the production of suspension-based LAIs (Fig. 5.3). This

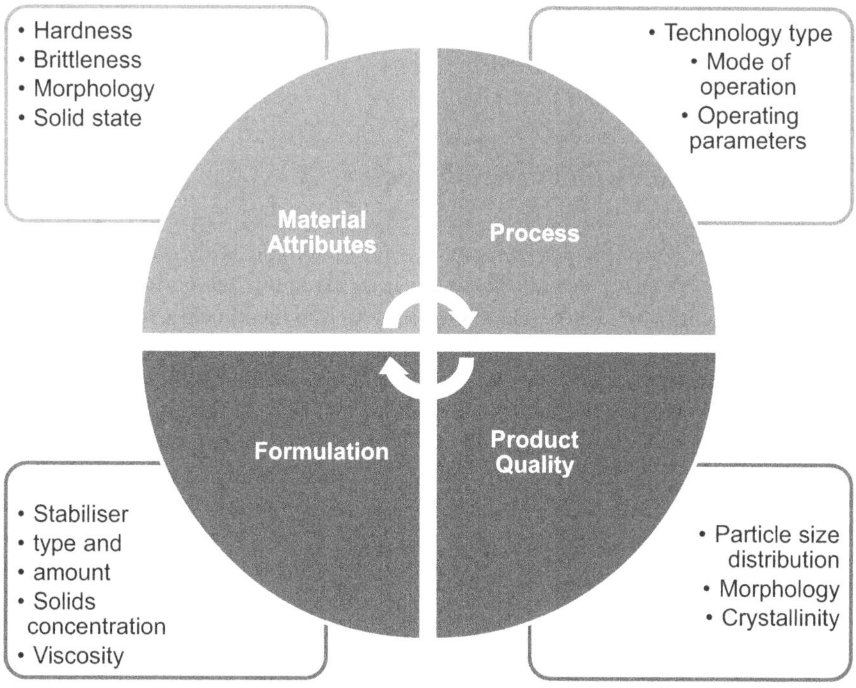

Fig. 5.3 Factors impacting nano-/microsuspension LAI process and product quality

section focuses on top-down comminution processes (wet stirred media milling and high-pressure homogenization (HPH)) as they are the most described in the literature, and will drill down into the main influencers towards product quality, mapping out their interdependence and the current state of the art in their understanding, challenges and opportunities.

Drug Molecule Properties

Inherent material properties govern the millability of a particle and the literature consensus is that the governing material parameters are crystal shape, crystallographic slip planes, intermolecular interactions or the arrangement of molecules in the crystal lattice and mechanical properties (Kaur et al. 2022; Knieke et al. 2009; Halme et al. 2019). It is commonly known that breakage occurs at the grain boundaries or slip planes, i.e. the weakest points in a polycrystalline material (Knieke et al. 2009). However, breakage can also be influenced by particle geometry as well as the presence of defects/cracks formed during its crystallization, where it is generally suggested that more elongated particles and the presence of cracks can result in more rapid particle size reduction (Halme et al. 2019). The crystal structure will result in certain mechanical properties (i.e. hardness and Young's modulus (or brittleness index)) and these play an essential role in the compound millability (Perkins et al. 2009; Salahshoor et al. 2024).

Formulation Screening in Small Scale

The most critical excipient in a suspension formulation is the stabilizer, which ensures the physical stability of the suspension, given that the particle size and its distribution define the release rate after injection. The theoretical understanding on stabilizer selection is relatively immature to define the stabilizer purely on physicochemical characterization of the compound and the stabilizers, with scarce efforts into harnessing existing datasets to derive formulation-induced parameters impacting processability (Nakach et al. 2014). Hence an empirical methodology (e.g. design of experiment, trial and error) is needed, which offers a relatively quick fix to the formulation definition, but unfortunately lacks robust scientific foundations. The inherent risk in an empirical screening approach lies in the potential of overlooking long-term stability issues or formulation robustness due to the absence of scientific comprehension, ultimately resulting in a more risk-full formulation development process of aqueous suspensions relative to other formulations that are better theoretically supported. The empirical solution often used incorporates a partial risk reduction by initial larger experimental screens followed by keeping parallel formulations in development until sufficient data has been generated to take a final decision.

The first surfactant screening typically runs with a defined API concentration, which could be determined by the anticipated desired concentration or compound availability, incorporating different types of surfactants, preferably approved for the intended administration route. Often the work is not restricted to these, and looks much broader (Lestari et al. 2015; Nakach et al. 2014). To minimize the variable factors at this early screening stage Nakach et al. (2014) proposed only to run with a stabilizer concentration of 3% (w/v) and later optimize this when the stabilizer and surfactant would be defined.

The intention with the initial formulation screening is to define stabilizers that can ensure a physical stable formulation, which can be easily resuspended and administered during the product's shelf life. Long-term stability studies are the best approach for that; however, as a balance between time and science, screening stability studies are often run over several weeks at elevated temperature, such as 40 °C, often using simple first response factors, e.g.:

- Particle size distribution
- Resuspendability
- Syringeability

These factors are typically determined just after the milling and for several weeks under stressed conditions, whereby a trend may be detected. The study does not provide an optimized formulation but gives clear indications of stabilizers that do not interact well with the solid compound, and ones that seem to stabilize the suspension at least on the short term, thereby providing guidance on which surfactants that are, and are not relevant, to include in further investigations.

At early stages of development there is often limited access to API; hence, small-scale milling may be relevant. Roller milling has historically been applied in this field (Lestari et al. 2015; Nakach et al. 2014), where the compound is placed in a laboratory test tube with a media containing surfactants and beads, the latter typically zirconium, and rotated at relatively low speeds for days to weeks. This method can be used with limited equipment investment and can easily be scaled from, e.g. 1–20 mL per batch and even more if needed. The method ensures continuous grinding of the API and has been reported capable of producing nanosuspensions (Lestari et al. 2015; Nakach et al. 2014) for brittle compounds. Another more recently described method for suspension formulation screening is the use of dual centrifugation (Hagedorn et al. 2017, 2019), see Fig. 5.4.

The milling in the dual centrifugation is of the order of several hours according to the procedure proposed by Hagedorn et al. (2017, 2019), owing to its high energy, which significantly shortens the surfactant screening phase relative to the low energy roller milling procedure. Zulbeari and Holm (2023, 2024) investigated the reproducibility of the roller mill and the dual centrifugation and reported higher variation when producing the same suspension by roller milling as compared to dual centrifugation.

Beyond definition of the stabilizer, an important element in the formulation work is the optimization of the API and surfactant concentration, experiments that may be run at small scale using the same response factors as the initial screening. At this

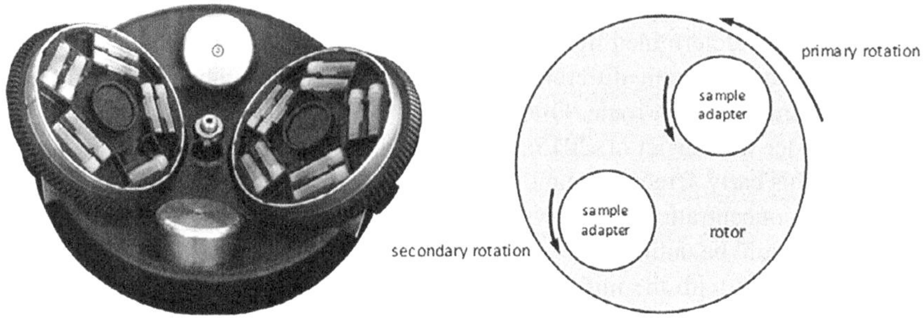

Fig. 5.4 Dual centrifugation rotor head (right) and a schematic picture of the principles (left) (Modified from Hagedorn et al. 2019 with permission)

stage, often more than one stabilizer may be kept in the investigational loop, due to the difficulties and uncertainty in predicting the long-term stability by short term stress studies. As a part of these studies, elements such as pH and the resuspending agent may also be explored; but for true stability studies according to the International Council for Harmonisation of Technical Requirements for Pharmaceuticals for Human Use (ICH) guidelines (ICH 2003) larger batches are normally needed. Hence, at this stage, transfer to a larger scale research manufacturing is normally needed.

Process Development and Manufacturing Scale Up

Process Analytical Technologies (PAT)

Implementing real-time process monitoring and control would allow non-invasive, continuous, rapid and real-time data acquisition, with much more process feedback than current offline measurements offer. There are particular benefits for real-time process analytical technologies (PAT) for suspension-based LAIs, as often sterile processing is needed (see Sect. 5.4.6), which puts a restriction on the number of samples that can be taken during processing and therefore limits the capability for in-process control via sampling and offline analysis. However, in-line analysis is complicated by several challenges: (1) high solids loadings resulting in: multiple scattering, possible particle interactions at high concentration, particle phenomena occurring under flow; (2) sub-micron particle size quantification; and (3) variations in API and drug product characteristics throughout the process. Expanding beyond measurements of particle size, real-time characterization of other properties (e.g. polymorphic behaviour and rheology) would significantly amplify our understanding of LAI processing and control and would be the next step towards real-time process evaluation. In terms of real-time (inline or online) particle size characterization, the following section will discuss the available techniques. Beyond the aforementioned challenges typical for LAI drug products, other challenges associated with PAT tools are:

- Presence of air bubbles, which would interfere with measurements of the API; this can be eliminated by using algorithms that discriminate according to the particle shape, utilizing that bubbles are uniquely perfectly spherical.
- The existence of high concentrations or multi-components in one system can negatively affect the ability to recognize and classify particles.
- The presence of low probability large particles and their appropriate detection. This requires good sample representation, i.e. careful consideration of the positioning of the tool, with knowledge of the flow profiles and flow cell design.
- Probe fouling.

The different PAT tools currently available for inline/online particle sizing of suspensions are presented next.

Focused Beam Reflectance Measurements (FBRM)

FBRM is based on the principle of laser diffraction, where a laser beam is transmitted through a probe; the optics rotate consistently at a particular speed, causing the laser beam to sweep through the particles as they move past the probe window. Particles, in turn, scatter the light back to the detector in the form of light pulses, which are counted, and the distance across each particle determined by multiplying the duration of each pulse by the scan speed (Mettler Toledo-b, USA). This method can extract 1D particle information in the form of a chord length distribution as well as other parameters (such as turbidity). Challenges associated with FBRM are that the movement (speed) and optical properties of the particles can affect the measurement, as well as its inability to detect submicron sized particles (Liu et al. 2022).

In-Line Microscopy

In-line microscopy, or imaging tools, is making significant advancements in particle tracking (Liu et al. 2022); they can extract 1D (size) or 2D (shape, aspect ratio, area, perimeter) information directly from an image (Fig. 5.5), which is less impacted by artefacts compared to other techniques (varying material properties during a process particle such as morphology or varying velocity of the particle due to changes in viscosity during a process). Combining high-quality microscopic images with advanced image processing technology, accurate information such as particle size distribution (PSD) can be quantified. Examples of tools commercially available are the EasyViewer (Mettler Toledo-a, U.S.A) and the Blaze probes (Blaze Metrics, USA). Typical outputs are particle count, chord length distribution and turbidity—the latter being useful for process profiling and is directly linked to PSD.

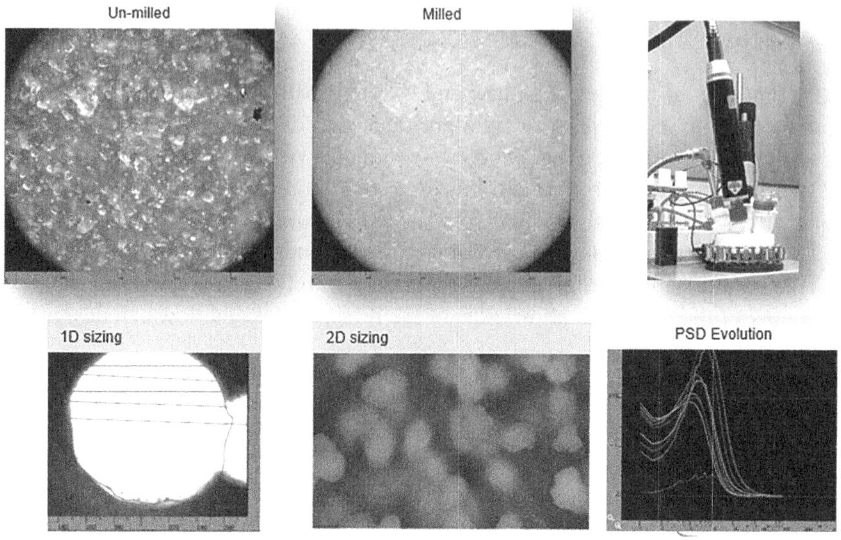

Fig. 5.5 Use of in-process microscopy to derive particle size information of a suspension-based LAI manufacturing process

Microscopy-based tools are restricted to micron sized particles and generally incapable of detecting sub-micron particles. In addition, they could be challenged when there are multi-components in one system with similar backscattering properties, which could negatively impact the ability to recognize and classify particles (limitations for FBRM too). Both the microscopy and the FBRM techniques have been more widely applied in crystallization processes (Sudhakar et al. 2023; Szilágyi et al. 2022), and less literature exists in the drug product space and suspension-based LAIs.

Dynamic Light Scattering (DLS)

DLS is a commonly used technique for analysing small particles ranging from 1 nm to a few micrometres in size, the major advantage here is having the ability to extend below into the nanometre range. DLS works by optically measuring the diffusion of particles; by analysing the fluctuations in scattering intensities over time, the diffusion coefficients and hydrodynamic radius of the particles can be calculated. There are fibre optical backscatter immersion probes available commercially that can perform DLS measurements, however, one limitation is that they cannot be used in moving fluids, which is a common requirement for a PAT. In highly concentrated systems typically experienced for LAIs, multiple scattering can occur further complicating the measurement.

Raman Spectroscopy

Particle size directly affects Raman measurements, which manifests in various spectral variations due to particle related effects, e.g. variation of the Raman shift or peak height (Chandramohan et al. 2011; Kirschner and Mitsos 2018). The advantages of Raman spectroscopy are that it is suited for measurements in highly concentrated suspension systems without the need for dilution and is potentially capable of extracting other physico-chemical information beyond particle size. However, to derive particle size from a Raman signal requires calibration with an orthogonal technique and as the Raman signal can be sensitive to the experimental conditions in the suspension (temperature, probe positioning), it can be challenging to ensure a reliable means to derive particle size from the signal.

Equipment Design from Lab to Pilot Scale

The choice of equipment for manufacturing during development should ensure representativeness to the commercial manufacturing process. In some cases, this could conflict with other priorities, e.g. having high throughput capabilities using very small amounts of material, which is a limiter particularly during early development and screening stages. To ensure scalability, material needs to be manufactured according to a similar processing methodology as the commercial line production; for top-down comminution processes, this means giving the product similar levels of specific energy as will be discussed later. In early development stages, the operation is often carried out in batch; the disadvantage of the screens at such small scales is that the representativeness of the process could be compromised, depending on the chosen equipment and the provided stress energies. Planetary, dual centrifuge and roller mills are used to obtain fundamental understanding of stabilization mechanisms and screening for empirical stabilizer identification in early development stages. This process is not sufficiently automated and conducted in batch-mode. Nonetheless, miniaturized milling assays guide the way towards quick screening approaches as discussed in Sect. 5.4.2.

Process Development

Wet Stirred Media Milling

Wet stirred media milling involves the use of a milling chamber that contains an agitator, grinding media, a dispersing medium (water), polymers/surfactants for stabilization and the drug to be reduced in size. The milling material typically consists of small beads made of stainless steel, glass, ceramic (such as yttrium stabilized zirconium dioxide) or highly cross-linked polystyrene resin. The latter two options are preferred to minimize contamination to the final product (Salazar et al. 2014). The particle size achieved during wet grinding is influenced by two opposing mechanisms: particle breakage on the one hand and the formation of agglomerates on the other. A plateau in the milling profile is common, also termed as 'grinding limit', which is a consequence of the crystallite size and any associated competing particle agglomeration phenomena, equipment capability for breakage and intensity of applied forces. The outcome of a bead milling process is determined by two key parameters. The first parameter is stress energy (SE) and represents the energy of a collision between two grinding media. This SE is proportional to the stress or kinetic energy of the grinding media, SE_{GM} (Arno Kwade 2003). The maximum stress energy of grinding media in the mill, $SE_{GM, max}$, is at the agitator tip, where the velocity is equal to the tip speed, v_t:

$$SE \propto SE_{GM,max} = d_{GM}^3 \rho_{GM} v_t^2 \tag{5.2}$$

where d_{GM}^3 is the grinding media diameter and ρ_{GM} is the grinding media true density.

The second parameter is the stress number (SN) (Kwade 2003), which represents the number of collisions that are required to break a product particle into fractions with the target size. SN is determined by the collision frequency, N_c, the probabilty that a particle is caught and sufficiently stressed at a media contact, P_S, and by the number of drug particles inside the mill, N_P:

$$SN \propto \frac{N_c P_S}{N_P} \tag{5.3}$$

While efforts have been made to calculate an average SE to account for the energy inhomogeneity within the mill, this is not a trivial matter and relies on assumptions on bead energy and occupancy distribution within a mill (Kwade 2004). The relative importance of SN and SE is dependent on the breakage behaviour of a particle as well as the combination of process and formulation parameters. Another common parameter used to describe the process performance of wet stirred media milling, is the specific energy $E_{m,\,BM}$ (Breitung-Faes 2017a; Stenger et al. 2005):

$$E_{m,BM} = \frac{\int_0^t (P - P_0)\,dt}{m_p} \tag{5.4}$$

Here, m_p is the mass of the solid product, P is the power inputted to carry out the comminution and P_0 is the no-load power.

Agitator speed has a direct impact on the kinetic energy of the grinding media where an increase in the stirrer speed leads to faster breakage due to the occurrence of more frequent bead-bead collisions with higher stress intensity (Parker et al. 2020). The optimal bead size required to effectively break a particle depends on the comminution goal and the properties of the API being milled; bead size influences the SN and SE in an opposing way, i.e. smaller grinding beads provide a higher number of stress events, but a lower stress energy, whereas larger beads exert higher stress energies, though at lower frequencies (Cerdeira et al. 2011). The bead load has a direct impact on the number of stresses: a larger number of beads will result in a higher frequency of stress events. The effect of suspension flow rate will depend on the mode of operation, where differing opinions exist on its impact on the process performance; most recent papers have shown that flow rate has a negligible impact on the process (Heidari et al. 2024). Flow rate, however, as it impacts the velocity in which the suspension passes through the chamber, in principle, could induce effects such as bypassing (particles that pass through insufficiently grinded) or dead zoning (particles that reside in dead zones in the chamber for prolonged periods of time). Thus, applying basic reaction engineering knowledge implies keeping the superficial velocity through the mill constant to maintain similar behaviour across equipment types.

High-Pressure Homogenization (HPH)

HPH relies on the principle of forcing fluid through a microscale gap thereby generating high pressures and velocities. Variations of this technology exist, essentially varying in the geometry of the homogenizer valve (also termed interaction chamber). The HPH valve design variations can be classified into different types (Fig. 5.6):

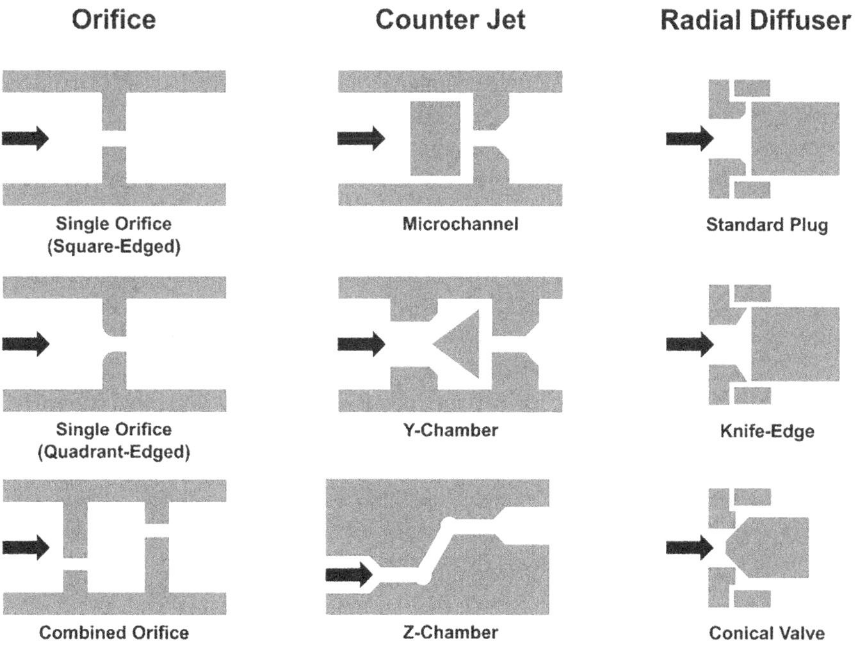

Fig. 5.6 Different high pressure homogenizer geometries: (orifice valves), counter jet valves (microfluidizers) and radial diffusing valves (piston gap homogenizers) (Inguva et al. 2024)

1. Orifice valves
2. Microfluidizers (also termed counter jet valves)
3. Piston-gap homogenizers (also termed radial diffusing valves)

Besides the above variations in valve design, it is common for HPH to consist of two serially mounted valves, where the second valve is used for creating a back-pressure for the first (thereby suppressing unwanted/exaggerated cavitation) or alternatively to carry out a form of initial deagglomeration prior to the particle size fine-tuning with the second valve.

The piston-gap homogenizers and microfluidizers are the most described HPH equipment in the literature for pharmaceutical applications. In both manufacturing methods, the suspension undergoes pressure-driven homogenization, where a suspension is delivered to an interaction chamber using a high-pressure pump through a small gap (several tens of microns), resulting in high velocities and thereby high pressures of the order of 1000–2000 bar. The valve geometry dictates the specifics of the fluid dynamics, and it is not uncommon to have different processing outcomes at the same pressures depending on the valve design. Forces such as cavitation and collisions of the particles with each other as well as with the interaction chamber result in particle comminution. The two main parameters influencing process performance in a HPH process is the homogenization pressure and the valve geometry. The number of homogenization cycles required to create a micro or nanosuspension depends on the applied energy, where in the context of HPH this is largely controlled by the homogenization pressure, which in itself is a function of the geometry of the equipment (homogenization gap) as well as the flow rate.

One limitation with HPH equipment is inability to process large input material sizes due to the small interaction chambers (tendency for blockage). Therefore, often pre-homogenization would be needed to circumvent this issue. In addition, feeds possessing properties such as high viscosity or large particles, typically have poor pumpability and cannot be reliably fed into a homogenizer (Inguva et al. 2024).

Similar particle sizes can be produced by both HPH and wet stirred media milling reaching comparable sub-micron particle size ranges (Kanthamneni et al. 2016; Nakach et al. 2018; Salazar et al. 2014; Shegokar et al. 2011; Chogale et al. 2020). While for particle shape, results differed, where drastic differences in morphology when comparing high-pressure homogenization with wet stirred media milling were attained, with more rounded particles being reported for the former (Nakach et al. 2018). These results are likely to be specific to the compound properties investigated in these studies, and could vary according to the compound properties and formulation (Keck and Müller 2006).

To provide an analogous parameter that provides a high-level description of the energy applied in a high-pressure homogenization process, specific energy E_m can be calculated for the HPH process (Méndez Cañellas et al. 2023a):

$$E_{m\,\mathrm{HPH}} = \frac{P \cdot Q \cdot t}{m_p}$$

where P is the milling pressure, Q is the flow rate, t is the milling time and m_p is the suspension mass.

Due to differences in geometry for different scale HPH equipment, large differences in the Reynolds and cavitation numbers often exist when operating at similar pressures across the scales (Håkansson 2017). Therefore, careful process design and evaluation of the process performance parameters is required—to attain similar processing behaviour on scale-up. It is worth to note that some HPH equipment suppliers offer numbering out the interaction chambers as a scale-out principle (as opposed to increasing the valve size) which could alleviate these scale-up differences.

Piston-Gap (Radial Diffusing Valve) Homogenizers

Two types of piston gap (also known as radial diffusing valve) homogenization are described in the literature for nanocrystals: Dissocubes® and Nanopure®, where the former utilizes water as the dispersion medium and the latter a non-aqueous/low vapour pressure dispersion solvent (extending the working range to lower temperatures thereby avoiding chemical degradation as well as reducing extensive cavitation where required) (Cornier et al. 2016; Wright and Burgess 2012). A schematic of the valve geometry is shown in Fig. 5.7. The fluid is fed axially into a passage head (or valve seat) and then accelerated radially into the gap in between the passage head and the impact head, further colliding with the impact ring before leaving the valve—it is at this point that the pressure drops and cavitation occurs due to bubble formation and collapse. The parameters that influence the pressure drop across the valve are the adjustable valve gap and the flow rate.

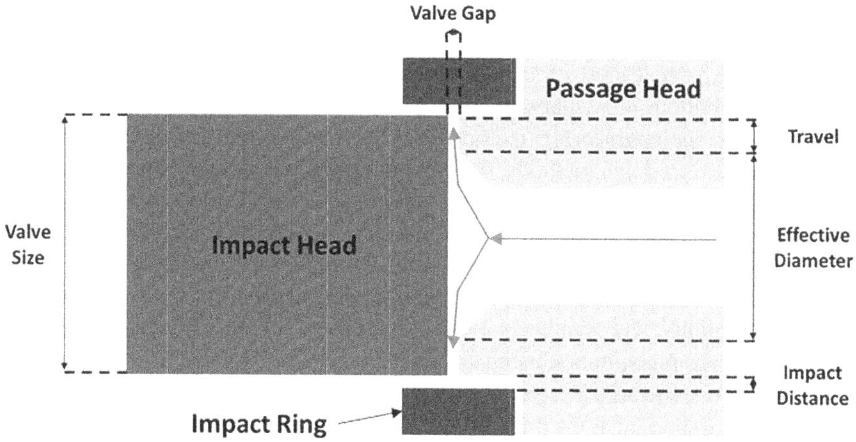

Fig. 5.7 Working principles of a piston-gap homogenizer (radial diffusing valve). (Adapted from Inguva et al. 2024)

Microfluidizers (Counter Jet Valve)

In a microfluidizer (or counter jet valve), a jet is impinged either with itself or against the microfluidic walls. In a 'Z' type chamber, the suspension changes its direction a few times leading to particle collision while in the 'Y' type, the suspension stream is split and recombined, where the two streams collide frontally.

Homogenization pressure along with number of passes are the most influential process parameters, as well as the common formulation parameters: stabilizer type and concentration (Hao et al. 2012; Verma et al. 2009). While the impact of pressure is clear, increasing the number of cycles increases the probability that larger particles undergo breakage. A study comparison between dry jet milling and microfluidization can be found in Méndez Cañellas (2023) showing that for indomethacin and naproxen, while both techniques can produce suspensions with a median size range of 1–10 μm, the dry jet milling was more compatible with a wider range of excipients and presented better stability in the median particle size and span. However, in some cases (certain stabilizer types) the microfluidization was able to produce smaller nanosized particles, while the dry jet mill was less capable. This could be due to the higher applied specific energies in the microfluidization (calculated to be 5–25 times higher) thereby providing higher energy intensity to produce smaller particles, potentially at the expense of generating high surface energies that could impact subsequent stability.

Bottom-Up Synthesis

The manufacturing challenges for LAIs typically focus on controlling the physicochemical properties of the drug, both during production and throughout the shelf life of the final product. Processes that offer enhanced control or can fine-tune the drug to achieve the desired effect in humans are highly sought after. For suspension-based LAIs, bottom-up synthesis of nanoparticles/suspensions presents significant opportunities for engineering nanoparticles to achieve specific product performance. Unlike top-down methods, bottom-up approaches generate nanoparticles by assembling them from drug molecules, allowing for the creation of particles with unique nano- or micro-structures and eliminating the need for additional top-down processing. This approach enables fine-tuning of particle size without high-energy size-reduction steps, potentially providing more precise control over drug characteristics and stability. The synthesis of drug nanoparticles can be achieved through methods such as precipitation using an antisolvent and nucleation of a desired crystal phase is done by rapidly creating a very high supersaturated solution. The choice of solvents is important as the drug has to be soluble in at least one solvent. Variations of this process include the Hydrosol® capable of generating crystalline drug particles and the Nanomorph® typically generating amorphous drug nanoparticles. A thorough review of the differing bottom-up nanoparticle synthesis methods can be found in Chan and Kwok (2011).

Bottom-up methods have the following challenges: stabilization of the nanoparticles after generation of the solid phase due to the fast timescales, i.e. preventing the flux of any further molecules to the newly generated nanocrystal surfaces and any subsequent growth. This can be achieved through a number of ways: either via cooling, immediate isolation or by application of a tailored surface poisoning. Another hurdle is the use of organic solvents, necessitating a solvent switching step to convert the medium to water after the particle generation, with the assurance that the required residual solvent levels are met.

Recently, LongActNow, an EU Commission–funded project investigated industrially viable methods of bottom-up nanosuspension production in a concerted effort. The project has demonstrated that bottom-up synthesis of suspensions with the desired quality attributes for LAIs is not only feasible but can open new avenues, e.g. towards particle engineering, while maintaining physical and chemical stability, potentially also reducing cost and increasing speed of batch deliveries for drug product manufacturing. The investigated methods were antisolvent crystallization, supercritical antisolvent spray drying and the associated downstream processing in the form of membrane separation to perform the solvent switching required for the final aqueous drug product (Anjum et al. 2024a; Hugo Silva et al. 2022; Méndez Cañellas et al. 2023b; Nandi et al. 2023a, b, 2024a, b; Anjum et al. 2024b).General solids suspension mixing guidance can be followed to achieve homogeneity during milling, crystallization and filling. This is covered extensively in the mixing handbooks such as (Kresta 2004). In regards to mixing process design, correlations (Grenville et al. 2016) or alternatively computational fluid dynamics can assist in selecting the tank and impeller geometry and speed.

Mechanistic Modelling

Modelling milling operations for pharmaceutical compounds relies on a fundamental understanding of breakage behaviour of organic materials. The three types of breakage phenomena relevant to organic material are (1) shear impacts, (2) frontal impacts and (3) cavitation forces. For high-pressure homogenization, particularly the radial diffusing piston gap, cavitation forces have been reported to dominate (Nakach et al. 2018), while for stirred media milling, shear and impact forces can both apply.

Various types of models exist for wet stirred media milling that range in complexity, including the stress model (Breitung-Faes 2017b; Breitung-Faes and Kwade 2019; Flach et al. 2016; Flach et al. 2018; Nöske et al. 2020), first-order breakage rate model (Nakach et al. 2017), population balance models (Clancy et al. 2024), the microhydrodynamic model (Eskin et al. 2005; Guner et al. 2021) and computational fluid dynamics—discrete element methods (CFD-DEM) (Beinert et al. 2014; de Carvalho et al. 2023; Jayasundara et al. 2009; Larsson et al. 2020), the latter providing insight into the grinding media velocities, occupancies, energies and in some cases, particle breakage kinetics. While there are a few modelling approaches available, there is no unanimous consensus on a single fully predictive model that can be generalized for all APIs. All the above models rely on some form of calibration based on experimental data, and to reach a fully predictive model requires incorporation of all relevant compound material parameters that can themselves be predicted or measured apriori.

For high-pressure homogenization, characterizing the process involves solving a set of equations that describe the pressure drop across the valve, the turbulence, shear rate/stress, cavitation and impact forces. CFD has also proved as a tool that can simulate flow and velocity profiles across the valve which can aid in equipment and process design (Inguva et al. 2024).

In the absence of a model, some general rules of thumb can aid in technology transfers and scale ups, such as maintaining geometric, dynamic and kinematic similarity, i.e. equipment type and mode of operation, stress intensity and/or specific energy.

Data-driven (statistical) models that assess the relationship between process parameters and process performance can be utilized in the event where an adequate mechanistic depiction of the process is not feasible (He et al. 2020). However, these models rely more heavily on inputs from experimentation, and are product and geometry specific; accuracy and uniformity in the experimental setups of the harnessed datasets can be challenging. The quality-by-design (QbD) approach towards process design promotes the use of process performance parameters that are equipment and scale agnostic (i.e. stress intensity rather than a culmination of process parameters). This approach has recently been exemplified (Bitterlich et al. 2024) potentially setting the path towards similar approaches and applications in the future.

Drug-Release Modulation and In Vitro–In Vivo Correlations

When developing a new LAI product, frequent changes occur in formulation, manufacturing processes, scales and sites. These changes pose challenges in demonstrating bioequivalence, which is the ability of two formulations of the same drug or two drug products to achieve the same therapeutic effect. Demonstrating bioequivalence typically requires in vivo studies in humans, increasing development costs and time to market. Establishing in vitro–in vivo correlations (IVIVC) could alleviate these challenges by potentially supporting biowaivers and reducing regulatory burdens, leading to time and cost savings (Suarez-Sharp et al. 2016; Emami 2006; Tosca et al. 2021). IVIVC involves developing predictive mathematical models that describe the relationship between in vitro drug dissolution and in vivo response of different formulations. It can guide formulation and/or process development changes throughout drug product development stages and serve as a surrogate for bioequivalence studies when pre-approval and post-approval changes are necessary. However, there is no regulatory guidance for establishing IVIVC for LAI products. This is most likely due to the lack of standardized in vitro dissolution methods and the complexity of capturing in vivo performance, including dissolution from particles, distribution at the depot, inflammatory response and capillary absorption, through in vitro methodologies.

The FDA has defined three levels of IVIVC: Level A, Level B and Level C (FDA 1997), representing different degrees of correlation between in vitro and in vivo data and providing varying levels of confidence in the predictions made. It is worth noting that the FDA does not require a specific level of IVIVC for drug approval. However, having a higher level of correlation (like Level A or Level B) can give more confidence in how well the drug works and how equivalent it is to other versions, which could mean fewer human studies are needed or support changes after approval. Among LAI aqueous suspension products on the market, only Invega Sustenna has reported a Level A IVIVC in clinical studies (Bao et al. 2023; FDA 2009). Since conducting clinical studies to show IVIVC can be resource-intensive, some groups have turned to non-clinical animal models for developing release methods in the lab and using computer models. These animal studies might help pave the way for successful IVIVC development in humans (Bao et al. 2023), but research is clearly lacking in the field.

Administration

Vehicle Selection for Suspension Preparation

The vehicle for LAI suspensions contains five main groups of excipients, where the individual class of excipient is added if needed (Ali et al. 2010; Kumar and Verma 2023):

1. Stabilizer(s)
2. Viscosity increasing excipient
3. Flocculating agent(s)
4. Isotonic agent
5. pH adjustment

The definition of the stabilizer is described in Sect. 5.4.2 and relies on an empirical screening process. Addition of some of the other excipients may be defined later in development as more experience is obtained with the formulation. As an example, excipients to support flocculation may be added if issues with resuspending the suspension is observed as a part of the stability study, which may explain the multiple commercial formulations containing both a classical non-ionic surfactant and PEG3350 (see Table 5.1).

Excipients to increase the viscosity, such as carboxymethyl cellulose, can be added to the formulation to slow the sedimentation rate. This may be relevant in case where the healthcare provider prefers to prepare the formulation before the patient comes into the consultation, e.g. if a lyophilized cake for resuspension. In this case, the resuspended suspension should be sufficiently stable for a period of time to allow for administration of an accurate dose.

Another critical component necessary for formulating an LAI suspension is a tonicity agent. Osmolality serves as a measure of the osmotic pressure exerted by all soluble species (colligative properties), both ionizable and non-ionizable, within the drug product (Wang 2015). Injection of hypertonic preparations has been linked to heightened local pain and discomfort and should hence be avoided if possible (Graven-Nielsen et al. 1997; Lei and You 2012). Osmolarity levels below 600 mOsms/kg are considered acceptable, with ideal levels falling between 285 and 295 mOsm/kg (close to the plasma isotonicity value of 300 mOsm/kg) (Taghizadeh et al. 2022). Common tonicity agents found in LAI products (see Table 5.1) include sodium chloride, mannitol and sucrose. In addition, other excipients that may be present in LAI suspensions include buffering agents, pH adjusters, viscosity modifiers, bulking agents, antioxidants and antimicrobial preservatives.

Syringeability and Injectability

Syringeability and injectability are essential considerations for injectable formulations. Despite their apparent similarity, they represent distinct parameters, where syringeability refers to the ease with which a solution can be withdrawn from its container into a syringe. This factor is influenced by variables such as viscosity, surface tension and potential interactions between the solution and its container. Injectability, on the other hand, covers not only a broader spectrum of elements and the ease of preparing and administering the dose, but also considerations such as patient comfort, the likelihood of adverse reactions at the injection site, and the overall ergonomic experience for both the patient and the healthcare provider (Watt et al. 2019).

Aqueous suspensions in general behave like water unless the included API has a morphology that prevents this. The syringeability and injectability of the suspension is therefore in general good, as limited amounts of excipients are added that could increase the viscosity of the formulation. One of the main elements related to both syringeability and injectability of suspensions is a good process for how to resuspend a lyophilized version of the suspension. For premade suspensions in a vial or a prefilled syringe, the suspension may sediment during storage, and may need to be properly resuspended to get both a good syringeability and injectability, as an unevenly distributed suspension can be harder to work with. A training device has hence been developed and marketed by Janssen to ensure sufficient training to the healthcare givers intended to handle the prefilled syringes containing a premade suspension (Shore 2024).

At the formulation development stage there is often a desire to have as high API concentration as possible in the formulation thereby reducing the needed injection volume, with the intention of inducing less pain for the patient upon injection as well as providing the possibility to contain a dose that providing the possibility to contain a dose that would allow for maximum possible duration of drug release. The amount of API added to the formulations are in general limited by the 'flowability' of the suspension, i.e. at some stage, the solid compound could dominate the formulation behaviour, and the system may not be injectable any longer. The precise concentration where this occurs is compound specific and there is, to the best of our knowledge, no guidance in the literature if this can be compensated, at least partially, by formulation design.

Impact of Needle Size and Geometry on Injectability

Aqueous suspensions for LAI purposes in general have mean particle sizes below 10 μm (Nguyen et al. 2022). These formulations are in general injected IM (see Table 5.1), for which the kit for Ability Maintaina suggest using a 21–23 G needle dependent upon injection size and patient (FDA 2015). At this needle size, the inner diameter is 510 to 227 μm; hence, issues with needle size when administering the aqueous LAI suspension with Ability Maintaina would not be anticipated in the clinical setting.

In general, the syringeability and injectability should be tested as a part of the formulation design, with a range of different needles. Beside the formulator's subjective opinion, a mechanical setup is normally also recommended where the injection force is determined on a texture analyser or similar as suggested in the ISO guideline (ISO 2022).

Sterilization Methods

The European Medicines Agency (EMA) provides guidance on selecting the correct sterilization methods, which include terminal sterilization, heat, sterile filtration or aseptic processing. The decision trees provided in the EMA guideline assist in choosing the optimal sterilization method while considering various potential issues and risks (European Medicines Agency 2019). It is generally recommended to select the first feasible option as the methods typically offer decreasing assurance of sterility as you move down the decision trees. Figure 5.8 depicts a decision tree for sterilizing aqueous suspension products. Drug products intended to be sterile should ideally undergo terminal sterilization in their final container and significant efforts should be made to enable this process. If terminal sterilization in the final container is not feasible, an alternative strategy combining drug substance sterilization (using dry heat or ionizing radiation; Fig. 5.9) and aseptic manufacturing should be used. In this case, all components in contact with the drug product throughout the manufacturing process should be sterilized using one of the methods suggested in the guideline.

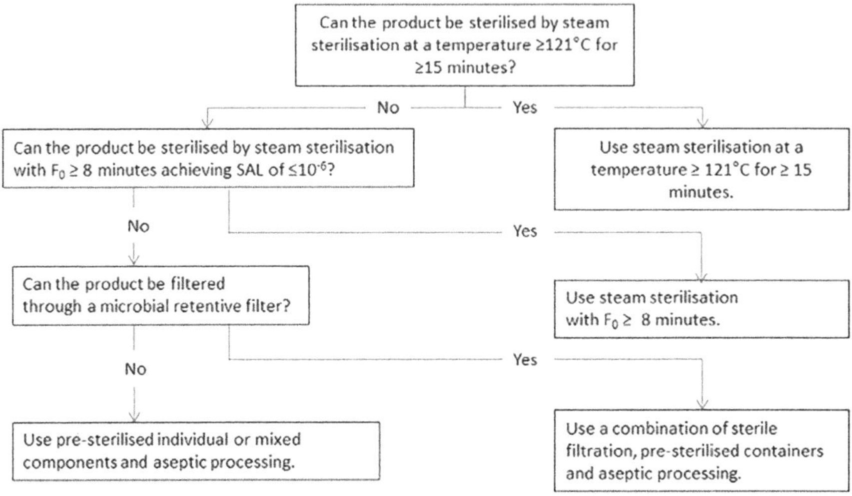

Fig. 5.8 Decision tree for sterilization choices for aqueous products (European Medicines Agency 2019)

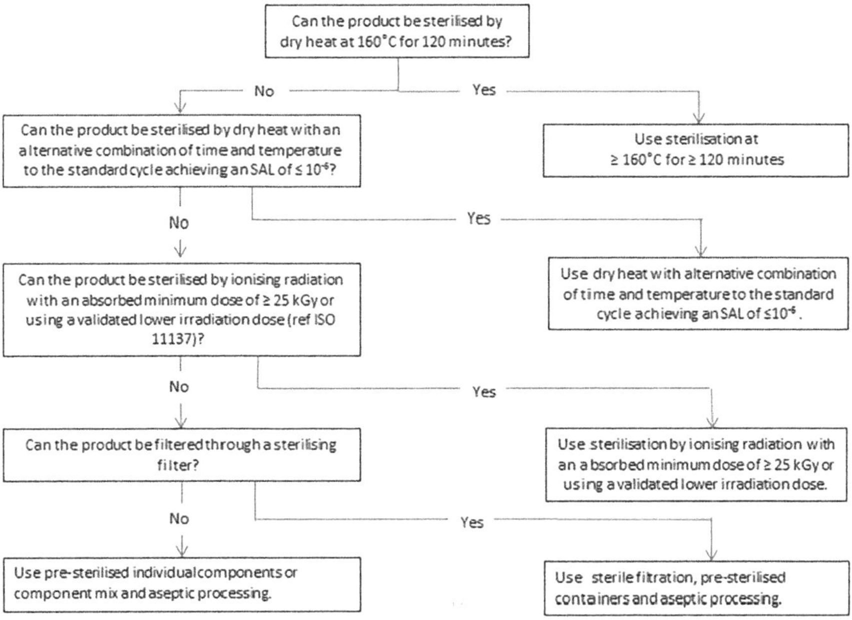

Fig. 5.9 Decision tree for sterilization choices for dry powder products, non-aqueous liquid or semi-solid products (European Medicines Agency 2019)

5.5 Discussion and Prospective

Development, Manufacturing and Commercialization Challenges

There are several challenges related to the development, manufacturing and commercialization of LAIs:

- Predictive modelling: developing models to predict the manufacturability, stability and pharmacokinetics of compounds. These models are useful for differentiating between drug candidates during early development and supporting scale-up during late development.
- Novel drug delivery platforms and manufacturing methods: exploring new ways to engineer and control the physical and chemical properties of LAIs. This includes aspects such as particle size distribution and surface properties, which directly affect the drug-release profile and shelf life stability in the body.
- Process monitoring and control: establishing non-invasive and rapid analytical methods to monitor and control the manufacturing process without compromising the sterility of the final product.

Manufacturability

Regulatory guidelines from the FDA (2004a, 2018) and the concept of quality by design (QbD) (FDA 2004b, 2005) have provided a strategic framework for the pharmaceutical industry to enhance its production processes and product quality (Van Buskirk et al. 2014). The International Council for Harmonisation of Technical Requirements for Pharmaceuticals for Human Use (ICH) guidelines also emphasize the importance of demonstrating a deeper understanding of pharmaceutical and manufacturing science, which facilitates more flexible regulatory filings and easier transfers between scales and equipment after drug approval. To this end, there have been scarce efforts to predict the processability of nanosuspensions using mechanistic modelling and machine learning. Further advancements are necessary to fully simulate and predict processing outcomes, taking into account material and formulation parameters specific to pharmaceutical compounds and drug products. Particularly, understanding the influence of API properties on the grinding limit is of great interest, as it could enable predictions on whether a nanosuspension can be successfully created from a given compound.

Drug Product Stability

A key element to successful drug product development is a stable formulation. Small particles in liquid formulations, such as nano- and micro-particles, are known to be thermodynamically unstable and tend to group together thereby potentially impacting the PSD critical quality attributed (CQA). This may also happen during processing, where a material's surface could undergo changes, including the formation of new facets/alterations in existing facets, which can affect the overall surface energy, potentially impacting the stability of the product during its shelf life. Understanding this behaviour requires time-consuming and resource-intensive trial-and-error experiments. Predicting the stability of these complex systems and enabling data science approaches requires a wealth of data, including physico-chemical information of each component, processing conditions and analytical results. In addition, the stability of the drug product is not solely dependent on the formulation, but it can also be influenced by process parameters. Molecular dynamics modelling offers potential in simulating the long-term stability effects by considering factors such as miscibility, solubility and other physico-chemical properties influencing product stability (Han et al. 2019).

Drug-Release Rate

Gaining a better understanding of how the physico-chemical properties of LAIs influence drug release is crucial in selecting and developing the right compounds and formulations early in the development process. When an injectable drug delivery system is administered, multiple processes occur, including dissolution, nucleation and permeation. These processes are influenced by various factors, including the drug's physico-chemical parameters (e.g. solubility, diffusivity, molecular weight) and the physiological environment (e.g. pH, fluid volume). However, the understanding of all these events is not yet complete. By comprehending and modelling these events, it becomes possible to facilitate the selection of suitable drug candidates, formulations, in vitro methods and animal models.

Novel Drug Delivery

The goal of long-acting therapeutics is to maintain the activity of drugs for extended periods while ensuring tolerability by the human body. The advancement of LAI drug delivery relies on strategies that include slow and controlled release, high efficacy and tolerability. To improve drug delivery platforms, it is essential to address issues like burst release, which can lead to undesirable side effects, if the therapeutic window of the drug is narrow, and limit the clinical application of LAIs. Traditional excipients used for tablet formulations can be unsafe or immunogenic when used for parenteral delivery, making further research into drug delivery platforms and excipients necessary to broaden their application. Leveraging device capabilities can offer additional benefits, such as self-administration options, reducing reliance on healthcare systems and improving market uptake of the LAI platform, particularly in developing countries. Examples include disposable pre-filled syringes or self-administrable micro-needle patches that provide safe and effective injections without the risk of cross-contamination or errors.

5.6 Conclusion

Suspension-based LAIs offer a promising approach to sustained drug delivery, providing benefits such as reduced dosing frequency, improved patient compliance and consistent therapeutic levels. On the commercial front, LAIs are finding applications across various therapeutic areas, with several marketed products already FDA-approved, primarily for treating schizophrenia and HIV. However, regulatory challenges remain, particularly in demonstrating bioequivalence and establishing IVIVC, which could lead to time and cost savings. Overall, although LAIs offer exciting prospects for sustained drug delivery, ongoing research and innovation are essential to overcome the industrial and regulatory hurdles that persist.

In conclusion, aqueous suspensions for LAIs present promising industrial applications, with significant advancements in formulation and manufacturing methods. Both top-down and bottom-up techniques offer viable paths for creating micro- and nanosuspensions, with bottom-up approaches showing potential for greater control over particle properties. Addressing manufacturing and scale-up challenges involves optimizing particle size, drug stability and production efficiency through careful selection of stabilizers, process parameters and equipment design. Ensuring high-quality products requires meticulous process monitoring and the adoption of non-invasive, real-time analytics. In addition, the importance of formulation excipients and sterilization methods for stability and safety, along with innovative approaches to process development and scale-up, is crucial for achieving consistent drug release and quality in LAIs.

References

Ali Y, Kimura A, Coffey MJ, Tyle P. Pharmaceutical development of suspension dosage form. In: Kulshreshtha AK, Singh ON, Wall GM, editors. Pharmaceutical suspensions. New York: Springer; 2010.

Alidori S, Subramanian R, Holm R. Patient-centric long-acting injectable and implantable platforms—an industrial perspective. Mol Pharm. 2024;21(9):4238–58.

Al-Rifai N, Holm R. Small but mighty. The Chemical Engineer. 2021. https://www.thechemicalengineer.com/features/small-but-mighty/

Anjum F, Viville T, Nandi S, Wessner M, De Witte B, Collas A, Sadowski G. Bottom-up production of injectable itraconazole suspensions using membrane technology. Int J Pharm. 2024a;654(March):123977. https://doi.org/10.1016/j.ijpharm.2024.123977.

Anjum F, Wessner M, De Witte B, Al-Rifai N, Collas A, Sadowski G. Tailoring the use of excipients in bottom-up production of naproxen crystal suspensions via membrane technology. Int J Pharm. 2024b;652(November 2023):123846. https://doi.org/10.1016/j.ijpharm.2024.123846.

Bao Q, Wang X, Wan B, Zou Y, Wang Y, Burgess DJ. Development of in vitro-in vivo correlations for long-acting injectable suspensions. Int J Pharm. 2023;634:122642. https://doi.org/10.1016/j.ijpharm.2023.122642.

Beinert S, Schilde C, Gronau G, Kwade A. CFD-discrete element method simulations combined with compression experiments to characterize stirred-media mills. Chem Eng Technol. 2014;37(5):770–8. https://doi.org/10.1002/ceat.201300716.

Bitterlich A, Laabs C, Busmann E, Grandeury A, Juhnke M, Bunjes H, Kwade A. Challenges in nanogrinding of active pharmaceutical ingredients. Chem Eng Technol. 2014;37(5):840–6. https://doi.org/10.1002/ceat.201300697.

Bitterlich A, Laabs C, Krautstrunk I, Dengler M, Juhnke M, Grandeury A, Bunjes H, Kwade A. Process parameter dependent growth phenomena of naproxen nanosuspension manufactured by wet media milling. Eur J Pharm Biopharm. 2015;92:171–9. https://doi.org/10.1016/j.ejpb.2015.02.031.

Bitterlich A, Mihorko A, Juhnke M. Design space and control strategy for the manufacturing of wet media milled drug nanocrystal suspensions by adopting mechanistic process modeling. Pharmaceutics. 2024;16(3) https://doi.org/10.3390/pharmaceutics16030328.

Blackwood C, Sanga P, Nuamah I, Keenan A, Singh A, Mathews M, Gopal S. Patients' preference for long-acting injectable versus oral antipsychotics in schizophrenia: results from the patient-reported medication preference questionnaire. Patient Prefer Adherence. 2020;14:1093–102. https://doi.org/10.2147/PPA.S251812. PMID: 32753849; PMCID: PMC7342487

Blaze Metrics. Blaze metrics. https://www.blazemetrics.com/. Accessed 5 May 2024.

Böttcher AC, Schilde C, Kwade A. Experimental assessment of grinding bead velocity distributions and stressing conditions in stirred media mills. Adv Powder Technol. 2021;32(2):413–23. https://doi.org/10.1016/j.apt.2020.12.022.

Breitung-Faes S. Estimation of product relating energy of wet operated stirred media mills in terms of process transfer to other mill geometries and sizes. Miner Eng. 2017;103–104:33–42. https://doi.org/10.1016/j.mineng.2016.08.008.

Breitung-Faes S, Kwade A. Mill, material, and process parameters—a mechanistic model for the set-up of wet-stirred media milling processes. Adv Powder Technol. 2019;30(8):1425–33. https://doi.org/10.1016/j.apt.2019.04.013.

Cerdeira AM, Mazzotti M, Gander B. Role of milling parameters and particle stabilization on nanogrinding of drug substances of similar mechanical properties. Chem Eng Technol. 2011;34(9):1427–38. https://doi.org/10.1002/ceat.201100155.

Chan HK, Kwok PCL. Production methods for nanodrug particles using the bottom-up approach. Adv Drug Deliv Rev. 2011;63(6):406–16. https://doi.org/10.1016/j.addr.2011.03.011.

Chandramohan P, Srinivasan MP, Velmurugan S, Narasimhan SV. Cation distribution and particle size effect on Raman spectrum of CoFe2O4. J Solid State Chem. 2011;184(1):89–96. https://journals.scholarsportal.info/details/00224596/v184i0001/89_cdapseorsoc.xml

Chogale M, Gite S, Patravale V. Comparison of media milling and microfluidization methods for engineering of nanocrystals: a case study. Drug Dev Ind Pharm. 2020;46(11):1763–75. https://doi.org/10.1080/03639045.2020.1821046.

Clancy DJ, Guner G, Chattoraj S, Yao H, Faith MC, Salahshoor Z, Martin KN, Bilgili E. Development of a semi-mechanistic modeling framework for wet bead milling of pharmaceutical nanosuspensions. Pharmaceutics. 2024;16(3):1–35. https://doi.org/10.3390/pharmaceutics16030394.

Cornier J, Owen A, Kwade A, Van de Voorde M, Finke JH, Juhnke M, Kwade A, Bunjes H. Pharmaceutical nanotechnology: innovation and production. New York: Wiley; 2016. https://doi.org/10.1002/9783527800681.ch9.

de Carvalho RM, Oliveira ALR, Petit HA, Tavares LM. Comparing modeling approaches in simulating a continuous pilot-scale wet vertical stirred mill using PBM-DEM-CFD. Adv Powder Technol. 2023;34(9):104135. https://doi.org/10.1016/j.apt.2023.104135.

Dokoumetzidis A, Macheras P. A century of dissolution research: from Noyes and Whitney to the biopharmaceutics classification system. Int J Pharm. 2006;321(1–2):1–11. https://doi.org/10.1016/j.ijpharm.2006.07.011. Epub 2006 Jul 15. PMID: 16920290

Emami J. In vitro - in vivo correlation: from theory to applications. J Pharm Pharm Sci. 2006;9(2):169–89. PMID: 16959187

Eskin D, Zhupanska O, Hamey R, Moudgil B, Scarlett B. Microhydrodynamic analysis of nanogrinding in stirred media mills. AICHE J. 2005;51(5):1346–58. https://doi.org/10.1002/aic.10392.

European Medicines Agency. Guideline on the sterilisation of the medicinal product , active substance , excipient and primary container. 2019. https://www.ema.europa.eu/en/documents/scientific-guideline/guideline-sterilisation-medicinal-product-active-substance-excipient-and-primary-container_en.pdf.

FDA. FDA document of Invega Sustenna. 2009. https://www.accessdata.fda.gov/drugsatfda_docs/nda/2009/022264s000clinpharmr.pdf. Accessed 16 May 2009.

FDA. FDA document of Abilify Maintena. 2015. https://www.accessdata.fda.gov/drugsatfda_docs/label/2016/202971s008lbl.pdf, Accessed 16 May 2009.

Flach F, Konnerth C, Peppersack C, Schmidt J, Damm C, Breitung-Faes S, Peukert W, Kwade A. Impact of formulation and operating parameters on particle size and grinding media wear in wet media milling of organic compounds—a case study for pyrene. Adv Powder Technol. 2016;27(6):2507–19. https://doi.org/10.1016/j.apt.2016.09.026.

Flach F, Breitung-Faes S, Kwade A. Model based process optimization of nanosuspension preparation via wet stirred media milling. Powder Technol. 2018;331:146–54. https://doi.org/10.1016/j.powtec.2018.03.011.

Göke K, Roese E, Arnold A, Kuntsche J, Bunjes H. Control over particle size distribution by autoclaving poloxamer-stabilized trimyristin nanodispersions. Mol Pharm. 2016;13(9):3187–95. https://doi.org/10.1021/acs.molpharmaceut.6b00395.

Graven-Nielsen T, Arendt-Nielsen L, Svensson P, Jensen TS. Quantification of local and referred muscle pain in humans after sequential i.m. injections of hypertonic saline. Pain. 1997;69(1–2):111–7. https://doi.org/10.1016/s0304-3959(96)03243-5. PMID: 9060020

Grenville RK, Giacomelli JJ, Brown DAR. Suspension of solid particles in vessels agitated by Rushton turbine imperllers. Chem Eng Res Des. 2016;109:730–3. https://doi.org/10.1016/j.cherd.2016.03.024.

Guner G, Yilmaz D, Bilgili E. Kinetic and microhydrodynamic modeling of fenofibrate nanosuspension production in a wet stirred media mill. Pharmaceutics. 2021;13(7):1–33. https://doi.org/10.3390/pharmaceutics13071055.

Hagedorn M, Bögershausen A, Rischer M, Schubert R, Massing U. Dual centrifugation - a new technique for nanomilling of poorly soluble drugs and formulation screening by an DoE-approach. Int J Pharm. 2017;530(1–2):79–88. https://doi.org/10.1016/j.ijpharm.2017.07.047. Epub 2017 Jul 18. PMID: 28733242

Hagedorn M, Liebich L, Bögershausen A, Massing U, Hoffmann S, Mende S, Rischer M. Rapid development of API nano-formulations from screening to production combining dual centrifugation and wet agitator bead milling. Int J Pharm. 2019;565:187–98. https://doi.org/10.1016/j.ijpharm.2019.04.082. Epub 2019 May 4. PMID: 31063837

Håkansson A. Scale-down failed—dissimilarities between high-pressure homogenizers of different scales due to failed mechanistic matching. J Food Eng. 2017;195:31–9. https://doi.org/10.1016/j.jfoodeng.2016.09.019.

Halme A, Quayle MJ, Nilsson Lill SO, Pettersen A, Fransson M, Boissier C. Utilizing crystal structures for predicting impact of mechanical and surface properties on particle fracture. Cryst Growth Des. 2019;19(7):3670–80. https://doi.org/10.1021/acs.cgd.8b01867.

Han R, Xiong H, Ye Z, Yang Y, Huang T, Jing Q, Lu J, Pan H, Ren F, Ouyang D. Predicting physical stability of solid dispersions by machine learning techniques. J Control Release. 2019;311–312(March):16–25. https://doi.org/10.1016/j.jconrel.2019.08.030.

Hao L, Wang X, Zhang D, Xu Q, Song S, Wang F, Li C, Guo H, Liu Y, Zheng D, Zhang Q. Studies on the preparation , characterization and pharmacokinetics of Amoitone B nanocrystals. Int J Pharm. 2012;433(1–2):157–64. https://doi.org/10.1016/j.ijpharm.2012.05.002.

He Y, Ye Z, Liu X, Wei Z, Qiu F, Li HF, Zheng Y, Ouyang D. Can machine learning predict drug nanocrystals? J Control Release. 2020;322(October 2019):274–85. https://doi.org/10.1016/j.jconrel.2020.03.043.

Heidari H, Muanpaopong N, Guner G, Yao HF, Clancy DJ, Bilgili E. A novel PBM for nanomilling of drugs in a recirculating wet stirred media mill: impacts of batch size, flow rate, and Back-mixing. Pharmaceutics. 2024;16(3) https://doi.org/10.3390/pharmaceutics16030353.

Hennart SLA, van Hee P, Drouet V, Domingues MC, Wildeboer WJ, Meesters GMH. Characterization and modeling of a sub-micron milling process limited by agglomeration phenomena. Chem Eng Sci. 2012;71:484–95. https://doi.org/10.1016/j.ces.2011.11.010.

Holm R, Lee RW, Glassco J, DiFranco N, Bao Q, Burgess DJ, Lukacova V, Alidori S. Long-acting injectable aqueous suspensions—summary from an AAPS workshop. AAPS J. 2023;25(3):1–19. https://doi.org/10.1208/s12248-023-00811-8.

Hugo Silva M, Kumar A, Hodnett BK, Tajber L, Holm R, Hudson SP. Impact of excipients and seeding on the solid-state form transformation of indomethacin during liquid Antisolvent precipitation. Cryst Growth Des. 2022;22(10):6056–69. https://doi.org/10.1021/acs.cgd.2c00678.

ICH. Stability testing of new drug substances and products. ICH Guideline. 2003;Q1A(R2)

Inguva P, Grasselli S, Heng PWS. High pressure homogenization—an update on its usage and understanding. Chem Eng Res Des. 2024;202(October 2023):284–302. https://doi.org/10.1016/j.cherd.2023.12.026.

ISO. ISO 11608-1:2022: needle-based injection systems for medical use—requirements and test methods. 2022. https://www.iso.org/standard/70733.html. Accessed 16 May 2024.

Jain KMH, Ho T, Hoe S, Wan B, Muthal A, Subramanian R, Foti C. Accelerated and biopredictive in vitro release testing strategy for single agent and combination long-acting injectables. J Pharm Sci. 2024;113(7):1885–97. https://doi.org/10.1016/j.xphs.2024.02.013.

Jayasundara CT, Yang RY, Guo BY, Yu AB, Rubenstein J. Effect of slurry properties on particle motion in IsaMills. Miner Eng. 2009;22(11):886–92. https://doi.org/10.1016/j.mineng.2009.04.009.

Kanthamneni N, Valiveti S, Patel M, Xia H, Tseng Y-C. Enhanced bioavailability of danazol nanosuspensions by wet milling and high-pressure homogenization. Int J Pharm Investig. 2016;6(4):218. https://doi.org/10.4103/2230-973x.195931.

Kaur A, Yadav JP, Sathe RY, Puri V, Bharatam PV, Bansal AK. Understanding poor milling behavior of voriconazole from crystal structure and intermolecular interactions. Mol Pharm. 2022;19(3):985–97. https://doi.org/10.1021/acs.molpharmaceut.1c00978.

Keck CM, Müller RH. Drug nanocrystals of poorly soluble drugs produced by high pressure homogenisation. Eur J Pharm Biopharm. 2006;62(1):3–16. https://doi.org/10.1016/j.ejpb.2005.05.009.

Kirschner JM, Mitsos A. Polymer particle sizing from Raman spectra by regression of hard model parameters. J Raman Spectrosc. 2018;49(8):1402–11. https://doi.org/10.1002/jrs.5387.

Knieke C, Sommer M, Peukert W. Identifying the apparent and true grinding limit. Powder Technol. 2009;195(1):25–30. https://doi.org/10.1016/j.powtec.2009.05.007.

Kresta SM, editor. Handbook of industrial mixing. 1st ed. Wiley; 2004. https://doi.org/10.1002/0471451452.

Kumar P, Verma M. Pharmaceutical suspensions: an updated patent review on novel suspending agents and recent advancement. Recent Adv Drug Deliv Formul. 2023;17(3):193–209. https://doi.org/10.2174/0126673878246149231010085610. PMID: 38050417

Kwade A. A stressing model for the description and optimization of grinding processes. Chem Eng Technol. 2003;26(2):199–205. https://doi.org/10.1002/ceat.200390029.

Kwade A. Mill selection and process optimization using a physical grinding model. Int J Miner Process. 2004;74(Suppl):93–101. https://doi.org/10.1016/j.minpro.2004.07.027.

Larsson S, Pålsson BI, Parian M, Jonsén P. A novel approach for modelling of physical interactions between slurry, grinding media and mill structure in wet stirred media mills. Miner Eng. 2020;148 https://doi.org/10.1016/j.mineng.2019.106180.

Lee M, Cho YW, Park JH, Chung H, Jeong SY, Choi K, Moon DH, Kim SY, Kim IS, Kwon IC. Size control of self-assembled nanoparticles by an emulsion/solvent evaporation method. Colloid Polym Sci. 2006;284(5):506–12. https://doi.org/10.1007/s00396-005-1413-3.

Lei J, You HJ. Variation of pain and vasomotor responses evoked by intramuscular infusion of hypertonic saline in human subjects: influence of gender and its potential neural mechanisms. Brain Res Bull. 2012;87(6):564–70. https://doi.org/10.1016/j.brainresbull.2011.11.003. Epub 2011 Nov 12. PMID: 22100333

Lestari ML, Müller RH, Möschwitzer JP. Systematic screening of different surface modifiers for the production of physically stable nanosuspensions. J Pharm Sci. 2015;104(3):1128–40. https://doi.org/10.1002/jps.24266. Epub 2015 Jan 28. PMID: 25630623

Li M, Alvarez P, Bilgili E. A microhydrodynamic rationale for selection of bead size in preparation of drug nanosuspensions via wet stirred media milling. Int J Pharm. 2017;524(1–2):178–92. https://doi.org/10.1016/j.ijpharm.2017.04.001.

Liu J, Kuang W, Liu J, Gao Z, Rohani S, Gong J. In-situ multi-phase flow imaging for particle dynamic tracking and characterization: advances and applications. Chem Eng J. 2022;438(January):135554. https://doi.org/10.1016/j.cej.2022.135554.

Margolis DA, Gonzalez-Garcia J, Stellbrink HJ, Eron JJ, Yazdanpanah Y, Podzamczer D, Lutz T, Angel JB, Richmond GJ, Clotet B, Gutierrez F, Sloan L, Clair MS, Murray M, Ford SL, Mrus J, Patel P, Crauwels H, Griffith SK, Sutton KC, Dorey D, Smith KY, Williams PE, Spreen WR. Long-acting intramuscular cabotegravir and rilpivirine in adults with HIV-1 infection (LATTE-2): 96-week results of a randomised, open-label, phase 2b, non-inferiority trial. Lancet. 2017;390(10101):1499–510. https://doi.org/10.1016/S0140-6736(17)31917-7. Epub 2017 Jul 24. PMID: 28750935

Méndez Cañellas F, Al-Rifai N, Padrela L, Tajber L, Khamiakova T, Otava M, Geertman R. A comparative study of dry and wet top-down milling approaches for the preparation of microparticle suspensions. Powder Technol. 2023a;428(June):118829. https://doi.org/10.1016/j.powtec.2023.118829.

Méndez Cañellas F, Geertman R, Tajber L, Padrela L. Sorbitol-coated indomethacin and naproxen particles produced by supercritical CO2-assisted spray drying. J Supercrit Fluids. 2023b;199(April):105969. https://doi.org/10.1016/j.supflu.2023.105969.

Mettler Toledo-a. EasyViewer with iC vision. https://www.mt.com/us/en/home/products/L1_AutochemProducts/particle-size-analyzers/easyviewer.html. Accessed 5 May 2024.

Mettler Toledo-b. ParticleTrack with FBRM technology. https://www.mt.com/int/en/home/products/L1_AutochemProducts/particle-size-analyzers/particletrack-fbrm.html. Accessed 5 May 2024.

Muller R, Al Shaal L, Shegokar R. SmartCrystal combination technology--scale up from lab to pilot scale and long term stability. Pharmazie. 2010;65:877–84.

Nakach M, Authelin JR, Tadros T, Galet L, Chamayou A. Engineering of nano-crystalline drug suspensions: employing a physico-chemistry based stabilizer selection methodology or approach. Int J Pharm. 2014;476(1–2):277–88. https://doi.org/10.1016/j.ijpharm.2014.09.048. Epub 2014 Sep 30. PMID: 25278040

Nakach M, Authelin JR, Agut C. New approach and practical modelling of bead milling process for the manufacturing of nanocrystalline suspensions. J Pharm Sci. 2017;106(7):1889–904. https://doi.org/10.1016/j.xphs.2017.02.036.

Nakach M, Authelin JR, Perrin MA, Lakkireddy HR. Comparison of high pressure homogenization and stirred bead milling for the production of nano-crystalline suspensions. Int J Pharm. 2018;547(1–2):61–71. https://doi.org/10.1016/j.ijpharm.2018.05.042.

Nandi S, Padrela L, Tajber L, Collas A. Development of long-acting injectable suspensions by continuous antisolvent crystallization: an integrated bottom-up process. Int J Pharm. 2023a;648(August):123550. https://doi.org/10.1016/j.ijpharm.2023.123550.

Nandi S, Padrela L, Tajber L, Collas A. Nucleation kinetics-based solvent selection for the liquid antisolvent crystallization of a lipophilic intermediate. J Mol Liq. 2023b;375:121306. https://doi.org/10.1016/j.molliq.2023.121306.

Nandi S, Collas A, Tajber L, Padrela L. Supercritical enhanced atomization induced particle entrapment for continuous production of injectable micro and nano-suspensions. J CO2 Util. 2024a;81(March):102732. https://doi.org/10.1016/j.jcou.2024.102732.

Nandi S, Verstrepen L, Hugo Silva M, Padrela L, Tajber L, Collas A. Continuous microfluidic antisolvent crystallization as a bottom-up solution for the development of long-acting injectable formulations. Pharmaceutics. 2024b;16(3):376. https://doi.org/10.3390/pharmaceutics16030376.

Nguyen VTT, Darville N, Vermeulen A. Pharmacokinetics of long-acting aqueous nano-/microsuspensions after intramuscular administration in different animal species and humans: a review. AAPS J. 2022;25(1):4. https://doi.org/10.1208/s12248-022-00771-5. PMID: 36456852

Nöske M, Li K, Brüning K, Breitung-Faes S, Kwade A. Impact of stress conditions on stirred media milling of a two component mixture. Miner Eng. 2020;146(November 2019):106100. https://doi.org/10.1016/j.mineng.2019.106100.

Osorio-Arias JC, Vega-Castro O, Martinez-Monteagudo SI. Fundamentals of high-pressure homogenization of foods. In: Innovative food processing technologies: a comprehensive review, vol. 3. Wiley; 2020. p. 244–73. https://doi.org/10.1016/b978-0-08-100596-5.23021-7.

Parker N, Rahman M, Bilgili E. European journal of pharmaceutics and biopharmaceutics impact of media material and process parameters on breakage kinetics—energy consumption during wet media milling of drugs. Eur J Pharm Biopharm. 2020;153(June):52–67. https://doi.org/10.1016/j.ejpb.2020.05.024.

Perkins MC, Bunker M, James J, Rigby-Singleton S, Ledru J, Madden-Smith C, Luk S, Patel N, Roberts CJ. Towards the understanding and prediction of material changes during micronisation using atomic force microscopy. Eur J Pharm Sci. 2009;38(1):1–8. https://doi.org/10.1016/j.ejps.2009.05.004.

Salahshoor Z, Desai PM, Chattoraj S, Lam S. Exploring the relationship between bulk Young's modulus of materials and milling efficiency during wet bead milling of pharmaceutical compounds. Int J Pharm. 2024;660(June):124365. https://doi.org/10.1016/j.ijpharm.2024.124365.

Salazar J, Müller RH, Möschwitzer JP. Combinative particle size reduction technologies for the production of drug nanocrystals. J Pharm. 2014;2014:265754. https://doi.org/10.1155/2014/265754.

Shegokar R, Nakach M. Large-scale manufacturing of nanoparticles—an industrial outlook. In: Drug delivery aspects: Volume 4: expectations and realities of multifunctional drug delivery systems. Elsevier; 2020. https://doi.org/10.1016/B978-0-12-821222-6.00004-X.

Shegokar R, Singh KK, Müller RH, College CUS. Nevirapine nanosuspension: comparative investigation of production methods. Nanotechnol Dev. 2011;1:e4. https://doi.org/10.4081/nd.2011.e4.

Shore. 2024. https://shore-group.com/work/multisensory-drug-preparation-trainer/. Accessed 14 April 2024.

Siepmann J, Siepmann F. Mathematical modeling of drug dissolution. Int J Pharm. 2013;453(1):12–24. https://doi.org/10.1016/j.ijpharm.2013.04.044. Epub 2013 Apr 22. PMID: 23618956

Stender HH, Kwade A, Schwedes J. Stress energy distribution in different stirred media mill geometries. Int J Miner Process. 2004;74(Suppl):103–17. https://doi.org/10.1016/j.minpro.2004.07.003.

Stenger F, Mende S, Schwedes J, Peukert W. Nanomilling in stirred media mills. Chem Eng Sci. 2005;60(16):4557–65. https://doi.org/10.1016/j.ces.2005.02.057.

Suarez-Sharp S, Li M, Duan J, Shah H, Seo P. Regulatory experience with in vivo in vitro correlations (IVIVC) in new drug applications. AAPS J. 2016;18(6):1379–90. https://doi.org/10.1208/s12248-016-9966-2. Epub 2016 Aug 1. PMID: 27480319

Sudhakar P, Kumari A, Kundu S, Sankar VR, Thella PK, Shah K, Bhargava SK. Design and optimization of antisolvent crystallization of L-aspartic acid using response surface model:

focused beam reflectance. Chem Eng Res Des. 2023;191:172–82. https://doi.org/10.1016/j.cherd.2023.01.020.

Szilágyi B, Quon JL, Papageorgiou CD, Nagy ZK. Monitoring and digital design of the cooling crystallization of a high-aspect ratio anticancer drug using a two-dimensional population balance model. Chem Eng Sci. 2022;257:117700. https://doi.org/10.1016/j.ces.2022.117700.

Taghizadeh B, Jaafari MR, Zarghami N. New insight into the importance of formulation variables on parenteral growth hormone preparations: potential effect on the injection-site pain. Front Endocrinol (Lausanne). 2022;13:963336. https://doi.org/10.3389/fendo.2022.963336. PMID: 36263321; PMCID: PMC9576007

Tosca EM, Rocchetti M, Pérez E, Nieto C, Bettica P, Moscoso Del Prado J, Magni P, De Nicolao G. In vitro-in vivo correlation (IVIVC) population modeling for the in silico bioequivalence of a long-acting release formulation of progesterone. Pharmaceutics. 2021;13(2):255. https://doi.org/10.3390/pharmaceutics13020255. PMID: 33673306; PMCID: PMC7918143

Van Buskirk GA, Asotra S, Balducci C, Basu P, Didonato G, Dorantes A, Eickhoff WM, Ghosh T, González MA, Henry T, Howard M, Kamm J, Laurenz S, Mackenzie R, Mannion R, Noonan PK, Ocheltree T, Pai U, Poska RP, et al. Best practices for the development, scale-up, and post-approval change control of IR and MR dosage forms in the current quality-by-design paradigm. AAPS PharmSciTech. 2014;15(3):665–93. https://doi.org/10.1208/s12249-014-0087-x.

Verma S, Lan Y, Gokhale R, Burgess DJ. Quality by design approach to understand the process of nanosuspension preparation. Int J Pharm. 2009;377(1–2):185–98. https://doi.org/10.1016/j.ijpharm.2009.05.006.

Wang W. Tolerability of hypertonic injectables. Int J Pharm. 2015;490(1–2):308–15. https://doi.org/10.1016/j.ijpharm.2015.05.069. Epub 2015 May 29. PMID: 26027488

Watt RP, Khatri H, Dibble ARG. Injectability as a function of viscosity and dosing materials for subcutaneous administration. Int J Pharm. 2019;554:376–86. https://doi.org/10.1016/j.ijpharm.2018.11.012. Epub 2018 Nov 7. PMID: 30414478

Wright J, Burgess D, editors. Long acting injections and implants. Springer; 2012. https://doi.org/10.1017/CBO9781107415324.004.

Zhang Y, Li Y, Zhao X, Zu Y, Wang W, Wu W. Preparation, characterization and bioavailability of oral puerarin nanoparticles by emulsion solvent evaporation method. RSC Adv. 2016;6:69889–901. https://doi.org/10.1039/c6ra08413c.

Zulbeari N, Holm R. Wet bead milling by dual centrifugation - an approach to obtain reproducible and differentiable suspensions. Int J Pharm. 2023;646:123455. https://doi.org/10.1016/j.ijpharm.2023.123455. Epub 2023 Sep 28. PMID: 37776963

Zulbeari N, Holm R. Is roller milling - the low energy wet bead media milling - a reproducible and robust milling method for formulation investigation of aqueous suspensions? Int J Pharm. 2024;651:123733. https://doi.org/10.1016/j.ijpharm.2023.123733. Epub 2023 Dec 22. PMID: 38142873

Chapter 6
Oil-Based Long-Acting Injectable Systems: Formulation Insights and Future Perspectives

Niranjan G. Kotla, Samaneh Validi, and Farshad Ramazani

List of Abbreviations

1D	One-dimensional
2D	Two-dimensional
3D	Three-dimensional
12-HSA	12-Hydroxystearic acid
CGC	Critical gelation concentration
*C*max	Maximum concentration
DNS	Dinalbuphine sebacate
DOX	Doxorubicin
DS	Drug substance
IM	Intramuscular
ISFOs	Lipid in situ forming oleogels
IVIVC	In vitro–in vivo correlation
LAI	Long-acting injectable
LLCs	Lyotropic liquid crystals
log *P*	Partition coefficient
NMP	N-methyl-pyrrolidone
PK–PD	Pharmacokinetic and pharmacodynamic
SC	Subcutaneous
SDE	Sebacoyl dinalbuphine ester
*T*max	Time to maximum concentration

N. G. Kotla
Novartis Institutes for Biomedical Research (NIBR), Novartis Pharma AG, Basel, Switzerland

Department of Health Technology, Section for Biotherapeutic Engineering and Drug Targeting, Technical University of Denmark, Lyngby, Denmark

S. Validi · F. Ramazani (✉)
Technical Research and Development, Novartis Pharma AG, Basel, Switzerland
e-mail: farshad.ramazani@novartis.com

F. Ramazani (ed.), *Biodegradable Long Acting Injectables and Implants*,
https://doi.org/10.1007/978-3-031-72302-5_6

6.1 Introduction

Long-acting injectable (LAI) formulations are designed to prolong the action of therapeutics in the body, resulting in fewer dosing errors and improving patient compliance. LAIs have been used for drug release either at the site of action locally or for systemic delivery through subcutaneous (SC) and intramuscular (IM) routes, overcoming the need for frequent/repeated dosing (Li et al. 2022). When compared to traditional oral administration, LAIs have numerous distinct advantages such as a long-lasting effect, avoidance of stomach enzymatic degradation, reduced dosing frequency, and improved adherence in patients with chronic conditions that require frequent repeated dosing (Kotla et al. 2023a, b; Nkanga et al. 2020).

LAI delivery systems have been shown to be effective in the treatment and management of chronic diseases such as diabetes, osteoarthritis, cancer, hormonal imbalances, and neurological disorders by reducing the dosing frequency while maintaining favorable pharmacokinetic and pharmacodynamic (PK–PD) profiles (Chaudhary et al. 2019; Kotla et al. 2023b; Li et al. 2022; Nkanga et al. 2020). Given recent advances and interest in LAI therapies, the number of research findings and the marketed products in this field is growing. Several proven LAI technology platforms that have entered the market include oil-based formulations, micro- and nano-crystal suspensions, polymeric particle-based products, and in situ depots/implants. For these LAIs, the selection of the delivery system depends largely on the drug substance's properties, mode of action, and site of therapeutic action, among others.

Oil-based lipophilic compound formulations consist of an active pharmaceutical ingredient, oil, and possibly a co-solvent (such as benzyl benzoate, benzyl alcohol, and ethyl lactate). Several vegetable or synthetic oils including castor oil, soybean oil, sesame oil, cottonseed oil, fractionated coconut oil, and triglyceride oils have been used as the vehicle oil in LAI products (Rahnfeld and Luciani 2020; Schwendeman et al. 2014; Wilkinson et al. 2022). Figure 6.1 depicts a schematic representation of several oil-based long-acting injectable formulations, such as oil-based solutions, suspensions, or prodrug forms, and oleogels of lipophilic drugs in an oil carrier. The advantages of oil-based formulations (ease of procedure, low scale-up cost) led to the successful development of oil-based LAI products. Although these strategies have several advantages, they also present some key limitations, including applicability limited only to lipophilic drugs, injection site tolerability issues, depot physical/mechanical properties (size, shape, and volume), altered drug-release behavior, and so on. This chapter specifically focuses on oil-based LAI system properties, drug-to-formulation considerations, and their release mechanisms. We also highlight the formulation strategies of oil-based delivery systems with a summary of various oil-based formulations in clinical trials and marketed products.

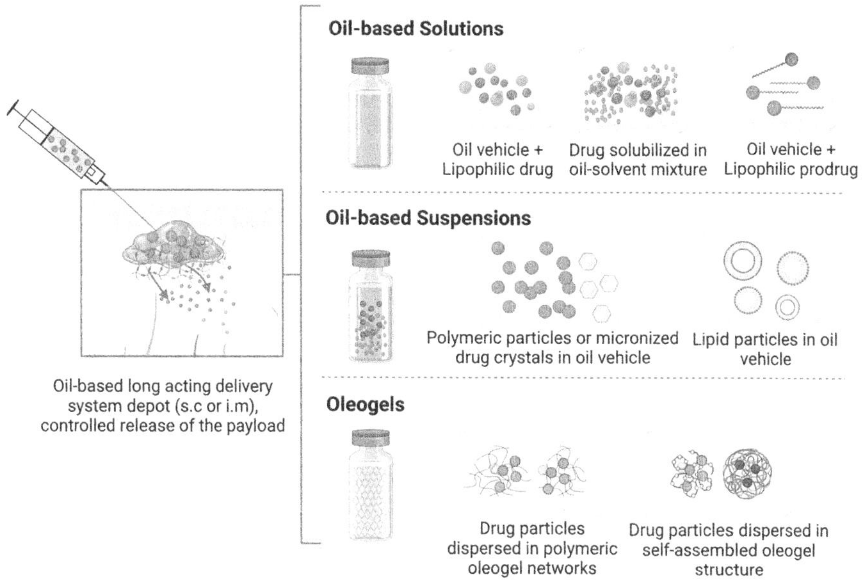

Fig. 6.1 Oil-based long-acting injectable delivery system administration via subcutaneous or intramuscular route to deliver the payload in a controlled-release manner, various formulation strategies/types, and composition

6.2 Oil-Based LAIs: Formulation Considerations, Composition, and Recent Developments

In this section, we will discuss various aspects of drug molecule from physicochemical characteristics to formulation composition in developing an ideal oil-based formulation for long-acting delivery applications. An ideal oil-based LAI formulation pharmacological effect(s) depends on multiple factors including (1) the drug molecule properties (log *P*, solubility, potency, and chemical stability); (2) formulation type (oil-based solutions, oil suspension, prodrug form, conjugated systems); (3) type of oil (composition and viscosity); and (4) formulation matrices behavior and local environment (depot degradation, passive diffusion, tissue fluid impact, drug partitioning between oil and tissues fluid, etc.).

Partition coefficient (log *P*) is an important physicochemical property of lipophilic agents that play a significant role in the oil-based formulation kinetics. Drugs having a high log *P* value, indicating high lipophilicity, are more likely to be accepted for oil-based formulations since they have a higher affinity for oils and can be easily dissolved or suspended in the vehicle. This feature may increase drug substance (DS) solubility, stability, and bioavailability.

From an analytical view, physicochemical characterization of the oil-based systems is simple. In vitro drug release can be studied using a "holder" for oil solution, commonly a dialysis membrane that is fitted in a conventional apparatus.

Factors Impacting Extended Release of Oil-Based Formulations

Several factors are important in modulating drug release from oil-based formulations including drug substance properties (such as log *P*, prodrug forms, fatty acid type, and molecular weight), formulation attributes (type of formulation, viscosity, usage of gelators, co-solvents, etc.), and the in vivo behavior depending on the site of injection. Oil-based products with high log *P* values (>3.8) are in general preferred for long-acting delivery (Nkanga et al. 2020). This can be obtained by increasing the length of the fatty-acid chain (Remenar 2014). Esterification of alcohol groups of drug molecules with fatty acids can reduce the drug's aqueous solubility and increase its oil partitioning. A drug molecule with a strong affinity for the oil phase could potentially release too slowly, whereas a molecule with a high aqueous solubility will partition out of the oil phase and release quickly. Literature evidence showed that decanoate tail (C10) seemed to have the optimum release rate for many compounds (Remenar 2014). The relationship between log *P* and pharmacokinetics is being studied by scientists (Han et al. 2021; Wilkinson et al. 2022).

The drug molecule's ester modification may be carried out using a wide variety of endogenous fatty acids to modify its physical characteristics, such as melting point, solubility, and partition coefficient (Wilkinson et al. 2022). As an example, an antipsychotic drug substance is esterified to a long-chain fatty acid, and the resulting molecule is dispersed in vegetable oils to extend the release (Park et al. 2013; Spanarello and La Ferla 2014). Besides, due to their potential benefits over synthetic oils, research on the use of vegetable oil vehicles for oil-based long-acting formulations is still underway. Vegetable oils have several advantages including biodegradability, renewability, and lesser toxicity over synthetic oils. Soybean oil has been used as a carrier for depot injections and transdermal patches. Other vegetable oils that have been explored for use in oil-based long-acting formulations include olive oil, sunflower oil, and corn oil.

Viscosity refers to a fluid's resistance to flow and is a measure of the internal friction as it flows. In simpler terms, it describes how "sticky" a fluid is. While developing formulations with oil, it is important to use oil that is physically or chemically stable and has the right viscosity. The viscosity of the oil affects how far it spreads and impacts syringeability of the depot formulation (Weng Larsen and Larsen 2009). Highly viscous oils are more common for depot formulations as they help maintain the shape of the depot after injection, enabling prolonged drug release and less spreading. On the other hand, low viscous oils are desirable for good injectability, to minimize pain upon injection and excellent processability, e.g., easy filtration. Generally, Newtonian oil-based solutions with viscosities up to approximately 60 mPa·s exhibit good syringeability, as commercial parenteral solutions typically fall within the range of 29–60 mPa·s (Larsen et al. 2012).

A study by Larsen et al. compared the clearance rates of different vegetable oils after IM and SC administration in pigs, demonstrating the impact of the oil's viscosity on the clearance rate in the body (Larsen et al. 2001). It was found that oils with high viscosity, such as castor oil with a viscosity of 283 mPa·s and sesame oil with a viscosity of 33 mPa·s, had a longer in vivo half-life compared to oil with a low viscosity of 12 mPa·s. However, high viscosity oils have difficulties with syringeability; corn oil was preferred for its good injectability and sustained drug release (Zhang et al. 2019). The release of drugs from oily solutions is affected by various factors, including drug partitioning between the oil and tissue fluid, injection volume, site of injection, lipid degradation, and the shape of the depot (O'Brien et al. 2021; Weng Larsen and Larsen 2009). This issue can be solved by adding co-solvents such as N-methyl-pyrrolidone (NMP) and ethanol and using low viscosity precursors. For example, Saklani et al. developed doxorubicin (DOX)-loaded monoolein, phosphatidylcholine (PC), d-α-tocopherol polyethylene glycol 1000 succinate (TPGS), and ethanol-based lyotropic liquid crystals (LLCs) for localized delivery of DOX in solid tumors (Saklani et al. 2022). Another approach to improve injectability is by using a microemulsion-based system (Li et al. 2019). Lipid-based depots face challenges with syringeability and injectability due to their large particle size and high viscosity (Berteau et al. 2015). Organic solvents can be employed to reduce the viscosity of depot systems, although there are concerns regarding toxicity. In-situ forming lipid depot systems have partially addressed deliverability issues; however, they may cause burst release. Furthermore, the extent of depot spreading at the injection site is highly variable and influenced by factors such as tissue structure, injection rate, and recipient physiology, which can impact drug release kinetics and contribute to variability (Kempe and Mäder 2012).

The gelators (e.g., aluminum stearate, silica, methyl cholate, cholesteryl stearate, and dextrin palmitate) have been used as an alternative to raise oil viscosity and improve the sustained drug release performance of oil-based solutions (Nippe and General 2015; Wilkinson et al. 2022). The oleogels containing saturated fatty acids are formulated by dissolving the saturated fats in the oil phase at an elevated temperature. Subsequently, the hot mixture is cooled to room temperature. During this cooling process, fat crystals begin to crystallize within the oil phase. If the concentration of the saturated fatty acid exceeds the critical gelation concentration (CGC), which is the maximum concentration required for inducing oil gelation, the saturated fats establish an interconnected colloidal structure. The growth of fat crystals can occur in one-dimension (1D), two-dimension (2D), three-dimension (3D), or a combination of these directions. In a recent study with steroid ZK28 delivery, it was observed that gelation of medium-chain triglycerides with methyl cholate resulted in a controlled release of the drug in vivo in Wistar rats, possibly due to low drug solubility in the vehicle, slower drug diffusivity, and a high vehicle viscosity (Nippe et al. 2013). A recent study conducted by Dümichen et al. (2023) highlighted that in situ forming implants (ISFI) based on PLGA often require a large volume of organic solvents such as N-methyl-pyrrolidone (NMP). As an alternative approach, authors sought to develop lipid in situ forming oleogels (ISFOs) as improved delivery strategy. 12-Hydroxystearic acid (12-HSA) was chosen as the oleogelling agent, and 12-HSA-based ISFOs offer potential alternatives to PLGA/NMP ISFI systems with advantages such as avoidance of PLGA acidic degradation byproducts, use of less chemical solvent with low toxicity (Dümichen et al. 2023). The numerous

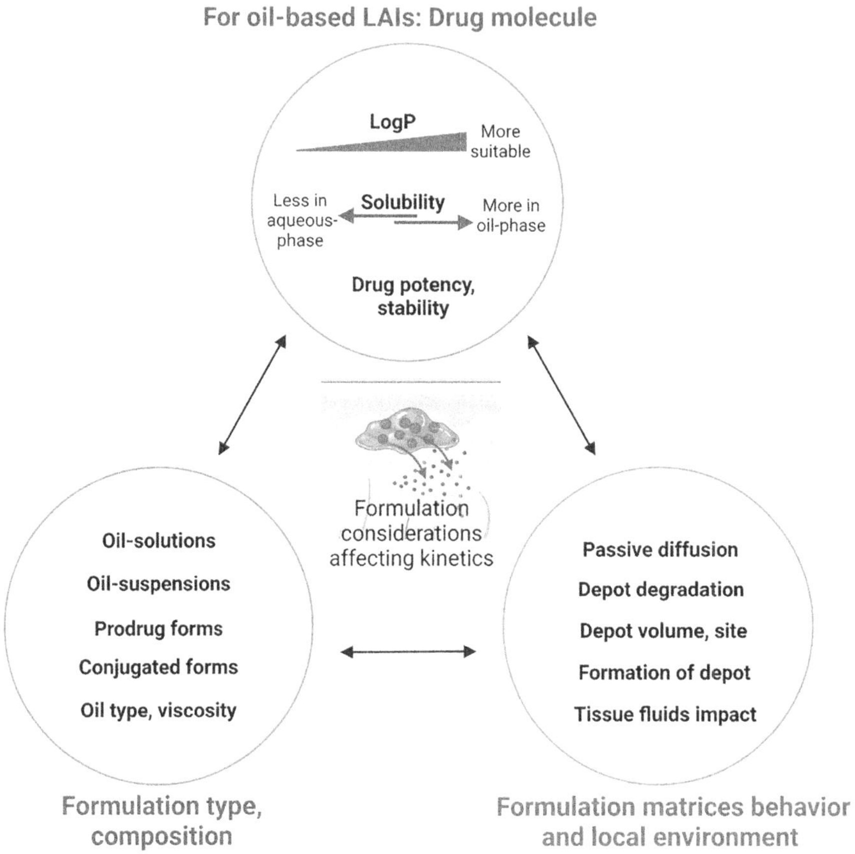

Fig. 6.2 Various factors contributing to the pharmacokinetics of the oil-based formulations

parameters that contribute to the pharmacokinetics of the payload in oil-based formulations are highlighted in Fig. 6.2.

Yeh et al. (2017) studied benzyl benzoate/sesame oil-based DNS (dinalbuphine sebacate) formulation impact on patients with hemorrhoidectomy in a clinical study (moderate to severe post-operative) discomfort. According to the pharmacokinetic profile of nalbuphine (an opioid analgesic) in individuals treated with the prolonged-release formulations, the oil-based injections showed a 7–10-day analgesic effect. Recent clinical findings showed that a single prodrug of sebacoyl dinalbuphine ester (SDE) IM injection (150 mg/2 mL)-based formulation showed promising results in patients who underwent hemorrhoidectomy for postoperative long-acting analgesia (Yeh et al. 2017). In a comparable investigation, the weight ratio of benzyl benzoate (a solubilizing agent) to sesame oil (the solvent/oil ratio) had a direct influence on dinalbuphine sebacate (DNS) release. Pharmacokinetic research in beagle dogs using IM injection elucidated the link between in vitro and in vivo release (Li et al. 2017). According to the pharmacokinetic analysis, the formulation with a higher benzyl benzoate to oil ratio had a longer drug-release profile with a lower maximum concentration (*C*max) and a longer time to peak blood concentration level (*T*max). By increasing the solvent/oil ratio in the formulation, they were able to extend the release duration; however, the authors did not explain the reasons behind this phenomenon. Yet, establishing an in vitro–in vivo correlation (IVIVC) is extremely beneficial in accelerating the development of lipophilic solutions and developing a suitable product quality control approach (Li et al. 2017). Novel simulation methods can minimize the number of experiments and predict IVIVC for such products.

6.3 Manufacturing Strategies of Oil-Based Formulations

Manufacturing process of oil-based formulation is reasonably simple and involves standard manufacturing procedures. In the case of oily solutions, the drug is dissolved in an oil vehicle, filtered via a sterilized 0.2-micron size filter to remove microbial and potential foreign visible particle contaminations, and filled into vials in a clean room as shown in Fig. 6.3 (Duvnjak et al. 2023). For suspensions, the drug is dispersed in the oil vehicle including proper suspending agent using an inline mixture. Since the drug particles are in submicron range (e.g., 1–50 micron), they cannot be filtered. In this case, a terminal sterilization approach such as autoclaving is needed. If the excipients or the active pharmaceutical ingredient (API) degrades during terminal sterilization process, the alternative method would be aseptic manufacturing. So far, all marketed products are oily solutions not suspension. Often, the viscosity of the final formulations plays a key role in drug development. Highly viscous oils pose several issues such as difficulty of homogenization, filtration, and eventually injectability issues. To lower the viscosity, co-solvent such as benzyl alcohol is used in the formulation; benzyl alcohol not only lowers viscosity of the solution but also increases the solubility of the active pharmaceutical ingredient (Duvnjak et al. 2023; Wilkinson et al. 2022).

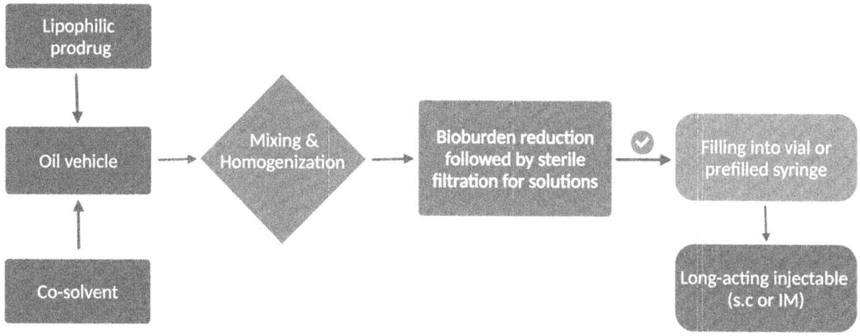

Fig. 6.3 Preparation of oily solutions. Lipophilic prodrug is added to the oil vehicle and, after dissolution, it is sterile filtered into vial or prefilled syringes

6.4 Current Landscape of Oil-Based LAI Formulations

Emerging evidence have showed the potential efficacy of the prodrug oil-based LAI formulations, which involve the covalent bonding of a fatty-acid chain and therapeutic agents to form prodrugs that are subsequently incorporated into the oil phase. In particular, decanoate, enanthate, and caproate are frequently employed to establish an ester linkage with the progenitor medication, thereby augmenting its solubility in the oil phase and improving its partitioning in the adipose tissues. Lyogen Depot® and Haldol Depot® are two parenteral formulations that belong to the class of typical antipsychotics. These oil-based formulations have been available in the United States as LAI formulations. Drug release from IM depots depends on the gradual release of prodrugs from fatty tissues into the circulation and prolonged hydrolysis of ester linkages to liberate the parent drug. Controlling drug release also requires the drug concentration in the oil phase, the oil depot's surface area, and the partition coefficient between the tissue fluid and the oil depot. Table 6.1 enlists some of the approved long-acting oil-based products.

Although several oily formulations have entered the market, but a recent trend in clinical trials shows a decreasing number. Some of the limitations of this formulation include high drug burst release and difficulty in tailoring drug release as compared to other delivery systems, such as PLGA microparticles, obligation of functional groups in parent drug to form lipophilic prodrugs, injection site pain, etc. (Duvnjak et al. 2023; Kotla et al. 2023b; Nkanga et al. 2020; Schwendeman et al. 2014).

6.5 Conclusion and Future Perspectives

This chapter delves into the sophisticated world of LAI formulations, with a specific focus on oil-based systems. The imperative need for sustained treatment of chronic illnesses has spurred the development of LAIs, offering enhanced pharmacokinetics and therapeutic efficacy. Within the realm of LAIs, oil-based formulations have emerged as a prominent category, utilizing various oils, lipids, and polymers in diverse combinations. The significance of oil-based formulations lies in their ability to sustain the release of lipophilic compounds, contributing to prolonged therapeutic effects. However, as discussed, these formulations come with their set of challenges, including applicability limited to lipophilic drugs, injection site tolerability issues, and variations in depot physical properties.

In the future, not only small molecules but also novel larger molecules such as therapeutic RNAs can be conjugated to a proper size of the fatty acid. The engineered molecule can be delivered into the body via intravenous injection, or it can be formulated in an oil-based depot to be administered via SC or IM injection. The fatty acid content not only prolongs the RNA circulation time but also increases its partitioning in the fatty tissues which, in turn, prolong its effect(s).

Table 6.1 List of approved long-acting oil-based products

Product name	Oil type	Therapeutic agent	Indication	Log *P*	Route of administration and dosing interval	Manufacturer
Faslodex®	Castor oil	Fulvestrant	Breast cancer	8.9	Intramuscular (250 mg/5 mL) Monthly (first dose), Bi week (after)	AstraZeneca
Clopixol Depot®	Fractionated coconut oil	Zuclopenthixol decanoate	Schizophrenia	9	Intramuscular (200 mg/mL) Every 2–4 weeks	Lundbeck Ltd
Nebido®	Castor oil	Testosterone Undecanoate	Testosterone deficiency	8.5	Intramuscular (250 mg/mL) Every 10–14 weeks	Bayer
Haldol Depot®	Sesame oil	Haloperidol decanoate	Tourette's syndrome Schizophrenia	4.3	Intramuscular (50 mg/mL) Every 4 weeks	Janssen Pharmaceuticals
Delatestryl®	Sesame oil	Testosterone enanthate	Breast cancer hypogonadism	7.0	Intramuscular (200 mg/mL) Every 2–4 weeks	NA
Fluanxol Depot®	Triglyceride oil solution	Flupentixol Decanoate	Schizophrenia	7.2	Intramuscular (20–100 mg/mL) Every 2–4 weeks	Lundbeck Ltd
Depo-Testosterone®	Cottonseed oil	Testosterone cypionate	Testosterone deficiency	6.4	Intramuscular (50–400 mg) Every 2–4 weeks	Pfizer
Noristerat®	Castor oil	Norethisterone enantate	Contraceptive	6	Intramuscular (200 mg/mL) Every 8 weeks	Bayer

(continued)

Table 6.1 (continued)

Product name	Oil type	Therapeutic agent	Indication	Log *P*	Route of administration and dosing interval	Manufacturer
Androcur Depot®	Castor oil	Cyproterone acetate	Prostate cancer	3.8	Intramuscular (100 mg/mL) Every 10–14 days	Bayer
Lyogen Depot®	Sesame oil	Fluphenazine Decanoate	Psychotic disorders	7.2	Intramuscular (25 mg/mL) Every 2–4 weeks	NA
Prima-Med®	NA	Methenolone enanthate	Androgen, anabolic steroid; anemia	7.5	Intramuscular (100 mg/mL) Weekly	Bioniche Pharma
Omegaven®	Fish oil triglycerides	Eicosapentaenoic and Docosahexaenoic acid	Parenteral nutrition-induced cholestasis	5.6, 6.2	Intravenous infusion (100 mg/mL) 4 weeks	Fresenius Kabi
Smoflipid ®	Soyabean, Olive, fish body oils with triglyceride mixture	NA	Parenteral nutrition	NA	IV infusion (200 mg/mL) 0.5–2 g/kg per day	Fresenius Kabi

From a manufacturing perspective, these systems are generally considered easy to scale up and commercialize. Although the existing literature provides a solid understanding of potential vehicles, their connection to release duration, and well-established in vitro methods, there are gaps in the in vivo data needed to truly define the IVIVC for the applied in vitro dissolution methods. Moreover, an in-depth comprehension of the effects of the route of administration (SC vs. IM and different sites of IM injection) on in vivo drug absorption is crucial. This knowledge is essential for constructing meaningful in-silico prediction models, facilitating rapid product development.

References

Berteau C, Filipe-Santos O, Wang T, Roja HE, Granger C, Schwarzenbach F. Evaluation of the impact of viscosity, injection volume, and injection flow rate on subcutaneous injection tolerance. Med Dev. 2015;8:473–84. https://doi.org/10.2147/MDER.S91019.

Chaudhary K, Patel MM, Mehta PJ. Long-acting injectables: current perspectives and future promise. Crit Rev Ther Drug Carrier Syst. 2019;36(2):137–81. https://doi.org/10.1615/CritRevTherDrugCarrierSyst.2018025649.

Dümichen A, Lucas H, Trutschel M-L, Mäder K. Development and characterization of novel in-situ-forming oleogels. Pharmaceutics. 2023;15(1):254. https://doi.org/10.3390/pharmaceutics15010254.

Duvnjak M, Villois A, Ramazani F. Biodegradable long-acting injectables: platform technology and industrial challenges. In: Handbook of experimental pharmacology. Berlin: Springer; 2023. p. 1–18. https://doi.org/10.1007/164_2023_651.

Han S, Mei L, Quach T, Porter C, Trevaskis N. Lipophilic conjugates of drugs: a tool to improve drug pharmacokinetic and therapeutic profiles. Pharm Res. 2021;38(9):1497–518. https://doi.org/10.1007/s11095-021-03093-x.

Kempe S, Mäder K. In situ forming implants - an attractive formulation principle for parenteral depot formulations. J Control Release. 2012;161(2):668–79. https://doi.org/10.1016/j.jconrel.2012.04.016.

Kotla NG, Langlois J-B, Fisch A, Kramer I, Halleux C. Surface modified cationic PLGA microparticles as long-acting injectable carriers for intra-articular small molecule drug delivery. Eur J Pharm Biopharm. 2023a;193:96–104. https://doi.org/10.1016/j.ejpb.2023.10.020.

Kotla NG, Pandey A, Vijaya Kumar Y, Ramazani F, Fisch A. Polyester-based long acting injectables: advancements in molecular dynamics simulation and technological insights. Drug Discov Today. 2023b;28(2):103463. https://doi.org/10.1016/j.drudis.2022.103463.

Larsen SW, Rinvar E, Svendsen O, Lykkesfeldt J, Friis GJ, Larsen C. Determination of the disappearance rate of iodine-125 labelled oils from the injection site after intramuscular and subcutaneous administration to pigs. Int J Pharm. 2001;230(1–2):67–75. https://doi.org/10.1016/s0378-5173(01)00860-2.

Larsen SW, Thing MA, Larsen C. Oily (lipophilic) solutions and suspensions. In: Wright JC, Burgess DJ, editors. Long acting injections and implants. Springer US; 2012. p. 113–35. https://doi.org/10.1007/978-1-4614-0554-2_7.

Li C-J, Ku M-Y, Lu C-Y, Tien Y-E, Chern WH, Huang J-D. In vitro and in vivo release of dinalbuphine sebacate extended release formulation: effect of the oil ratio on drug release. Int J Pharm. 2017;531(1):306–12. https://doi.org/10.1016/j.ijpharm.2017.08.083.

Li Y, Angelova A, Liu J, Garamus VM, Li N, Drechsler M, Gong Y, Zou A. In situ phase transition of microemulsions for parenteral injection yielding lyotropic liquid crystalline carriers of the antitumor drug bufalin. Colloids Surf B. 2019;173:217–25. https://doi.org/10.1016/j.colsurfb.2018.09.023.

Li W, Tang J, Lee D, Tice TR, Schwendeman SP, Prausnitz MR. Clinical translation of long-acting drug delivery formulations. Nat Rev Mater. 2022;7(5):406–20. https://doi.org/10.1038/s41578-021-00405-w.

Nippe S, General S. Investigation of injectable drospirenone organogels with regard to their rheology and comparison to non-stabilized oil-based drospirenone suspensions. Drug Dev Ind Pharm. 2015;41(4):681–91. https://doi.org/10.3109/03639045.2014.895375.

Nippe S, Preuße C, General S. Evaluation of the in vitro release and pharmacokinetics of parenteral injectable formulations for steroids. Eur J Pharm Biopharm. 2013;83(2):253–65. https://doi.org/10.1016/j.ejpb.2012.09.006.

Nkanga CI, Fisch A, Rad-Malekshahi M, Romic MD, Kittel B, Ullrich T, Wang J, Krause RWM, Adler S, Lammers T, Hennink WE, Ramazani F. Clinically established biodegradable long acting injectables: an industry perspective. Adv Drug Deliv Rev. 2020;167:19–46. https://doi.org/10.1016/j.addr.2020.11.008.

O'Brien MN, Jiang W, Wang Y, Loffredo DM. Challenges and opportunities in the development of complex generic long-acting injectable drug products. J Control Release. 2021;336:144–58. https://doi.org/10.1016/j.jconrel.2021.06.017.

Park EJ, Amatya S, Kim MS, Park JH, Seol E, Lee H, Shin Y-H, Na DH. Long-acting injectable formulations of antipsychotic drugs for the treatment of schizophrenia. Arch Pharm Res. 2013;36(6):651–9. https://doi.org/10.1007/s12272-013-0105-7.

Rahnfeld L, Luciani P. Injectable lipid-based depot formulations: where do we stand? Pharmaceutics. 2020;12(6) https://doi.org/10.3390/pharmaceutics12060567.

Remenar JF. Making the leap from daily oral dosing to long-acting injectables: lessons from the antipsychotics. Mol Pharm. 2014;11(6):1739–49. https://doi.org/10.1021/mp500070m.

Saklani R, Yadav PK, Nengroo MA, Gawali SL, Hassan PA, Datta D, Mishra DP, Dierking I, Chourasia MK. An injectable in situ depot-forming lipidic lyotropic liquid crystal system for localized Intratumoral drug delivery. Mol Pharm. 2022;19(3):831–42. https://doi.org/10.1021/acs.molpharmaceut.1c00782.

Schwendeman SP, Shah RB, Bailey BA, Schwendeman AS. Injectable controlled release depots for large molecules. J Control Release. 2014;190:240–53. https://doi.org/10.1016/j.jconrel.2014.05.057.

Spanarello S, La Ferla T. The pharmacokinetics of long-acting antipsychotic medications. Curr Clin Pharmacol. 2014;9(3):310–7. https://doi.org/10.2174/15748847113089990051.

Weng Larsen S, Larsen C. Critical factors influencing the in vivo performance of long-acting lipophilic solutions—impact on in vitro release method design. AAPS J. 2009;11(4):762–70. https://doi.org/10.1208/s12248-009-9153-9.

Wilkinson J, Ajulo D, Tamburrini V, Le Gall G, Kimpe K, Holm R, Belton P, Qi S. Lipid based intramuscular long-acting injectables: current state of the art. Eur J Pharm Sci. 2022;178:106253. https://doi.org/10.1016/j.ejps.2022.106253.

Yeh C-Y, Jao S-W, Chen J-S, Fan C-W, Chen H-H, Hsieh P-S, Wu C-C, Lee C-C, Kuo Y-H, Hsieh M-C, Huang W-S, Chung Y-C, Liou T-Y, Chiu H-H, Tseng W-K, Lee K-C, Wang J-Y. Sebacoyl Dinalbuphine Ester extended-release injection for long-acting analgesia: a multicenter, randomized, double-blind, and placebo-controlled study in hemorrhoidectomy patients. Clin J Pain. 2017;33(5):429–34. https://doi.org/10.1097/AJP.0000000000000417.

Zhang S, Wan Q, Xu X, Xing Y, Ding J, Yang S, Sun W, Lu M, Pan B. A novel oil-based suspension of a micro-environmental, pH-modifying solid dispersion for parenteral delivery: formulation and stability evaluation. Colloids Surf B Biointerfaces. 2019;179:382–92. https://doi.org/10.1016/j.colsurfb.2019.04.001.

Chapter 7
Clinical Applications of Long Acting Injectables: Systemic and Local Drug Delivery

Bwalya A. Witika, Madan S. Poka, Pedzisai A. Makoni, Farshad Ramazani, and Christian I. Nkanga

List of Abbreviations

ABSSTIs	Acute bacterial skin and soft tissue infections
APIs	Active pharmaceutical ingredients
ART	Antiretroviral treatment
aSAH	Aneurysmal subarachnoid haemorrhage
BLT	Bone marrow–liver–thymus
BPG	Benzathine penicillin G
BSI	Bloodstream infections
CAB	Cabotegravir
CSII	Continuous subcutaneous insulin infusion
CVDs	Cardiovascular diseases
DBV	Dalbavancin
DM	Diabetes mellitus
ER	Estrogen receptor
FDA	Food and Drug Administration
FPG	Fasting plasma glucose
GAS	Group A streptococcal

B. A. Witika (✉) · M. S. Poka · P. A. Makoni
Department of Pharmaceutical Sciences, School of Pharmacy, Sefako Makgatho Health Sciences University, Pretoria, South Africa
e-mail: Bwalya.witika@smu.ac.za

F. Ramazani
Technical Research and Development, Novartis Pharma AG, Basel, Switzerland

C. I. Nkanga (✉)
Department of Medicinal Chemistry and Pharmacognosy, Faculty of Pharmaceutical Sciences, University of Kinshasa, Kinshasa, Democratic Republic of the Congo
e-mail: christian.nkanga@unikin.ac.cd

F. Ramazani (ed.), *Biodegradable Long Acting Injectables and Implants*,
https://doi.org/10.1007/978-3-031-72302-5_7

GLP-1	Glucagon-like peptide-1
GnRH	Gonadotropin-releasing hormone
HAART	Highly active antiretroviral therapy
HIV	Human immunodeficiency virus
IE	Infective endocarditis
IM	Intramuscular
INSTI	Integrase strand transfer inhibitor
IOP	Intraocular pressure
LA-ART	Long-acting antiretroviral therapy
LAIs	Long-acting injectables
LAR	Long-acting release
LARCs	Long-acting reversible contraceptives
MIC_{50}	Minimum inhibitory concentration
MSM	Men who have sex with men
NMP	N-Methylpyrrolidone
NNRTI	Non-nucleoside reverse transcriptase inhibitor
NPH	Neutral protamine Hagedorn
NSTI	Integrase strand transfer inhibitor (I)
OA	Osteoarticular
OGTT	Oral glucose tolerance test
ORI	Oritavancin
PEP	Post-exposure prophylaxis
PG	Plasma glucose
PJIs	Prosthetic-joint infections
PrEP	Pre-exposure prophylaxis
RHD	Rheumatic heart disease
RPV	Rilpivirine
RZF	Rezafungin
SRLs	Somatostatin receptor ligands
T1DM	Type 1 diabetes mellitus
T2DM	Type 2 diabetes mellitus
TGW	Transgender women

7.1 Introduction

Long-acting injectables (LAIs) are designed to provide continuous release of active pharmaceutical ingredients (APIs) after administration. They may reduce the side effects owing to plasma fluctuation of immediate-release drugs (Lachman et al. 2003). LAIs are distinguished by their reduced frequency of administration, which improves patient compliance while reducing relapse and adverse effects. As a result of the successful commercialization of several LAI drug products, research and development of LAI drug formulations have seen a boost in recent decades (Li et al. 2022; Shi et al. 2021).

LAIs were initially used to maintain the systemic drug concentration at a therapeutic level for a longer time once injected into the body. Nowadays, in addition to systemic delivery, LAIs are also used for local drug delivery by injecting them directly into the site of action wherever possible (Ramazani et al. 2016). For example, they are injected into the knee space for the management of pain and osteoarthritis, or they are implanted at the back of the eye. In this respect, local drug delivery has opened a new avenue for the management of different chronic diseases as they ensure that the drug reaches its target tissue without the need to bypass many biological barriers (Bassand et al. 2022). This chapter provides an overview of the available marketed long-acting injectables, with particular attention to psychiatric disorders, hormonal applications, neoplastic diseases and infectious diseases. In addition, the ongoing clinical trials on LAIs in the management of different disease classes are exemplified.

7.2 Psychiatric Disorders

After the introduction of antipsychotics in the 1950s, therapeutic success was hampered by poor patient adherence to oral dosage preparations. This is mostly because some patients failed to recall instructions, even after healthcare professionals stressed the importance of these directives (Johnson 2009). Owing to this, fluphenazine enanthate, the first long-acting antipsychotic injectable, was developed in 1966. Nonetheless, this LAI antipsychotic exhibited side effects, such as tardive dyskinesia, which led to the development of fluphenazine decanoate, with reduced side effects, approximately 18 months later in 1967 (Brissos et al. 2014; Gottfries and Green 1974; Johnson 2009). A clinical trial on the use of fluphenazine decanoate, among others, in the treatment of outpatients (after their discharge from hospital) showed a reduced relapse frequency compared to the group prior to treatment onset (Gottfries and Green 1974). It was therefore evident that maintenance therapy was essential for adequate psychiatric treatment; thus, measures for improved medication adherence needed to be in place (Johnson and Freeman 1973). Consequently, in 1981 and 1986, haloperidol decanoate was introduced in Europe and the United States, respectively (Brissos et al. 2014).

Claims of better drug tolerance and fewer side effects than their predecessors accompanied the introduction of second-generation oral antipsychotics. Further, these agents were shown to have the potential to prevent or alleviate accelerated frontotemporal cortical grey matter decline, as well as impart more neuroprotection than the first-generation antipsychotics (Brissos et al. 2014; Keefe et al. 2004; Robinson et al. 2006). Owing to these advantages, a decline in the use of LAI antipsychotics occurred, with prescribers opting for second-generation oral agents. However, their clinical applications are limited by metabolic syndrome, and improved adherence has not been associated with their use (Brissos et al. 2014; Meyer and Stahl 2009). To this end, Marcus et al. reported better patient compliance from the use of LAIs than oral agents, with 51.8 and 67.7% of patients, respectively, being non-compliant (Marcus et al. 2015). Consequently, in 2003, the extended-release risperidone microspheres were the first second-generation antipsychotics to be formulated into an LAI (Greer et al. 2020). Various second-generation LAIs have been developed and approved by the Food and Drug Administration (FDA) and other regulatory global authorities for use in psychiatry (Berk et al. 2004; Greer et al. 2020; LactMed 2022a, b). Healthcare professionals have thus been able to harness the therapeutic advantages of second-generation LAIs while improving patient adherence (Johnson and Freeman 1973).

Clinical Applications of LAIs in Schizophrenia Management

Schizophrenia, a chronic condition characterized by a frequently disabling mental disorder, affects approximately 1% of the global population. It presents as signs and symptoms of unknown aetiology, with predominant psychosis, mostly as auditory hallucinations, paranoid delusions and bizarre behaviours (Insel 2010). There is no cure for schizophrenia, with clinicians generally adopting a combined approach involving pharmacologic therapy and psychological interventions to manage psychotic episodes and prevent relapse (Maroney 2020). A recent retrospective study compared the clinical outcomes in patients with schizophrenia following diverse event-driven initiation strategies of LAIs. This study found that administering an LAI before non-adherence from oral antipsychotics or occurrence of a schizophrenia-related inpatient/emergency room visit led to reduced all-cause inpatient days of inpatient stay and emergency visits (Correll et al. 2023). Further, other studies that compared the use of LAIs to oral agents for first-episode or newly diagnosed schizophrenia reported reduced relapse rates (Subotnik et al. 2015), improved adherence (Pilon et al. 2017) and better patient satisfaction (Parellada et al. 2005) to regimens that include the former. The reports provide a positive trend for the application of LAIs in the management of schizophrenia from a therapeutic (patient adherence and reduced relapse) and acceptability (improved satisfaction) perspective. The efficacy and safety, as well as patient selection and dosing considerations of LAIs, are discussed in the following sections from a real-world perspective.

Efficacy and Safety of LAIs in Schizophrenia

Although under prescribed and/or administered, LAIs are among the most effective treatments in psychiatry. Typically, they are used as adherence maintenance therapy in chronic schizophrenia; however, recent studies have shown their efficacy in early-phase or first-episode disease following the failure by most patients to attain long-lasting symptom alleviation or functional retrieval even after responding positively to initial oral antipsychotic treatment (Correll et al. 2016; Robinson et al.

1999). The safety and efficacy of LAIs have been investigated in various placebo-controlled, randomized controlled trials, with researchers reporting that they have a good therapeutic profile and reduce hospital admissions. Nonetheless, the incidence of extrapyramidal effects was found to increase as the duration of therapy increased (Rossi et al. 2012), which may be attributed to the increased occupancy of dopamine-2 receptors with time (Takada et al. 2018). Further, the efficacy and safety of LAIs compared to oral antipsychotics and among LAIs have been investigated by various authors (Correll et al. 2016). The findings of these aspects in relation to specific LAIs compared to their oral counterparts are summarized in Table 7.1.

Table 7.1 LAIs versus their corresponding oral agents

Drug agent	Study description and findings	References
Risperidone	A double-blind study of LAI and oral risperidone in patients with schizophrenia; the two treatments showed comparable efficacy; efficacy and safety of LAI risperidone were demonstrated	Chue et al. (2005)
Olanzapine	A randomized, double-blind trial of maintenance treatment in patients with schizophrenia; olanzapine LAI had comparable efficacy and safety to oral olanzapine; however, injection-related adverse events are common, including post-injection delirium/sedation syndrome, which can be managed by supportive care and treatment (Seebaluck et al. 2023)	Kane et al. (2010)
Paliperidone palmitate 1-month	Post hoc analysis on pooled data from 10 randomized controlled long-term clinical trials (≥6 months) involving patients with schizophrenia or bipolar disorder taking once-monthly paliperidone palmitate LAI and once-daily paliperidone palmitate extended-release tablets; the risk of tardive dyskinesia was similar irrespective of the route of administration	Gopal et al. (2014)
Aripiprazole	Two double-blind randomized controlled trials comparing LAI to oral aripiprazole in patients with schizophrenia; LAI aripiprazole was non-inferior to the oral agent in terms of relapse risk; both formulations had a comparable safety profile	Fleischhacker et al. (2014), Ishigooka et al. (2015)
Zuclopenthixol	Randomized clinical trial comparing oral versus LAI zuclopenthixol in patients with schizophrenia and previous violence; violent acts were less in the LAI group; irrespective of the formulation, non-adherence to therapy was the best predictor of violence	Arango et al. (2006)
Fluphenazine enanthate	Controlled comparative study of oral fluphenazine and LAI fluphenazine enanthate in patients with acute and chronic psychosis; the LAI had equal therapeutic potency to the oral agent	Van Praag et al. (1970)

Patient Selection and Dosing Considerations

Various clinicians concur that the most obvious reason for prescribing LAIs is patient non-compliance; however, initiation of these agents is not the 'silver bullet' to solving the problem. Adopting therapeutic alliance has been shown to play a vital role in the management of these patients following better insight into the disease by the patients and their caregivers, in turn, yielding better attitude to therapy, improved compliance and less relapses and hospitalizations (Heres et al. 2008; Jaeger and Rossler 2010; Kikkert et al. 2006; McEvoy et al. 1989). Furthermore, clinicians should systemically guide patients' decisions regarding treatment with LAIs in relation to various factors, such as the incidence of positive symptoms, history of relapse, poor compliance, history of non-adherence, suicide risk and previous hospitalizations. To this end, the choice of LAI to prescribe/administer, as well as the dose, must always consider clinical judgement (safety and efficacy) and the preferences of the patients and those who care for them (Rossi et al. 2012). As an example, in an expert consensus survey on identifying patients and clinical scenarios for the use of LAI antipsychotics, for a patient with schizophrenia who has been on an antipsychotic therapy for ≥2 years, regardless of adherence, LAI therapy was considered very or usually appropriate in the presence of these factors: poor understanding of illness and importance of treatment, homeless or unstable housing circumstances, history of more than two hospitalizations for psychotic relapses, history of violence, living alone, suicide attempt, substance abuse, prominent psychotic symptoms, young adult (18–25 years) and cognitive impairment (Sajatovic et al. 2018b).

LAIs in the Management of Bipolar Disorder

Characterized by recurrent mood fluctuations and instability, bipolar disorder affects approximately 1–2% of the global population (Bartoli et al. 2023; McIntyre et al. 2020, 2022). Patients with bipolar disorder have a poor quality of life, present with various addictive behaviours that co-occur as well as physical comorbidities and have a high mortality rate (Bartoli et al. 2023; Kessing et al. 2015; Schoepf and Heun 2014). Depressive and manic relapses of bipolar disorder have been reported to often be a consequence of poor treatment adherence to oral agents (Chakrabarti 2019; MacDonald et al. 2016). Owing to this, LAIs used for the management of schizophrenia have been proposed as an option for bipolar disorder therapeutics (Boyce et al. 2018; Keramatian et al. 2019) and have been shown to improve core clinical outcomes, including hospitalizations and emergency department visits, and prevent manic relapses (Bartoli et al. 2023). In clinical application, considerations for using LAIs in this patient population include poor adherence or risk thereof, multiple or infrequent but serious episodes, residual symptoms from multiple oral agent therapy or dosage form preference (injectable versus oral) (Sajatovic et al. 2018a). Consequently, LAIs are widely used for bipolar disorder off-label;

nonetheless, only risperidone microspheres and aripiprazole monohydrate are approved for maintenance therapy of bipolar disorder by the FDA (Bartoli et al. 2023; D'Agostino et al. 2022).

Effectiveness of LAIs in Bipolar Disorder Treatment

In a study that aimed to determine the efficacy of LAI risperidone in maintenance therapy of bipolar I disorder, eligible patients with current or recent manic or mixed episodes were administered either oral or LAI risperidone. Following the treatment period, those who maintained response were randomly allocated to receive placebo of LA risperidone injections at a ratio of 1:1 for up to 24 months. The authors found that the time to recurrence for any mood episode was significantly longer in the LAI compared with the placebo group. Further, the time to recurrence of elevated-mood episodes was different between the groups (Quiroz et al. 2010). Including the study, various studies (e.g., non-randomized, randomized, open-label, adequately powered randomized, double-blind trials and studies), prior to its approval for use in bipolar disorder, showed that risperidone LAI was effective in preventing relapse or hospitalization during bipolar maintenance therapy (Bobo and Shelton 2010).

In a study that aimed to evaluate the efficacy, safety and tolerability of aripiprazole LAI for bipolar I disorder, patients with the condition and presently experiencing a manic episode were stabilized sequentially using oral and LAI aripiprazole and then randomized to the LAI or placebo groups (1:1 ratio to 52 weeks of double-blind treatment). LAI aripiprazole significantly delayed the time to recurrence of any mood episode compared with the placebo group. In addition, significantly fewer patients in the LAI group experienced recurrence of any mood episode than those in the placebo group (Calabrese et al. 2017).

Future Directions and Research in LAIs for Bipolar Disorder

LAIs boast practical advantages of guaranteeing the delivery of medication at the required or prescribed dose, as well as ensuring ease in monitoring adherence to the particular injectable. However, these agents have some shortcomings in the management of patients with bipolar disorder, including the risk of extrapyramidal side effects and tardive dyskinesia, which may surpass those of patients with schizophrenia, the development of possible treatment-related exacerbation of depressive symptoms and lack of response to therapy (Bobo and Shelton 2010). In an attempt to provide adequate maintenance therapy while preventing the aforementioned limitations, psychiatrists often prescribe both oral and LAI second-generation antipsychotics on an off-label basis because of positive data in published literature (Aguglia et al. 2021; Salvi et al. 2019), which reports the effectiveness of LAIs for long-term maintenance therapy of bipolar disorder as follows: aripiprazole > olanzapine > risperidone > paliperidone (Aguglia et al. 2021). It is evident that there is a dearth of information on what influences the clinicians to prescribe a certain LAI for bipolar disorder in everyday clinical practice. Consequently, real-world data that elucidate prescriptive patterns of clinicians, safety and efficacy of the prescribed LAIs, and their approved application in bipolar disorder (beyond their use in schizophrenia), are warranted to guide future recommendations.

7.3 LAIs of Antibiotics and Antifungals

Introduced in 1950s, benzathine penicillin G (BPG) is the first-ever FDA-approved long-acting injectable antibiotic used for the treatment and prevention of group A streptococcal (GAS) infections (Wyber et al. 2013). It is the preferred secondary prophylactic agent for preventing rheumatic heart disease (RHD) associated with rheumatic fever (Markowitz 1985; Wyber et al. 2013). In addition, BPG is the first-line agent in the treatment of early syphilis, particularly in pregnant women to prevent mother-to-child transmission and people living with HIV (Clement et al. 2014; WHO 2016). Other indications of BPG include the treatment of yaws, Lyme disease and management of hyposplenism in sickle cell disease (Ali et al. 2019; Wyber et al. 2013). BPG is made up of two penicillin G molecules reacting with diphenylethylene diamine, resulting in a salt form that has an aqueous solubility of 0.02%. As a result of low solubility, when administered intramuscularly (IM), it is slowly absorbed, producing prolonged therapeutic serum concentration (Pedroso et al. 2001). BPG is typically administered as a single-dose three to four timesweekly, with adherence rates of >80% among patients with RHD, where the treatment duration lasts for 10 years or more (Mekonen et al. 2020; Musoke et al. 2013; Wyber et al. 2013). Pain and swelling at the injection site are the common adverse effects reported for the available BPG formulations. In some cases, severe allergic (3.2%) and anaphylactic (0.2%) reactions (Marantelli et al. 2019; Wyber et al. 2013) are reported. Both the lyophilized and ready-to-use formulations were reported as having challenges with administering due to needle blockages and aggregation of particles. Further, the lack of cold-chain facilities in under-resourced countries has a negative impact on the disease burden (Hand et al. 2020; Wyber et al. 2015). Hence, better formulation strategies such as an implantable device or in situ gel forming systems that can maintain the plasma concentrations over multiple months are needed for improving access to BPG and compliance.

Dalbavancin (DBV) and Oritavancin (ORI) are novel semi-synthetic, second-generation lipoglycopeptide antibiotics that have been marketed since 2014 (Durante-Mangoni et al. 2020; Rosenthal et al. 2018); they have longest half-lives of 14.4 and 16 days, respectively, than any commercially available antibiotics (Baiardi et al. 2023; Cain et al. 2022). These broad-spectrum antibiotics are primarily approved for the treatment of acute bacterial skin and soft tissue infections (ABSSTIs) caused by gram-positive bacteria, where they act by inhibiting the peptidoglycan cross-linking in the bacterial cell wall (Dimopoulou et al. 2023; Rosenthal et al. 2018). DBV and ORI are available as lyophilized powder for injection, administered as a single infusion of 1500 mg and 1200 mg respectively (Lovatti et al. 2023; Rosenthal et al. 2018). The promising clinical outcomes of DBV against multi-drug resistant infections have led to several off-label clinical applications such as osteoarticular (OA) infections, bloodstream infections (BSI), endocarditis and prosthetic-joint infections (PJIs) (Lovatti et al. 2023). A systematic review conducted by Lovatti et al. (2023) studied the efficacy of the DBV against OA infections, where the results indicated high success rate of 92.3%, when three doses of DBV are administered on a weekly basis. Similar results were reported for infective endocarditis (IE) and prosthetic-joint infections (PJIs) by several small studies conducted at single centres (Durante-Mangoni et al. 2020; Ruiz-Sancho et al. 2023).

A systematic review and meta-analysis conducted to evaluate the efficacy and safety of ORI for ABSSTIs in comparison with the other comparators proved to have similar or equivalent effects. In addition, there was a reduced risk of readmission and emergency admissions (Zhang et al. 2021). Like DBV, ORI also reported for off-label use in the treatment of BSI, IE (Texidor et al. 2024) and osteomyelitis (Scoble et al. 2020), where the drug was found to be effective against a wide range of gram-positive bacteria with less adverse events. It is important to note that a significant reduction in the hospital stays resulted in reduced treatment costs. From the perspective of anti-microbial stewardship and pharmacoeconomics, both the DBV and ORI can reduce the risk of resistance, length of hospital stay and emergency admissions and improve patient quality of life. However, there is still a need for more clinical trials to determine the optimal dosing for off-label indications, long-term safety and efficacy.

Rezafungin (RZF), available as Rezzayo™, is the only long-acting injectable antifungal drug approved in 2023 by the US FDA for the treatment of candidemia and invasive candidiasis in adults, and as prophylaxis in preventing the invasive fungal diseases in blood and marrow transplant recipients (Garcia-Effron 2020; Syed 2023). The novel once-weekly echinocandin has a terminal half-life of approximately 130 h, allowing once-a-week dose of 400 mg as intravenous infusion (Lepak et al. 2018; Syed 2023). Like other first-generation echinocandins, RZF inhibits cell wall formation by inhibiting 1,3-β-D-glucan synthesis (Garcia-Effron 2020). The pooled analysis data from multi-centre, double-blind, double-dummy, randomized phase 2 (STRIVE) and phase 3 (ReSTORE) trials indicated the potential early treatment benefit and similar safety and efficacy profiles to caspofungin for 30-day all-cause mortality (Thompson et al. 2023). RZF is a structural analogue of anidulafungin with choline aminal ether replacing the hemiaminal region at the C5 ornithine position, resulting in a significant reduction in the non-enzymatic chemical degradation (Krishnan et al. 2017; Syed 2023). Unlike several chemical instability issues reported for the first-generation echinocandins, the structural change offered increased solubility and exceptional stability in plasma, aqueous and buffered solutions, allowing more flexible dosing and storage of reconstituted preparations (Krishnan et al. 2017). In addition, RZF has shown to have better tissue penetration, similar minimum inhibitory concentration (MIC_{50}) and the same mechanisms of resistance (Garcia-Effron 2020).

7.4 Human Immunodeficiency Virus (HIV) Infection

Human immunodeficiency virus (HIV) infection has become a chronic manageable condition in recent decades (Günthard et al. 2016; Saag et al. 2018). It is treated with highly active antiretroviral therapy (HAART), which consists of three antiretroviral drugs taken daily, typically in pill form (Saag et al. 2018). The use of two antiretroviral drugs, tenofovir and emtricitabine, for pre-exposure prophylaxis in prevention among high-risk individuals has also been approved (Chou et al. 2019; Jotwani et al. 2018). HAART is effective in suppressing viral replication in more than 90% of patients, but this suppression requires daily use (Altice et al. 2019; Clay et al. 2018; Nachega et al. 2014).

Oral antiretroviral drug administration has been shown to be a burden for patients, causing distress and leading to non-adherence in a significant number (Altice et al. 2019). Treatment failure, the emergence of drug-resistant HIV strains and the advancement of the disease can all be predicted by non-adherence (Engler et al. 2019; Iacob et al. 2017; Ma et al. 2016). As a result, efforts have been made to make HIV treatment and prevention easier by using long-acting injectable antiretroviral medications that can be administered for longer durations (Augustine et al. 2018; Gao et al. 2018; Jackson et al. 2014; Margolis et al. 2017; Müller et al. 2011; Witika et al. 2021).

Development and Formulation of Long-Acting Antiretroviral Therapy

There are only two available long-acting antiretroviral therapy (LA-ART) formulations, namely cabotegravir (CAB), an HIV-1 integrase strand transfer inhibitor (INSTI), and rilpivirine (RPV), a non-nucleoside reverse transcriptase inhibitor (NNRTI).

CAB is effective when administered via quarterly intramuscular (IM) injection, according to dose selection studies in phase IIa clinical trials. Similarly, clinical findings show that RPV is effective in the treatment of HIV maintenance and is suitable for long-acting injection due to its good physicochemical properties. An intramuscular nanosuspension of RPV can last 4–8 weeks (once-monthly dose) (Andrews and Heneine 2015; Trezza et al. 2015).

According to the results from phase II prevention trials, injectable CAB was safe and well tolerated by participants. A separate phase III non-inferiority trial investigated the use of intramuscular CAB to prevent HIV infection in men who have sex with men (MSM) and transgender women (TGW). A preliminary report from this trial supports the efficacy of cabotegravir in preventing HIV infection when compared to an oral tenofovir/emtricitabine combination (Landovitz et al. 2021; Markowitz et al. 2017; National Institute of Allergy and Infectious Diseases 2016).

Clinical trials for injectable cabotegravir for pre-exposure prophylaxis (PrEP) are still ongoing (Landovitz et al. 2023; Markowitz et al. 2017; World Health Organization 2020). A study of over 3000 women in seven African countries concluded unequivocally that long-acting injectable CAB effectively protected women from HIV acquisition when used as PrEP (National Institute of Allergy and Infectious Diseases 2016; World Health Organization 2020).

RPV is a second-generation NNRTI with an oral formulation. For treatment-naive HIV-1-infected individuals with a viral load (HIV-1 RNA) of 100,000 copies/mL or less, this form is combined with other antiretroviral drugs (Cohen et al. 2011; Molina et al. 2011). It has been developed as a long-acting injectable form of RPV with a half-life of 30–90 days. This form allows for intramuscular administration every 4–8 weeks (Jackson et al. 2014; McGowan et al. 2016; Verloes et al. 2015; Williams et al. 2015). In contrast, the oral form has a half-life of 24–45 h and must thus be taken daily to be effective (Aouri et al. 2017). In early clinical studies (Jackson et al. 2014; Verloes et al. 2015), a comparable drug concentration to that of the oral form was detected in the plasma up to 84 days after administration of the long-acting injectable.

Another INSTI being developed as a long-acting injectable is raltegravir (Kovarova et al. 2016). It is approved for HIV treatment and prevention (post-exposure prophylaxis (PEP)) in its oral form (Günthard et al. 2016). The development of a long-acting agent is still in the early stages. Humanized bone marrow–liver–thymus (BLT) mice were given 7.5 mg of subcutaneous long-acting raltegravir. The pharmacokinetic properties of 400 mg oral raltegravir were similar in this study, and the treatment was effective in protecting BLT mice from HIV vaginal challenge for up to 4 weeks after a single dose, whereas control mice were not protected from HIV infection. The drug's plasma level was also found to be comparable to that of the oral form (Kovarova et al. 2016). This study also revealed the presence of the active drug in the female genital tracts of mice, a finding that, if replicated in humans, could indicate usefulness as a preventive intervention, as sexual intercourse is one of the most common routes of HIV transmission.

7.5 Diabetes Mellitus

Diabetes mellitus (DM) is a complex chronic illness characterized by hyperglycaemia or high blood glucose levels caused by deficiencies in insulin secretion, action or both. The chronic metabolic imbalance associated with this disease puts the patients at high risk for long-term macro- and microvascular complications, which can lead to frequent hospitalization and complications, including an increased risk for cardiovascular diseases (CVDs) (1).

Diabetes is clinically diagnosed using one of four plasma glucose (PG) criteria viz. elevated fasting plasma glucose (FPG) (>126 mg/dL), 2 h PG during a 75 g oral glucose tolerance test (OGTT) (>200 mg/dL), random PG (>200 mg/dL) with classic signs and symptoms of hyperglycaemia or haemoglobin A1C level > 6.5% (Chaudhury et al. 2017).

Diabetes is generally classified as type 1 (T1DM) or type 2 (T2DM) based on the aetiology of disease. Type 1 diabetes is caused primarily by autoimmune-mediated destruction of pancreatic beta cells, which results in insulin deficiency (Mishra and Ndisang 2014; Ndisang and Jadhav 2014; Petersen et al. 2017). Changes in lipid metabolism, increased hyperglycaemia-mediated oxidative stress, endothelial cell dysfunction and apoptosis are usually associated with this (Mishra and Ndisang 2014; Ndisang and Jadhav 2014; Petersen et al. 2017). Similarly, in type 2 diabetes, increased glucotoxicity, lipotoxicity, endoplasmic reticulum-induced stress and apoptosis cause beta cell loss (Hossain et al. 2007; Mishra and Ndisang 2014; Ndisang et al. 2014; Ndisang and Jadhav 2014; Petersen et al. 2017). Whereas type 1 diabetes is characterized by the presence of beta cell autoantibodies, type 2 diabetes is thought to be caused by a combination of peripheral insulin resistance and dysfunctional insulin secretion by pancreatic beta cells (Mishra and Ndisang 2014; Ndisang and Jadhav 2014; Petersen et al. 2017). Both types of diabetes, however, are associated with a wide range of complications, including cardiomyopathy, nephropathy and neuropathy.

One of the mainstays of the management of type 1 and 2 diabetes is the use of insulin injectables. Management of diabetes and indeed the outcomes of treatment have greatly been improved using LAI insulin.

Glargine

Insulin glargine (glargine) was the first long-acting insulin analogue to enter clinical use in 2000. Glargine is a human insulin analogue created using recombinant DNA technology. The insulin molecule has been modified in two ways: the C-terminus of the B chain has been elongated with the addition of two arginine residues inserted at position B30, and asparagine has been replaced with glycine at position (Rosskamp and Park 1999).

The main goal for the development of glargine was to circumvent the strong peak action profile and poor reproducibility of neutral protamine Hagedorn (NPH). It also offered a longer duration of action, which better met basal insulin needs throughout the day (Becker et al. 2005; Lepore et al. 2000; Porcellati et al. 2007a). With an end-of-action time of up to 24 h, glargine showed a nearly flat time-action profile and reduced variability when compared to NPH (Heinemann et al. 2000; Porcellati et al. 2007a). It is also longer acting than NPH insulin in people with type 1 diabetes and who are dose-dependent (Heise et al. 2004; Klein et al. 2007; Lepore et al. 2000; Porcellati et al. 2007a, b).

The strongest evidence supporting the suitability of glargine as basal insulin came from studies that contrasted continuous subcutaneous insulin infusion (CSII) with once-daily glargine in a multiple daily injection regimen (Bruttomesso et al.

2008; Garg et al. 2004; Lepore et al. 2004). There were no differences observed in the frequency of hypoglycaemia or the level of glycaemic control (A1C) attained between the two regimens. In addition, there was no difference in the plasma glucose variability between the two regimes in the Bolli et al. study (Bolli et al. 2009).

Overall, glargine provides improved convenience with a once-daily dose, flexibility in the timing of injections (morning, pre-dinner or pre-bedtime) and lifestyle benefits for people with type 1 diabetes (Ashwell et al. 2006). Eating at set times is not necessary, and fasting for up to 18 h is acceptable (Mucha et al. 2004). Moreover, it has been shown that exercise has no effect on glargine's subcutaneous absorption (Peter et al. 2005).

When attempting to achieve a target A1C level of 7% (achieved in more than 50% of subjects), glargine also offers improved safety (reduced risk of nocturnal hypoglycaemia) and convenience (once-daily administration) (Riddle et al. 2003). Although the level of glycaemic control achieved is the same as with NPH (Riddle et al. 2003), the lower risk of hypoglycaemia allows for more aggressive dose titration and, as a result, a greater likelihood of reaching and maintaining a lower glycaemic target value over time.

Insulin Detemir

Insulin detemir (detemir) was the second clinically available basal insulin analogue. Detemir is also made using recombinant DNA technology, with the amino acid threonine at position B30 of the human insulin molecule deleted and a myristic fatty acid residue added to the e-amino group of the lysine residue at position B29 (Barlocco 2003). The acylation of the insulin molecule with a fatty acid residue increases its binding affinity to albumin in the subcutaneous, intravascular and extracellular compartments, which delays absorption and reduces the variability of its biological activity (Havelund et al. 2004).

Subsequent research performed by Kurtzhals (2004) suggested that albumin binding enhances the reproducibility of the time-action profile of acylated insulins while also delaying absorption. Moreover, it has been suggested that passive diffusion through a paracellular or transcellular pathway, which delays the time it takes for the insulin to reach its receptor, is the main non-saturable process responsible for the delayed transcapillary transport of insulin detemir in skeletal muscle (Hamilton-Wessler et al. 2002). Since acylated insulin analogues have a much lower affinity for the insulin receptor than human insulin preparations, higher concentrations of active principle are required to achieve equivalent biopotency (Kurtzhals et al. 2000). For equivalent metabolic effects, temir must therefore be administered subcutaneously at five times the concentration of other basal insulins (Hordern et al. 2005) and intravenously at roughly eight times the concentration of human recombinant insulin (Rossetti et al. 2008).

Comparative Outcomes Between Glargine and Detemir

The efficacy and safety of glargine and detemir as an add-on therapy to oral diabetes medications have been compared in people with type 2 diabetes (Rosenstock et al. 2008). In a 52-week, open-label, parallel, multinational trial of 582 insulin-naive individuals, glargine and detemir were compared. Basal insulin was started once daily in the evening at a dose of 12 units, and detemir was added in the morning if predinner plasma glucose exceeded 7.0 mmol/L (126 mg/dL) (Rosenstock et al. 2008).

Detemir and glargine both improved A1C (from 8.6% at baseline to 7.2% and 7.1%, respectively; difference, 0.05% [95% CI, −0.11 to 0.21%]), with similar FPG in both groups. Furthermore, the proportion of subjects who achieved A1C 7.0% without experiencing hypoglycaemia was comparable (33% vs. 35%, respectively). In completers, however, weight gain was less with detemir than with glargine (3.0 vs. 3.9 kg; difference, −0.9 kg [95% confidence interval, −1.6 to −0.2 kg]; $P = 0.01$). In the detemir-treated group, 55% required a twice-daily dose of 1.0 U/kg, while the remaining 45% only required a once-daily dose of 0.52 U/kg (the overall mean daily dose of detemir was 0.78 U/kg). In comparison, the mean daily dose of glargine (once daily in all subjects) was 0.44 U/kg. This study found that glargine and detemir may be equivalent as basal insulins in type 2 diabetes in terms of glycaemic goals (lowering A1C and risk of hypoglycaemia), but that detemir required a higher dose with a twice-daily regimen in more than 50% of cases with slightly higher risk of hypoglycaemia.

These preliminary findings were confirmed in another 24-week, randomized, 'treat-to-target' trial comparing glargine (once daily) with detemir (twice daily) in 973 insulin-naive individuals with type 2 diabetes who were not adequately controlled on oral diabetes medications. Glargine achieved comparable glycaemic control to detemir at much lower doses, but with more weight gain (Swinnen et al. 2010). Further research has shown that detemir (once or twice daily) and glargine (once daily) have similar glycaemic efficacy and safety when used in basal-bolus regimens with short-acting insulin analogues (Hollander et al. 2008; Raskin et al. 2009).

Bydureon®

Bydureon® is a formulation of exenatide, a glucagon-like peptide-1 (GLP-1) agonist, which is used for the treatment of type II diabetes. It is composed of Poly(lactic-co-glycolic acid) (PLGA)-based microspheres containing 5% (w/w) exenatide. Multiple studies have shown that Bydureon® exhibits superior hypoglycaemic efficacy and a reduced incidence of adverse effects when compared to the free drug, administered in solution through subcutaneous injection twice daily at a dosage of 5–10 μg (Li et al. 2013; Painter et al. 2013). Bydureon® can be utilized as adjunctive therapy in combination with established oral anti-diabetic medications, such as metformin and/or sulfonylurea (Al-Mrayat et al. 2012; Painter et al. 2013).

On administration of Bydureon®, exenatide is released gradually over a period of approximately 10 weeks. The release of exenatide initially occurs due to the presence of exenatide on the surface of the microparticles. This is followed by a gradual release of exenatide from the MP matrix because of the hydration and erosion of PLGA. The gradual release phase exhibits two consecutive peaks of exenatide in the bloodstream, occurring around week 2 and week 6 to 7, respectively. The plasma concentration of exenatide gradually increases and remains stable after 6 to 7 weeks at approximately 300 pg/mL when administered on a weekly basis (Cai et al. 2013).

Symlin®

Symlin is a subcutaneously administered formulation of pramlintide acetate. Pramlintide is a synthetic version of the naturally occurring pancreatic peptide amylin. Amylin and pramlintide have similar effects on lowering postprandial glucose and glucagon and slowing gastric emptying. Pramlintide use in type 1 and insulin-dependent T2DM is associated with modest reductions in HbA1c, often accompanied by weight loss (Hoogwerf et al. 2008).

7.6 Long-Acting Injectables (LAI) for Hormone Disorders

LAI Contraceptives for Birth Control

LAIs have a long history of use as contraceptives. Since 1960, long-acting reversible contraceptives (LARCs) like medroxyprogesterone acetate (three monthly injections) have been available. Long-acting hormonal contraceptives benefit women who want to plan their families or space their births, particularly in developing countries. Though LAI contraceptives have side effects such as irregular menstrual cycles, they are extremely effective and simple to use (Benagiano et al. 2015). Microcrystal suspension, esterification, drug-loaded microsphere suspension, implants and in situ forming depots are all injectable hormonal contraceptive formulations. These formulations result in a single injection that, in some cases, lasts for over 6 months (Halpern et al. 2015; Wu et al. 2015).

LARC methods include hormonal and non-hormonal contraception; hormonal contraception includes injectables and implants, which are very effective, safe, reversible and cost-effective. Subcutaneous implants are either rods or capsules that provide contraception for up to 5 years. Steroids are incorporated within the polymeric matrix (dimethylpolysiloxane; DPS) of contraceptive rods. Steroid crystals are encapsulated in polymeric shells (hollow polymer tubes) in contraceptive capsules (Alkema et al. 2013; Benagiano et al. 2015; Joshi et al. 2015; Winner et al. 2012).

Male hormonal contraceptives based on synthetic androgen or a combination of androgen and progesterone are being researched. There are already testosterone hormonal depots on the market, but they are mostly used for hormonal replacement therapy (Chao and Page 2016; Wang and Swerdloff 2010). Norethisterone enantate, etonogestrel, levonorgestrel, medroxyprogesterone acetate (DMPA) and hydroxy progesterone caproate are currently available as LAIs for contraception.

LAIs for Acromegaly

Acromegaly primarily results from excessive secretion of growth hormones. In general, growth hormone antagonists are employed to regulate the secretion of growth hormone (Davies et al. 1998; McKeage et al. 2003). A sustained-release formulation of a somatostatin analogue is becoming increasingly popular for the management of acromegaly. There are several long-acting injectable (LAI) medications available. These include octreotide long-acting release (LAR), which is given into the muscle every 4 weeks; lanreotide LAR, which is injected under the skin every 4–6 weeks; and pasireotide LAR, which is administered into the muscle every 4 weeks. Octreotide and lanreotide are classified as first-generation somatostatin receptor ligands (SRLs). These ligands have a success rate of approximately 70% in patients (Chanson et al. 2000). In 2014, the FDA granted approval to pasireotide LAR (Signifor® LAR), primarily due to its utilization of encapsulation technology using microspheres. Recent advancements in long-acting medications include octreotide implants and CAM2029, which is a long-acting formulation of octreotide. These medications use a fluid crystal delivery system, where the drug is held within a liquid crystal matrix. Somatoprim, also known as DG3173, is a highly specific ligand for somatostatin receptors that is currently being evaluated in clinical trials, specifically in phase II or phase III (Chanson et al. 2000; Cuevas-Ramos and Fleseriu 2016).

7.7 LAIs for Cancers

Systemic chemotherapy holds a crucial role in cancer treatment, but its use often leads to significant toxicities due to the high doses required and the exposure of healthy tissues to anticancer drugs (Palumbo et al. 2013). This approach can also inadvertently foster the development of drug-resistant cancer cells, as tumours may be exposed to suboptimal drug concentrations, complicating effective treatment (Zhang et al. 2023). Surgical removal is generally the preferred method for treating solid tumours. However, the challenge arises when complete removal is not achievable, often due to undetected tumour nodules, leading to the risk of the tumour re-emerging (Lv et al. 2023). This issue of local or regional tumour recurrence post-surgery is a common concern across various cancer types (Hu et al. 2017). In response to these challenges, the adoption of LAIs and implants is gaining attention. They offer a more targeted approach than traditional chemotherapy and help in reducing the risk of tumour recurrence after surgical procedures.

LAIs offer significant advantages over conventional cancer therapies, including maintaining consistent drug levels in the body, improving patient compliance and reducing the frequency of hospital visits (Saltzstein et al. 2018). This form of treatment enhances patients' quality of life by minimizing the disruption to their daily routines and reducing exposure to healthcare-associated infections. In essence, LAIs represent a significant advancement in oncology, offering a more patient-friendly, targeted and effective approach to cancer treatment.

In the subsequent sections, we present a few examples that illustrate the clinical application of LAIs in the treatment of various types of cancer, such as prostate, brain, liver and breast cancers. For those seeking in-depth information, there are comprehensive reviews available that extensively discuss the clinical applications of LAIs in cancer therapy (Pacheco et al. 2023; Padmakumar and Amiji 2023; Ramazani et al. 2016).

Prostate Cancer

Long-acting injectables (LAIs) have shown considerable promise in prostate cancer treatment, notably for their capacity to sustain consistent drug levels over extended periods. A notable instance is the use of an oil-based intramuscular (IM) formulation of 300 mg cyproterone acetate (Androcur Depot®). This formulation demonstrated equivalent efficacy to a daily oral regimen of 50 mg cyproterone acetate tablets, taken twice, in alleviating hot flush symptoms, a common adverse effect associated with antiandrogen therapy in prostate cancer patients (Chovan et al. 2000). Further research indicated that weekly IM injections of 300 mg Androcur Depot® achieved comparable blood concentrations of cyproterone acetate to an oral dose of 50 mg tablets taken 4 to 8 times a day (totalling 200–400 mg/day) (Rost et al. 1981). This suggests that a single IM injection of Androcur Depot® (300 mg) is just about 21% of the minimum oral dose required for effective weekly treatment of prostate cancer.

LAI formulations have shown comparable effectiveness to both oral and intravenous approaches. In a clinical study, it was shown that monthly subcutaneous insertions of a PLGA-based implant with goserelin acetate (Zoladex® 3.6 mg), an analogue of luteinizing hormone-releasing hormone, successfully managed testosterone and luteinizing hormone levels (Ahmed 1985). These results were comparable to those obtained from a continuous 60-day infusion of goserelin acetate. In addition, in a randomized, multi-centre trial focusing on patients with advanced prostate cancer, Zoladex® 3.6 mg was found to be more tolerable than the daily 3 mg injections of diethylstilbestrol, another analogue of luteinizing hormone-releasing hormone (Citrin et al. 1991).

Several other studies further highlight the potential of LAIs (Suprefact® Depot; Firmagon®; Eligard®) in providing sustained and effective prostate cancer treatment. For instance, Pettersson et al. (2006) conducted a study where they observed that a single subcutaneous injection of a PLGA implant containing the luteinizing hormone-releasing hormone analogue buserelin (Suprefact® Depot 9.45 mg) was effective in reducing serum testosterone levels to below the lower castration limit for approximately 6 months. This finding is particularly promising for the neoadjuvant therapy or long-term management of androgen deprivation. In a similar vein, clinical studies involving degarelix acetate (Firmagon®) in prostate cancer patients have demonstrated remarkable efficacy. These studies showed that degarelix acetate (a gonadotropin-releasing hormone (GnRH) receptor antagonist) effectively suppresses testosterone secretion and maintains this suppression at castration levels (testosterone ≤50 ng/dL) for up to 12 months of treatment (Firmagon 2020). The controlled-release mechanism of these formulations plays a vital role, notably enhancing patient quality of life. This is achieved by reducing the frequency of administration and minimizing side effects, thereby improving adherence to the treatment regimen.

Brain Cancer

To effectively combat blood-borne cancers, therapeutic agents need to penetrate the blood vessel walls in sufficient amounts to reach the cancer cells and counteract the resistance posed by the local microenvironment around these cells. The efficacy of drugs against primary brain tumours and brain metastases is often hindered by the brain's unique microenvironment, particularly the blood-brain barrier (Arvanitis et al. 2020). Current treatments for malignant gliomas include surgical resection, radiotherapy and chemotherapy (Ohka et al. 2012). However, resection often proves inefficient for long-term tumour control, as recurrences are common near the original resection site, indicating that the tumour cells have deeply infiltrated the surrounding brain tissue (Franzini et al. 2019).

LAIs have been developed for targeted delivery of anticancer drugs to the brain, featuring customized release profiles to enhance treatment effectiveness and minimize side effects. For instance, Gliadel® Wafer, a polymeric implant formulation of carmustine, was evaluated in a phase III randomized, placebo-controlled, double-blind trial involving 240 patients with newly diagnosed malignant glioma (Westphal et al. 2003). The study found that implanting Gliadel® Wafers during primary surgical resection was well tolerated, caused no systemic toxicities or adverse effects and eliminated the need for additional surgery. This localized chemotherapy approach reduced the death risk by 29% in the treated group. Another phase III clinical trial demonstrated that Gliadel® Wafers implanted in the resection cavity delivered carmustine concentrations 100 times higher locally than systemic administration, offering superior site-specific efficacy and reduced systemic toxicity, as evidenced by undetectable drug levels in plasma (Giese et al. 2004).

Benny et al. provide another example with their study on intracranially administered imatinib-loaded PLGA microspheres in mice implanted with human U87-MG cells (Benny et al. 2009). The results showed a 79% reduction in tumour size 14 days post-injection, with no toxicity observed in major organs (liver, kidney, brain, heart, muscle, spleen). Compared to Kilic et al. (2000), who found that 30 repeated doses of 1.5 mg systemic imatinib were needed for similar effects in glioma models, Benny et al.'s study suggests that a single dose of 1.25 mg imatinib in LAI microspheres could achieve comparable antitumoural effects with a significantly lower total dose. These studies collectively indicate that LAIs drug delivery to brain can notably reduce potential systemic toxicity while providing similar therapeutic outcome.

Breast Cancer
Around 80% of breast cancers are characterized by the expression of the estrogen receptor (ER), which is essential for their growth and survival, facilitated by estrogen binding to the ER (Nathan and Schmid 2017). This binding leads to the movement of the ER into the cell nucleus, initiating a series of signals that encourage the growth of breast cancer cells. The ER, as a nuclear receptor, is key in this progression. Fulvestrant, a drug targeting this process, acts as a selective estrogen receptor degrader. It binds to the ER, obstructs its function and breaks it down, thereby completely blocking the estrogen-signalling pathway through the ER. In an open, parallel-group, multi-centre trial involving 451 postmenopausal women with advanced breast cancer, the efficacy and tolerability of fulvestrant were assessed (Howell et al. 2002). The study was conducted in comparison with anastrozole, another endocrine treatment. Fulvestrant was tested as a long-acting oil-based injectable (Faslodex®), formulated with 96% ethanol, benzyl alcohol, benzyl benzoate and castor oil. Patients were randomized to receive either a once-monthly intramuscular injection of 250 mg fulvestrant (one × 5 mL) or a daily oral dose of 1 mg anastrozole. The results indicated that monthly intramuscular administrations of 250 mg/mL Faslodex® were as effective as the daily oral administrations of 1 mg anastrozole. Both treatments yielded comparable clinical benefits; the median times to progression were 5.5 months for fulvestrant and 5.1 months for anastrozole, with objective response rates of 20.7% and 15.7%, respectively. Faslodex® demonstrated its ability to consistently deliver fulvestrant directly into the systemic circulation. This trial serves as further evidence of the potential of LAIs to effectively substitute oral medications, which often require frequent dosing and are hindered by variable absorption and patient compliance issues.

7.8 LAIs for Allergic Inflammation

Allergic inflammation occurs from an intricate interconnectedness between various inflammatory cells, including dendritic cells, mast cells, lymphocytes, basophils, eosinophils and, in some cases, neutrophils, leading to the production of a myriad of inflammatory mediators. During allergic inflammation, sensory nerves are sensitized and activated, leading to symptoms. The exacerbation of symptoms may be a consequence of defective endogenous anti-inflammatory mechanisms. Corticosteroids are highly effective anti-inflammatory agents with broad-spectrum activity, including epigenetic modulation of the inflammatory response and suppression of a transcription factor (GATA3) responsible for allergic inflammatory responses (Barnes and Barnes 2011).

Ozurdex®, a sterile sustained-release rod-shaped ocular implant of dexamethasone, is indicated for the treatment of diabetic macular oedema, macular oedema following branch retinal vein occlusion or central retinal vein occlusion and ocular posterior segment inflammation caused by non-infectious uveitis (Clemente-Tomás et al. 2019). In a retrospective study of patients (20 eyes of 15 patients; mean age = 39.8 years) with non-infectious intermediate uveitis (with either cystoid macular edema or vitritis and non-responsive to standard therapy) who received Ozurdex®, ocular inflammation was found to be controlled in the majority of the patients. However, cataracts and raised intraocular pressure (IOP) were noted as adverse effects. Interestingly, Ozurdex® use during anterior segment surgery in patients with chronic, recurrent uveitis effectively controlled inflammation and IOP up to 3 months post-surgery (Ragam et al. 2014). Further, Ozurdex® administration was found to provide anatomic and functional improvements without serious ocular and systemic complications in patients with macular edema secondary to retinitis pigmentosa (Srour et al. 2013).

The PROPEL® steroid-releasing implant, a mometasone furoate controlled-release implant, prevents obstruction of the ethmoid sinus following chronic rhinosinusitis surgery. Different PROPEL® implants are available for specific sinus conditions (PROPEL®, ethmoid sinus; PROPEL Mini®, ethmoid sinus and frontal sinus ostia; and PROPEL Contour®, frontal sinus ostia and maxillary sinus ostia) (Kennedy 2012; Nkanga et al. 2020; Smith et al. 2020). Improved postoperative healing was reported with use of the implants, in turn minimizing additional surgical procedures, including adhesion lysis and the administration of systemic steroids (Kennedy 2012). An evidence-based review on drug-eluting implants in chronic rhinosinusitis recommended the use of SINUVA®, a mometasone furoate–releasing implant indicated for therapy in nasal polyps, for recurrent nasal polyps following endoscopic sinus surgery after positive clinical findings, including improved symptom scores, reduced polyp size and reduction in the need for revision surgery (Smith et al. 2020). Triamcinolone acetonide injectable suspension, marketed as Trivaris® or Triesence®, is indicated for various ocular inflammatory diseases such as uveitis, temporal arteritis, sympathetic ophthalmia and ocular inflammatory diseases refractory to topical corticosteroids. It is simply administered via needle injection (every 3–4 weeks) without requiring surgical placement (Gaballa et al. 2020). Betason L.A®, a long-acting injectable suspension of betamethasone, is indicated for inflammatory and allergic states among other disorders, and is administered via the intramuscular, intra-articular, intrabursal or intradermal routes (Nkanga et al. 2020). Kenalog® (triamcinolone acetonide injectable suspension) is used off-label for the treatment of refractory uveitis. However, when compared to Trivaris®, its intravitreal administration has been associated with adverse effects, including endophthalmitis, eye inflammation, elevated IOP and visual disturbances/loss (Nkanga et al. 2020; Parrish and Sternberg 2007), possibly due to the presence of benzyl alcohol (Lang et al. 2007; Nkanga et al. 2020).

7.9 Recent LAI Products in Clinical Trials

Several LAIs are undergoing clinical trials for the management of many chronic diseases. In the following sections, few examples are provided to show the importance of LAI as a patient-centric drug-delivery system for either local or systemic treatment of diseases.

Osteoarthritis and Articular Cartilage Defects

The management of articular cartilage defects is one of the most challenging clinical problems due to the limited ability of articular cartilage to regenerate. About 36% of athletes and 12% of the general population have cartilage injuries. At present, surgery (microfracture) is the only approved treatment for cartilage injuries. Therefore, developing new drug molecules or drug formulations which can promote local stem cells to regenerate the defect site without surgery would address the unmet medical need. After intra-articular injection, both large and small molecules rapidly exit the joint, via the lymphatic system and small blood vessels, respectively, resulting in low bioavailability of the drug in its site of action. To circumvent this issue, LAI formulations can be injected directly into the joint space and release the drug for an extended period. The long-acting formulation developed by Novartis for intra-articular injection to treat cartilage injuries in articulating joints is in phase 2 clinical trials (Oo and Hunter 2022). It aims to stimulate stem cells to repair acute/chronic joint injuries. Moreover, in a phase 3 study, Lorecivivint, a Wnt pathway inhibitor, microcrystal suspension developed by Samumed, significantly improved discomfort, function and joint space width in osteoarthritis patients. It is projected to be an effective treatment option for osteoarthritis in the near future (Kou et al. 2023).

Tendon Injuries

Injuries to tendons are widespread throughout the human body, often accompanied by tissue inflammation and degeneration. Despite numerous proposed surgical, rehabilitation, graft and drug therapies, tendon healing and persistent pain remain significant unmet medical needs. To address these challenges, we at Novartis suggested using local delivery of a long-acting injectable formulation based on PLGA microparticles to offer sustained delivery of drug therapeutics to tendon tissue. The available medication is now undergoing clinical trial phase 2 (Badry Bursulaya et al. 2018).

Schizophrenia

Teva Pharmaceuticals announced initiation of a phase 3 clinical trial on BEPO technology, which is an in situ forming gel based on amphiphilic co-polymer of lactic acid and polyethylene glycol dissolved in N-methylpyrrolidone (NMP). In this study, they are evaluating different doses of a monthly depot formulation of olanzapine, an antipsychotic drug (Roberge et al. 2020). The active ingredient of olanzapine was brought to market by Eli Lilly in the 1990s as tablet. It is an antagonist of serotonin–dopamine receptor that is used for management of schizophrenia and bipolar disorder. This is a good example for developing generic drugs that are capable of long-acting formulation, e.g. high potency, good stability and solubility in relevant solvent such as NMP.

Prostate Cancer

Camurus is developing a long-acting formulation of leuprorelin for prostate cancer treatment which is in phase 2 clinical trial. FluidCrystal® (or liquid crystal) consists of a non-aqueous liquid that transforms into a liquid crystal matrix at the administration site (Rahnfeld and Luciani 2020). The depot can be customized to release the drug gradually over a period ranging from days to months. The product is delivered through a convenient ready-to-use prefilled syringe with a 27-gauge needle and is designed as a low-volume injection.

Aneurysmal Subarachnoid Haemorrhage

A subarachnoid haemorrhage denotes bleeding in the space surrounding the brain. Typically, it occurs when a vulnerable region in a blood vessel (aneurysm) on the brain's surface ruptures, leading to leakage. Subsequently, the blood accumulates around the brain and within the skull, intensifying pressure on the brain. Nimodipine is an approved medication for treating aneurysmal subarachnoid haemorrhage (aSAH). An extended-release PLGA microparticle formulation of nimodipine has been designed for intraventricular or intracisternal administration. This investigation aimed to examine the hypothesis that delivering elevated sustained concentrations of nimodipine directly to the cerebrospinal fluid would yield superior efficacy in comparison to systemic administration (Hänggi et al. 2019). However, Edge Therapeutics, the inventor company, mentioned that they discontinued the clinical trial as the medication was unlikely to meet the primary endpoint.

Human immunodeficiency virus (HIV)

ViiV Healthcare, a global HIV specialist company, has announced positive results from the interim analysis of the LATITUDE phase III trial. The trial evaluated Cabenuva, a long-acting injectable antiretroviral treatment (ART) based on nanocrystal suspension for HIV (cabotegravir + rilpivirine). The study showed superior efficacy in maintaining viral load suppression compared to daily oral therapy for individuals facing challenges with adherence to daily medication. The LATITUDE study is ongoing across 31 sites in the United States, and participants were provided comprehensive adherence support while initially taking oral ART. The interim analysis revealed the effectiveness of long-acting injectable ART over daily oral standard care, leading to a recommendation for eligible participants to be offered the injectable treatment. Cabenuva combines cabotegravir and rilpivirine, offering a complete regimen for the treatment of HIV-1 infection in adults who are virologically suppressed (Bassand et al. 2022). The treatment aims to address challenges such as daily adherence barriers in people living with HIV.

Lumbosacral Radiculopathy or Sciatica

Lumbosacral radiculopathy also known as sciatica refers to a condition characterized by pain resulting from the compression or irritation of nerve roots located in the lower back. This syndrome may arise from various factors such as lumbar disc herniation, degenerative changes in the spinal vertebrae and constriction of the openings through which nerves exit the spinal canal. Clonidine exhibits properties that combat neuropathic pain and inflammation. The utilization of a new slow release–based on biodegradable polymer, PLGA–clonidine micropellets, ensures the continuous delivery of clonidine. The micropellets are administered via an applicator, conducted by fluoroscopy for accurate placement, close to the impacted nerve. In phase 2 clinical study, the peak plasma was noted 24 h post-injection, with an average elimination half-life spanning from 13 to 20 days. Throughout the study, for 1 year, there was a consistent and statistically significant decrease in both leg and back pain compared to baseline across all participant groups (Mekhail 2020).

A phase 3 clinical trial of the formulation has been planned to study the efficacy and safety in adults with lumbosacral radiculopathy. This method holds promise for extended pain alleviation while minimizing systemic side effects.

7.10 Conclusion

Several products have already been marketed for many years to manage chronic diseases, such as psychiatric disorders, diabetes, hormone therapy, different types of cancer and infectious diseases. In addition, the recent surge in clinical trials for long-acting injectable (LAI) products demonstrates their versatility and the potential to address unmet medical needs across diverse fields. From treating osteoarthritis without surgery to managing mental health conditions and addressing challenges in HIV therapy, LAI formulations demonstrate promise in providing sustained and effective treatments. While some trials face challenges, overall, these developments highlight the growing importance of LAI technology in reshaping the landscape of chronic disease management.

Acknowledgements This document has been produced with the financial assistance from the European Union (Grant no. DCI-PANAF/2020/420-028), through the African Research Initiative for Scientific Excellence (ARISE), pilot programme. ARISE is implemented by the African Academy of Sciences with support from the European Commission and the African Union Commission. The contents of this document are the sole responsibility of the author(s) and can under no circumstances be regarded as reflecting the position of the European Union, the African Academy of Sciences and the African Union Commission.

References

Aguglia A, Serafini G, Nebbia J, Salvi V, Martinotti G, Corbo M, Signorelli MS, Mineo L, Mencacci C, Di Sciascio G, Biggio G, Aguglia E, Amore M. Off-label use of second-generation antipsychotics in borderline personality disorder: a survey of Italian psychiatrists. J Pers Disord. 2021;35(3):321–35. https://doi.org/10.1521/PEDI_2019_33_445.

Ahmed. Preliminary report on use of depot formulation of LHRH analogue ICI 118630 (Zoladex) in patients with prostatic cancer. Br Med J (Clin Res Ed). 1985;290(6463):185–7.

Ali S, Long A, Nikiema JB, Madeira G, Wyber R. Availability and administration of benzathine penicillin G for the prevention of rheumatic fever in Africa: report of the working group on Penicillin, Pan-African Society of Cardiology Task Force on Rheumatic Heart Disease. Cardiovasc J Afr. 2019;30(6):369–72. https://doi.org/10.5830/cvja-2019-042.

Alkema L, Kantorova V, Menozzi C, Biddlecom A. National, regional, and global rates and trends in contraceptive prevalence and unmet need for family planning between 1990 and 2015: a systematic and comprehensive analysis. Lancet. 2013;381(9878):1642–52. https://doi.org/10.1016/S0140-6736(12)62204-1.

Al-Mrayat M, Lawrence V, Vloemans M. Prolonged release exenatide gets the go-ahead from NICE. Pract Diab. 2012;29(3):86–8. https://doi.org/10.1002/pdi.1664.

Altice F, Evuarherhe O, Shina S, Carter G, Beaubrun AC. Adherence to HIV treatment regimens: systematic literature review and meta-analysis. Patient Prefer Adherence. 2019;13:475–90. https://doi.org/10.2147/PPA.S192735.

Andrews CD, Heneine W. Cabotegravir long-acting for HIV-1 prevention. Curr Opin HIV AIDS. 2015;10(4):258–63. https://doi.org/10.1097/COH.0000000000000161.

Aouri M, Barcelo C, Guidi M, Rotger M, Cavassini M, Hizrel C, Buclin T, Decosterd LA, Csajk C. Population pharmacokinetics and pharmacogenetics analysis of rilpivirine in HIV-1-infected individuals. Antimicrob Agents Chemother. 2017;61(1) https://doi.org/10.1128/AAC.00899-16.

Arango C, Bombín I, González-Salvador T, García-Cabeza I, Bobes J. Randomised clinical trial comparing oral versus depot formulations of zuclopenthixol in patients with schizophrenia and previous violence. Eur Psychiatry. 2006;21(1):34–40. https://doi.org/10.1016/J.EURPSY.2005.07.006.

Arvanitis CD, Ferraro GB, Jain RK. The blood–brain barrier and blood–tumour barrier in brain tumours and metastases. Nat Rev Cancer. 2020;20(1):26–41. https://doi.org/10.1038/s41568-019-0205-x.The.

Ashwell SG, Gebbie J, Home PD. Optimal timing of injection of once-daily insulin glargine in people with type 1 diabetes using insulin lispro at meal-times. Diabet Med. 2006;23(1):46–52. https://doi.org/10.1111/j.1464-5491.2005.01726.x.

Augustine R, Ashkenazi DL, Arzi RS, Zlobin V, Shofti R, Sosnik A. Nanoparticle-in-microparticle oral drug delivery system of a clinically relevant darunavir/ritonavir antiretroviral combination. Acta Biomater. 2018;74:344–59. https://doi.org/10.1016/j.actbio.2018.04.045.

Baiardi G, Cameran Caviglia M, Piras F, Sacco F, Prinapori R, Cristina ML, Mattioli F, Sartini M, Pontali E. The clinical efficacy of multidose oritavancin: a systematic review. Antibiotics. 2023;12(10):1–14. https://doi.org/10.3390/antibiotics12101498.

Barlocco D. Insulin detemir. Novo Nordisk. Curr Opin Investig Drugs. 2003;4(4):449–54.

Barnes PJ, Barnes PJ. Pathophysiology of allergic inflammation. Immunol Rev. 2011;242(1):31–50. https://doi.org/10.1111/J.1600-065X.2011.01020.X.

Bartoli F, Cavaleri D, Nasti C, Palpella D, Guzzi P, Riboldi I, Crocamo C, Pappa S, Carrà G. Long-acting injectable antipsychotics for the treatment of bipolar disorder: evidence from mirror-image studies. Therap Adv Psychopharmacol. 2023;13:20451253231163636. https://doi.org/10.1177/20451253231163682.

Bassand C, Villois A, Gianola L, Laue G, Ramazani F, Riebesehl B, Sanchez-Felix M, Sedo K, Ullrich T, Duvnjak Romic M. Smart design of patient-centric long-acting products: from preclinical to marketed pipeline trends and opportunities. Expert Opin Drug Deliv. 2022;19(10) https://doi.org/10.1080/17425247.2022.2106213.

Becker RHA, Frick AD, Burger F, Potgieter JH, Scholtz H. Insulin glulisine, a new rapid-acting insulin analogue, displays a rapid time-action profile in obese non-diabetic subjects. Exp Clin Endocrinol Diabetes. 2005;113(8):435–43. https://doi.org/10.1055/s-2005-865806.

Benagiano G, Gabelnick H, Brosens I. Long-acting hormonal contraception. Womens Health. 2015;11(6):749–57. https://doi.org/10.2217/whe.15.68.

Benny O, Menon LG, Ariel G, et al. Local delivery of poly lactic-co-glycolic acid microspheres containing imatinib mesylate inhibits intracranial xenograft glioma growth. Clin Cancer Res. 2009;15(4):1222–31.

Berk M, Rathbone J, Mandriota-Carpenter SL. Clotiapine for acute psychotic illnesses. Cochrane Database Syst Rev. 2004;2004(4) https://doi.org/10.1002/14651858.CD002304.PUB2.

Bobo WV, Shelton RC. Risperidone long-acting injectable (Risperdal Consta®) for maintenance treatment in patients with bipolar disorder. Expert Rev Neurother. 2010;10(11):1637–58. https://doi.org/10.1586/ERN.10.143.

Bolli GB, Kerr D, Thomas R, Torlone E, Sola-Gazagnes A, Vitacolonna E, Selam JL, Home PD. Comparison of a multiple daily insulin injection regimen (basal once-daily glargine plus mealtime lispro) and continuous subcutaneous insulin infusion (lispro) in type 1 diabetes: a randomized open parallel multicenter study. Diabetes Care. 2009;32(7):1170–6. https://doi.org/10.2337/dc08-1874.

Boyce P, Irwin L, Morris G, Hamilton A, Mulder R, Malhi GS, Porter RJ. Long-acting injectable antipsychotics as maintenance treatments for bipolar disorder—a critical review of the evidence. Bipolar Disord. 2018;20:25–36. https://doi.org/10.1111/BDI.12698.

Brissos S, Veguilla MR, Taylor D, Balanzá-Martinez V. The role of long-acting injectable antipsychotics in schizophrenia: a critical appraisal. Therap Adv Psychopharmacol. 2014;4(5):198. https://doi.org/10.1177/2045125314540297.

Bruttomesso D, Crazzolara D, Maran A, Costa S, Dal Pos M, Girelli A, Lepore G, Aragona M, Iori E, Valentini U, Del Prato S, Tiengo A, Buhr A, Trevisan R, Baritussio A. In Type 1 diabetic patients with good glycaemic control, blood glucose variability is lower during continuous subcutaneous insulin infusion than during multiple daily injections with insulin glargine. Diabetic Med. 2008;25(3):326–32. https://doi.org/10.1111/j.1464-5491.2007.02365.x.

Bursulaya B, Fisch A, Lajiness JP, Lajiness P, Machauer R, Malekar S, Petrassi HMJ, Ramazani F, Remond A-C, Ullrich T, Usselmann P, Vangrevelinghe E. Substituted indazoles for treating tendon and/or ligament injuries. 2018.

Cai Y, Wei L, Ma L, Huang X, Tao A, Liu Z, Yuan W. Long-acting preparations of exenatide. Drug Des Devel Ther. 2013;7:963–70. https://doi.org/10.2147/DDDT.S46970.

Cain AR, Bremmer DN, Carr DR, Buchanan C, Jacobs M, Walsh TL, Moffa MA, Shively NR, Trienski TL. Effectiveness of Dalbavancin compared with standard of care for the treatment of osteomyelitis: a real-world analysis. Open Forum Infect Dis. 2022;9(2):1–6. https://doi.org/10.1093/ofid/ofab589.

Calabrese JR, Sanchez R, Jin N, Amatniek J, Cox K, Johnson B, Perry P, Hertel P, Such P, Salzman PM, McQuade RD, Nyilas M, Carson WH. Efficacy and safety of aripiprazole once-monthly in the maintenance treatment of bipolar I disorder: a double-blind, placebo-controlled, 52-week randomized withdrawal study. J Clin Psychiatry. 2017;78(3):324–31. https://doi.org/10.4088/JCP.16M11201.

Chakrabarti S. Treatment attitudes and adherence among patients with bipolar disorder: a systematic review of quantitative and qualitative studies. Harv Rev Psychiatry. 2019;27(5):290–302. https://doi.org/10.1097/HRP.0000000000000228.

Chanson P, Boerlin V, Ajzenberg C, Bachelot Y, Benito P, Bringer J, Caron P, Charbonnel B, Cortet C, Delemer B, Escobar-Jimenez F, Foubert L, Gaztambide S, Jockenhoevel F, Kuhn JM, Leclere J, Lorcy Y, Perlemuter L, Prestele H, et al. Comparison of octreotide acetate LAR and lanreotide SR in patients with acromegaly. Clin Endocrinol. 2000;53(5):577–86. https://doi.org/10.1046/j.1365-2265.2000.01134.x.

Chao JH, Page ST. The current state of male hormonal contraception. Pharmacol Ther. 2016;163:109–17. https://doi.org/10.1016/j.pharmthera.2016.03.012.

Chaudhury A, Duvoor C, Reddy Dendi VS, Kraleti S, Chada A, Ravilla R, Marco A, Shekhawat NS, Montales MT, Kuriakose K, Sasapu A, Beebe A, Patil N, Musham CK, Lohani GP, Mirza W. Clinical review of antidiabetic drugs: implications for type 2 diabetes mellitus management. Front Endocrinol. 2017;8:6. https://doi.org/10.3389/fendo.2017.00006.

Chou R, Evans C, Hoverman A, Sun C, Dana T, Bougatsos C, Grusing S, Korthuis PT. Preexposure prophylaxis for the prevention of HIV infection: evidence report and systematic review for the US preventive services task force. JAMA. 2019;321(22):2214–30. https://doi.org/10.1001/jama.2019.2591.

Chovan D, Mal M, Cerven I. 'Hot flush', an unpleasant symptom accompanying antiandrogen therapy of prostatic cancer and its treatment by cyproterone acetate. Int Urol Nephrol. 2000;32:77–9.

Chue P, Eerdekens M, Augustyns I, Lachaux B, Eriksson L, Pretorius H, David AS, Molc P. Comparative efficacy and safety of long-acting risperidone and risperidone oral tablets. Eur Neuropsychopharmacol. 2005;15:111–7. https://doi.org/10.1016/j.euroneuro.2004.07.003.

Citrin DL, Resnick M, Guinan P, Al-bussam N, Gau TC, Kennealey GT. A comparison of Zoladex and DES in the treatment of advanced prostate cancer: results of a randomized, multicenter trial. Prostate. 1991;18(2):139–46.

Clay PG, Yuet WC, Moecklinghoff CH, Duchesne I, Tronczyski KL, Shah S, Shao D. A meta-analysis comparing 48-week treatment outcomes of single and multi-tablet antiretroviral regimens for the treatment of people living with HIV. AIDS Res Ther. 2018;15(1):1–10. https://doi.org/10.1186/s12981-018-0204-0.

Clement ME, Okeke NL, Hicks CB. Treatment of syphilis: a systematic review. J Am Med Assoc. 2014;312(18):1905–17. https://doi.org/10.1001/jama.2014.13259.

Clemente-Tomás R, Hernández-Pérez D, Neira-Ibáñez P, Farías-Rozas F, Torrecillas-Picazo R, Osorio-Alayo V, Duch-Samper AM. Intracrystalline Ozurdex®: therapeutic effect maintained for 18 months. Int Ophthalmol. 2019;39(1):207–11. https://doi.org/10.1007/S10792-017-0780-3/FIGURES/4.

Cohen CJ, Andrade-Villanueva J, Clotet B, Fourie J, Johnson MA, Ruxrungtham K, Wu H, Zorrilla C, Crauwels H, Rimsky LT, Vanveggel S, Boven K. Rilpivirine versus efavirenz with two background nucleoside or nucleotide reverse transcriptase inhibitors in treatment-naive adults infected with HIV-1 (THRIVE): a phase 3, randomised, non-inferiority trial. Lancet. 2011;378(9787):229–37. https://doi.org/10.1016/S0140-6736(11)60983-5.

Correll CU, Citrome L, Haddad PM, Lauriello J, Olfson M, Calloway SM, Kane JM. The use of long-acting injectable antipsychotics in schizophrenia: evaluating the evidence. J Clin Psychiatry. 2016;77(suppl 3):21984. https://doi.org/10.4088/JCP.15032SU1.

Correll CU, Benson C, Emond B, Patel C, Lafeuille MH, Lin D, Morrison L, Ghelerter I, Lefebvre P, Mavros P. Comparison of clinical outcomes in patients with schizophrenia following different long-acting injectable event-driven initiation strategies. Schizophrenia. 2023;9(1):1–9. https://doi.org/10.1038/s41537-023-00334-3.

Cuevas-Ramos D, Fleseriu M. Pasireotide: a novel treatment for patients with acromegaly. Drug Des Devel Ther. 2016;10:227–39. https://doi.org/10.2147/DDDT.S77999.

D'Agostino A, Aguglia A, Barbui C, Bartoli F, Carrà G, Cavallotti S, Chirico M, Ostinelli EG, Zangani C, Martinotti G, Ostuzzi G, Nosè M, Purgato M, Turrini G, Mazzi MA, Papola D, Gastaldon C, Terlizzi S, Bertolini F, et al. Off-label long acting injectable antipsychotics in real–world clinical practice: a cross-sectional analysis of prescriptive patterns from the STAR Network DEPOT study. BMC Psychiatry. 2022;22(1):1–9. https://doi.org/10.1186/S12888-022-04071-2/FIGURES/2.

Davies PH, Stewart SE, Lancranjan I, Sheppard MC, Stewart PM. Long-term therapy with long-acting octreotide (Sandostatin-LAR®) for the management of acromegaly. Clin Endocrinol. 1998;48(3):311–6. https://doi.org/10.1046/j.1365-2265.1998.00389.x.

Dimopoulou D, Mantadakis E, Koutserimpas C, Samonis G. A narrative review on the role of Dalbavancin in the treatment of bone and joint infections. Antibiotics. 2023;12(10) https://doi.org/10.3390/antibiotics12101492.

Durante-Mangoni E, Gambardella M, Iula VD, De Stefano GF, Corrado MF, Esposito V, Gentile I, Coppola N. Current trends in the real-life use of dalbavancin: report of a study panel. Int J Antimicrob Agents. 2020;56(4) https://doi.org/10.1016/j.ijantimicag.2020.106107.

Engler K, Toupin I, Vicente S, Ahmed S, Lebouché B. A review of HIV-specific patient-reported measures of perceived barriers to antiretroviral therapy adherence: what themes are they covering? J Patient Rep Outcomes. 2019;3(1):1–8. https://doi.org/10.1186/s41687-019-0124-3.

Firmagon. Firmagon ® [Product Leaflet]. Parsippany, NJ. Approval in 2008, revised in 2020: Ferring Pharmaceuticals; 2020.

Fleischhacker WW, Sanchez R, Perry PP, Jin N, Peters-Strickland T, Johnson BR, Baker RA, Eramo A, McQuade RD, Carson WH, Walling D, Kane JM. Aripiprazole once-monthly for treatment of schizophrenia: double-blind, randomised, non-inferiority study. Br J Psychiatry. 2014;205(2):135–44. https://doi.org/10.1192/BJP.BP.113.134213.

Franzini A, Moosa S, Servello D, Small I, DiMeco F, Xu Z, Elias WJ, Franzini A, Prada F. Ablative brain surgery: an overview. Int J Hyperth. 2019;36(2):64–80. https://doi.org/10.1080/02656736.2019.1616833.

Gaballa SA, Kompella UB, Elgarhy O, Alqahtani AM, Pierscionek B, Alany RG, Abdelkader H. Corticosteroids in ophthalmology: drug delivery innovations, pharmacology, clinical applications, and future perspectives. Drug Deliv Transl Res. 2020;11(3):866–93. https://doi.org/10.1007/S13346-020-00843-Z.

Gao Y, Kraft JC, Yu D, Ho RJY. Recent developments of nanotherapeutics for targeted and long-acting, combination HIV chemotherapy. Eur J Pharm Biopharm. 2018;138:75–91. https://doi.org/10.1016/j.ejpb.2018.04.014.

Garcia-Effron G. Rezafungin—mechanisms of action, susceptibility and resistance: similarities and differences with the other echinocandins. J Fungi. 2020;6(4):1–23. https://doi.org/10.3390/jof6040262.

Garg SK, Walker AJ, Hoff HK, D'Souza AO, Gottlieb PA, Chase HP. Glycemic parameters with multiple daily injections using insulin glargine versus insulin pump. Diabetes Technol Ther. 2004;6(1):9–15. https://doi.org/10.1089/152091504322783350.

Giese A, Kucinski T, Knopp U, Goldbrunner R, Hamel W, Mehdorn HM. Pattern of recurrence following local chemotherapy with biodegradable carmustine (BCNU) implants in patients with glioblastoma. J Neurooncol. 2004;66(3):351–60.

Gopal S, Xu H, Bossie C, Burón JA, Fu DJ, Savitz A, Nuamah I, Hough D. Incidence of tardive dyskinesia: a comparison of long-acting injectable and oral paliperidone clinical trial databases. Int J Clin Pract. 2014;68(12):1514–22. https://doi.org/10.1111/IJCP.12493.

Gottfries CG, Green L. Flupenthixol decanoate--in treatment of out-patients. Acta Psychiatr Scand. 1974;50:15–24. https://doi.org/10.1111/J.1600-0447.1974.TB08890.X.

Greer D, Pasquale J, Wahrenberger JT. Objective and subjective benefits of a psychiatric pharmacist–led long-acting injectable medication training at a large, multisite organization. Mental Health Clin. 2020;10(5):264–9. https://doi.org/10.9740/MHC.2020.09.264.

Günthard HF, Saag MS, Benson CA, Del Rio C, Eron JJ, Gallant JE, Hoy JF, Mugavero MJ, Sax PE, Thompson MA, Gandhi RT, Landovitz RJ, Smith DM, Jacobsen DM, Volberding PA. Antiretroviral drugs for treatment and prevention of HIV infection in adults: 2016 recommendations of the international antiviral society-USA panel. JAMA. 2016;316(2):191–210. https://doi.org/10.1001/jama.2016.8900.

Halpern V, Stalter RM, Owen DH, Dorflinger LJ, Lendvay A, Rademacher KH. Towards the development of a longer-acting injectable contraceptive: past research and current trends. Contraception. 2015;92(1):3–9. https://doi.org/10.1016/j.contraception.2015.02.014.

Hamilton-Wessler M, Ader M, Dea MK, Moore D, Loftager M, Markussen J, Bergman RN. Mode of transcapillary transport of insulin and insulin analog NN304 in dog hindlimb: evidence for passive diffusion. Diabetes. 2002;51(3):574–82. https://doi.org/10.2337/diabetes.51.3.574.

Hand RM, Senarathna SMDKG, Page-Sharp M, Gray K, Sika-Paotonu D, Sheel M, Chuang VTG, Martinez J, Luna G, Manning L, Wyber R, Carapetis JR, Batty KT. Quality of benzathine penicillin G: a multinational cross-sectional study. Pharmacol Res Perspect. 2020;8(6):1–7. https://doi.org/10.1002/prp2.668.

Hänggi D, Etminan N, Mayer SA, Aldrich EF, Diringer MN, Schmutzhard E, Faleck HJ, Ng D, Saville BR, Macdonald RL. Clinical trial protocol: phase 3, multicenter, randomized, double-blind, placebo-controlled, parallel-group, efficacy, and safety study comparing EG-1962 to standard of care oral nimodipine in adults with aneurysmal subarachnoid hemorrhage [NEWTON-2 (Nimodipine microparticles to enhance recovery while reducing toxicity after subarachnoid hemorrhage)]. Neurocrit Care. 2019;30(1):88–97. https://doi.org/10.1007/s12028-018-0575-z.

Havelund S, Plum A, Ribel U, Jonassen I, Vølund A, Markussen J, Kurtzhals P. The mechanism of protraction of insulin detemir, a long-acting, acylated analog of human insulin. Pharm Res. 2004;21(8):1498–504. https://doi.org/10.1023/B:PHAM.0000036926.54824.37/METRICS.

Heinemann L, Linkeschova R, Rave K, Hompesch B, Sedlak M, Heise T. Time-action profile of the long-acting insulin analog insulin glargine (HOE901) in comparison with those of NPH insulin and placebo. Diabetes Care. 2000;23(5):644–9. https://doi.org/10.2337/diacare.23.5.644.

Heise T, Nosek L, Rønn BB, Endahl L, Heinemann L, Kapitza C, Draeger E. Lower within-subject variability of insulin detemir in comparison to NPH insulin and insulin glargine in people with type 1 diabetes. Diabetes. 2004;53(6):1614–20. https://doi.org/10.2337/diabetes.53.6.1614.

Heres S, Hamann J, Mendel R, Wickelmaier F, Pajonk FG, Leucht S, Kissling W. Identifying the profile of optimal candidates for antipsychotic depot therapy: a cluster analysis. Prog Neuro-Psychopharmacol Biol Psychiatry. 2008;32(8):1987–93. https://doi.org/10.1016/J.PNPBP.2008.09.025.

Hollander P, Cooper J, Bregnhøj J, Pedersen CB. A 52-week, multinational, open-label, parallel-group, noninferiority, treat-to-target trial comparing insulin detemir with insulin glargine in a basal-bolus regimen with mealtime insulin aspart in patients with type 2 diabetes. Clin Ther. 2008;30(11):1976–87. https://doi.org/10.1016/j.clinthera.2008.11.001.

Hoogwerf BJ, Doshi KB, Diab D. Pramlintide, the synthetic analogue of amylin: physiology, pathophysiology, and effects on glycemic control, body weight, and selected biomarkers of vascular risk. Vasc Health Risk Manag. 2008;4(2):355. https://doi.org/10.2147/VHRM.S1978.

Hordern SVM, Wright JE, Umpleby AM, Shojaee-Moradie F, Amiss J, Russell-Jones DL. Comparison of the effects on glucose and lipid metabolism of equipotent doses of insulin detemir and NPH insulin with a 16-h euglycaemic clamp. Diabetologia. 2005;48(3):420–6. https://doi.org/10.1007/s00125-005-1670-1.

Hossain P, Kawar B, El Nahas M. Obesity and diabetes in the developing world--a growing challenge. N Engl J Med. 2007;356(3):213–5. https://doi.org/10.1056/NEJMP068177.

Howell A, Robertson JFR, Albano JQ, Aschermannova A, Mauriac L, Kleeberg UR, Vergote I, Erikstein B, Webster A, Morris C. Fulvestrant, formerly ICI 182,780, is as effective as anastrozole in postmenopausal women with advanced breast cancer progressing after prior endocrine treatment. J Clin Oncol. 2002;20(16):3396–403. https://doi.org/10.1200/JCO.2002.10.057.

Hu W, Yang Y, Zhang L, Yin J, Huang J, Huang L, Gu H, Jiang G, Fang J. Post surgery circulating free tumor DNA is a predictive biomarker for relapse of lung cancer. Cancer Med. 2017;6(5):962–74. https://doi.org/10.1002/cam4.980.

Iacob SA, Iacob DG, Jugulete G. Improving the adherence to antiretroviral therapy, a difficult but essential task for a successful HIV treatment-clinical points of view and practical considerations. Front Pharmacol. 2017;8(Nov):299234. https://doi.org/10.3389/fphar.2017.00831.

Insel TR. Rethinking schizophrenia. Nature. 2010;468(7321):187–93. https://doi.org/10.1038/nature09552.

Ishigooka J, Nakamura J, Fujii Y, Iwata N, Kishimoto T, Iyo M, Uchimura N, Nishimura R, Shimizu N. Efficacy and safety of aripiprazole once-monthly in Asian patients with schizophrenia: a multicenter, randomized, double-blind, non-inferiority study versus oral aripiprazole. Schizophr Res. 2015;161(2–3):421–8. https://doi.org/10.1016/J.SCHRES.2014.12.013.

Jackson AGA, Else LJ, Mesquita PMM, Egan D, Back DJ, Karolia Z, Ringner-Nackter L, Higgs CJ, Herold BC, Gazzard BG, Boffito M. A compartmental pharmacokinetic evaluation of long-acting rilpivirine in HIV-negative volunteers for pre-exposure prophylaxis. Clin Pharmacol Ther. 2014;96(3):314–23. https://doi.org/10.1038/clpt.2014.118.

Jaeger M, Rossler W. Attitudes towards long-acting depot antipsychotics: a survey of patients, relatives and psychiatrists. Psychiatry Res. 2010;175(1–2):58–62. https://doi.org/10.1016/J.PSYCHRES.2008.11.003.

Johnson DAW. Historical perspective on antipsychotic long-acting injections. Br J Psychiatry. 2009;195(Suppl. 52):s7–s12. https://doi.org/10.1192/bjp.195.52.s7.

Johnson DAW, Freeman H. Drug defaulting by patients on long-acting phenothiazines. Psychol Med. 1973;3(1):115–9. https://doi.org/10.1017/S0033291700046407.

Joshi R, Khadilkar S, Patel M. Global trends in use of long-acting reversible and permanent methods of contraception: seeking a balance. Int J Gynecol Obstet. 2015;131:S60–3. https://doi.org/10.1016/j.ijgo.2015.04.024.

Jotwani V, Scherzer R, Glidden DV, Mehrotra M, Defechereux P, Liu A, Gandhi M, Bennett M, Coca SG, Parikh CR, Grant RM, Shlipak MG. Pre-exposure prophylaxis with Tenofovir Disoproxil fumarate/Emtricitabine and kidney tubular dysfunction in HIV-uninfected individuals. J Acquir Immune Defic Syndr. 2018;78(2):169–74. https://doi.org/10.1097/QAI.0000000000001654.

Kane JM, Detke HC, Naber D, Sethuraman G, Lin DY, Bergstrom RF, McDonnell D. Olanzapine long-acting injection: a 24-week, randomized, double-blind trial of maintenance treatment in patients with schizophrenia. Am J Psychiatry. 2010;167(2):181–9. https://doi.org/10.1176/APPI.AJP.2009.07081221.

Keefe RSE, Seidman LJ, Christensen BK, Hamer RM, Sharma T, Sitskoorn MM, Lewine RRJ, Yurgelun-Todd DA, Gur RC, Tohen M, Tollefson GD, Sanger TM, Lieberman JA. Comparative effect of atypical and conventional antipsychotic drugs on neurocognition in first-episode psychosis: a randomized, double-blind trial of olanzapine versus low doses of haloperidol. Am J Psychiatry. 2004;161(6):985–95. https://doi.org/10.1176/APPI.AJP.161.6.985.

Kennedy DW. The PROPEL™ steroid-releasing bioabsorbable implant to improve outcomes of sinus surgery. Expert Rev Respir Med. 2012;6(5):493–8. https://doi.org/10.1586/ERS.12.53.

Keramatian K, Chakrabarty T, Yatham LN. Long-acting injectable second-generation/atypical antipsychotics for the management of bipolar disorder: a systematic review. CNS Drugs. 2019;33(5):431–56. https://doi.org/10.1007/S40263-019-00629-Z/TABLES/6.

Kessing LV, Vradi E, Andersen PK. Life expectancy in bipolar disorder. Bipolar Disord. 2015;17(5):543–8. https://doi.org/10.1111/BDI.12296.

Kikkert MJ, Schene AH, Koeter MWJ, Robson D, Born A, Helm H, Nose M, Goss C, Thornicroft G, Gray RJ. Medication adherence in schizophrenia: exploring patients', carers' and professionals' views. Schizophr Bull. 2006;32(4):786–94. https://doi.org/10.1093/SCHBUL/SBL011.

Kilic T, Alberta JA, Zdunek PR, Acar M, Iannarelli P, O'Reilly T, Buchdunger E, Black PM, Stiles CD. Intracranial inhibition of platelet-derived growth factor-mediated glioblastoma cell growth by an orally active kinase inhibitor of the 2-phenylaminopyrimidine class. Cancer Res. 2000;60(18):5143–50.

Klein O, Lynge J, Endahl L, Damholt B, Nosek L, Heise T. Albumin-bound basal insulin analogues (insulin detemir and NN344): comparable time-action profiles but less variability than insulin glargine in type 2 diabetes. Diabetes Obes Metab. 2007;9(3):290–9. https://doi.org/10.1111/j.1463-1326.2006.00685.x.

Kou H, Qing Z, Zhao G, Sun X, Zhi L, Wang J, Chen X, Guo H, Zhang R, Ma J. Effect of lorecivivint on osteoarthritis: a systematic review and meta-analysis. Heliyon. 2023;9(8) https://doi.org/10.1016/j.heliyon.2023.e18682.

Kovarova M, Swanson MD, Sanchez RI, Baker CE, Steve J, Spagnuolo RA, Howell BJ, Hazuda DJ, Victor Garcia J. A long-acting formulation of the integrase inhibitor raltegravir protects humanized BLT mice from repeated high-dose vaginal HIV challenges. J Antimicrob Chemother. 2016;71(6):1586–96. https://doi.org/10.1093/jac/dkw042.

Krishnan BR, James KD, Polowy K, Bryant BJ, Vaidya A, Smith S, Laudeman CP. CD101, a novel echinocandin with exceptional stability properties and enhanced aqueous solubility. J Antibiot. 2017;70(2):130–5. https://doi.org/10.1038/ja.2016.89.

Kurtzhals P. Engineering predictability and protraction in a basal insulin analogue: the pharmacology of insulin detemir. Int J Obes Relat Metab Disord. 2004;28(2):S23–8. https://doi.org/10.1038/sj.ijo.0802746.

Kurtzhals P, Schäffer L, Sørensen A, Kristensen C, Jonassen I, Schmid C, Trüb T. Correlations of receptor binding and metabolic and mitogenic potencies of insulin analogs designed for clinical use. Diabetes. 2000;49(6):999–1005. https://doi.org/10.2337/diabetes.49.6.999.

Lachman L, Lieberman HA, Kanig JL. The theory and practice of industrial pharmacy. 3rd ed. Bombay: Varghese Publishing House; 2003. https://worldcat.org/title/148231775

LactMed. Flupenthixol. Drugs and Lactation Database; 2022a.

LactMed. Zuclopenthixol. Drugs and Lactation Database; 2022b.

Landovitz RJ, Donnell D, Clement ME, Hanscom B, Cottle L, Coelho L, Cabello R, Chariyalertsak S, Dunne EF, Frank I, Gallardo-Cartagena JA, Gaur AH, Gonzales P, Tran HV, Hinojosa JC, Kallas EG, Kelley CF, Losso MH, Madruga JV, et al. Cabotegravir for HIV prevention in cisgender men and transgender women. N Engl J Med. 2021;385(7):595–608. https://doi.org/10.1056/nejmoa2101016.

Landovitz RJ, Hanscom BS, Clement ME, Tran HV, Kallas EG, Magnus M, Sued O, Sanchez J, Scott H, Eron JJ, del Rio C, Fields SD, Marzinke MA, Eshleman SH, Donnell D, Spinelli MA, Kofron RM, Berman R, Piwowar-Manning EM, et al. Efficacy and safety of long-acting cabotegravir compared with daily oral tenofovir disoproxil fumarate plus emtricitabine to prevent HIV infection in cisgender men and transgender women who have sex with men 1 year after study unblinding: a secondary anal. Lancet HIV. 2023;10(12):e767–78. https://doi.org/10.1016/s2352-3018(23)00261-8.

Lang Y, Zemel E, Miller B, Perlman I. Retinal toxicity of intravitreal kenalog in albino rabbits. Retina. 2007;27(6):778–88. https://doi.org/10.1097/IAE.0B013E318030C517.

Lepak AJ, Zhao M, Andes DR. Candida auris in the neutropenic mouse invasive candidiasis. Antimicrob Agents Chemother. 2018;62(11):1–4.

Lepore M, Pampanelli S, Fanelli C, Porcellati F, Bartocci L, Di Vincenzo A, Cordoni C, Costa E, Brunetti P, Bolli GB. Pharmacokinetics and pharmacodynamics of subcutaneous injection of long-acting human insulin analog glargine, NPH insulin, and ultralente human insulin and continuous subcutaneous infusion of insulin lispro. Diabetes. 2000;49(12):2142–8. https://doi.org/10.2337/diabetes.49.12.2142.

Lepore G, Dodesini AR, Nosari I, Trevisan R. Effect of continuous subcutaneous insulin infusion vs multiple daily insulin injection with glargine as basal insulin: an open parallel long-term study. Diabetes Nutr Metab. 2004;17(2):84–9.

Li K, Yu L, Liu X, Chen C, Chen Q, Ding J. A long-acting formulation of a polypeptide drug exenatide in treatment of diabetes using an injectable block copolymer hydrogel. Biomaterials. 2013;34(11):2834–42. https://doi.org/10.1016/J.BIOMATERIALS.2013.01.013.

Li W, Tang J, Lee D, Tice TR, Schwendeman SP, Prausnitz MR. Clinical translation of long-acting drug delivery formulations. Nat Rev Mater. 2022;7(5):406–20. https://doi.org/10.1038/s41578-021-00405-w.

Lovatti S, Tiecco G, Mulé A, Rossi L, Sforza A, Salvi M, Signorini L, Castelli F, Quiros-Roldan E. Dalbavancin in bone and joint infections: a systematic review. Pharmaceuticals. 2023;16(7):1–18. https://doi.org/10.3390/ph16071005.

Lv Y, Li S, Liu Z, Ren Z, Zhao J, Tao G, Zheng Z, Han Y, Ye B. Impact of surgery and adjuvant chemotherapy on the survival of stage I lung adenocarcinoma patients with tumor spread through air spaces. Lung Cancer. 2023;177(December 2022):51–8. https://doi.org/10.1016/j.lungcan.2023.01.009.

Ma Q, Tso LS, Rich ZC, Hall BJ, Beanland R, Li H, Lackey M, Hu F, Cai W, Doherty M, Tucker JD. Barriers and facilitators of interventions for improving antiretroviral therapy adherence: a systematic review of global qualitative evidence. J Int AIDS Soc. 2016;19(1):21166. https://doi.org/10.7448/IAS.19.1.21166.

MacDonald L, Chapman S, Syrett M, Bowskill R, Horne R. Improving medication adherence in bipolar disorder: a systematic review and meta-analysis of 30 years of intervention trials. J Affect Disord. 2016;194:202–21. https://doi.org/10.1016/J.JAD.2016.01.002.

Marantelli S, Hand R, Carapetis J, Beaton A, Wyber R. Severe adverse events following benzathine penicillin G injection for rheumatic heart disease prophylaxis: cardiac compromise more likely than anaphylaxis. Heart Asia. 2019;11(2):1–9. https://doi.org/10.1136/heartasia-2019-011191.

Marcus SC, Zummo J, Pettit AR, Stoddard J, Doshi JA. Antipsychotic adherence and rehospitalization in schizophrenia patients receiving oral versus long-acting injectable antipsychotics following hospital discharge. J Manag Care Spec Pharm. 2015;21(9):754–68. https://doi.org/10.18553/JMCP.2015.21.9.754.

Margolis DA, Gonzalez-Garcia J, Stellbrink HJ, Eron JJ, Yazdanpanah Y, Podzamczer D, Lutz T, Angel JB, Richmond GJ, Clotet B, Gutierrez F, Sloan L, Clair MS, Murray M, Ford SL, Mrus J, Patel P, Crauwels H, Griffith SK, et al. Long-acting intramuscular cabotegravir and rilpivirine in adults with HIV-1 infection (LATTE-2): 96-week results of a randomised, open-label, phase 2b, non-inferiority trial. Lancet. 2017;390(10101):1499–510. https://doi.org/10.1016/S0140-6736(17)31917-7.

Markowitz M. Long-acting penicillins: historical perspectives. Pediatr Infect Dis. 1985;4(5):570–3. https://doi.org/10.1097/00006454-198509000-00054.

Markowitz M, Frank I, Grant RM, Mayer KH, Elion R, Goldstein D, Fisher C, Sobieszczyk ME, Gallant JE, Van Tieu H, Weinberg W, Margolis DA, Hudson KJ, Stancil BS, Ford SL, Patel P, Gould E, Rinehart AR, Smith KY, Spreen WR. Safety and tolerability of long-acting cabotegravir injections in HIV-uninfected men (ECLAIR): a multicentre, double-blind, randomised, placebo-controlled, phase 2a trial. Lancet HIV. 2017;4(8):e331–40. https://doi.org/10.1016/S2352-3018(17)30068-1.

Maroney M. An update on current treatment strategies and emerging agents for the management of schizophrenia. Am J Manag Care. 2020;25:S55–61.

McEvoy JP, Freter S, Everett G, Geller JL, Appelbaum P, Joy Apperson L, Roth L. Insight and the clinical outcome of schizophrenic patients. J Nerv Ment Dis. 1989;177(1):48–51. https://doi.org/10.1097/00005053-198901000-00008.

McGowan I, Dezzutti CS, Siegel A, Engstrom J, Nikiforov A, Duffill K, Shetler C, Richardson-Harman N, Abebe K, Back D, Else L, Egan D, Khoo S, Egan JE, Stall R, Williams PE, Rehman KK, Adler A, Brand RM, et al. Long-acting rilpivirine as potential pre-exposure prophylaxis for HIV-1 prevention (the MWRI-01 study): an open-label, phase 1, compartmental, pharmacokinetic and pharmacodynamic assessment. Lancet HIV. 2016;3(12):e569–78. https://doi.org/10.1016/S2352-3018(16)30113-8.

McIntyre RS, Berk M, Brietzke E, Goldstein BI, López-Jaramillo C, Kessing LV, Malhi GS, Nierenberg AA, Rosenblat JD, Majeed A, Vieta E, Vinberg M, Young AH, Mansur RB. Bipolar disorders. Lancet. 2020;396(10265):1841–56. https://doi.org/10.1016/S0140-6736(20)31544-0.

McIntyre RS, Alda M, Baldessarini RJ, Bauer M, Berk M, Correll CU, Fagiolini A, Fountoulakis K, Frye MA, Grunze H, Kessing LV, Miklowitz DJ, Parker G, Post RM, Swann AC, Suppes T, Vieta E, Young A, Maj M. The clinical characterization of the adult patient with bipolar disorder aimed at personalization of management. World Psychiatry. 2022;21(3):364–87. https://doi.org/10.1002/WPS.20997.

McKeage K, Cheer S, Wagstaff AJ. Octreotide long-acting release (LAR): a review of its use in the management of acromegaly. Drugs. 2003;63(22):2473–99. https://doi.org/10.2165/00003495-200363220-00014.

Mekhail NA. Safety and efficacy of novel epidural clonidine micropellets for lumbosacral radiculopathy. J Clin Neurol Neurosurg. 2020;3:2. https://doi.org/10.37421/jcnn.2020.3.109.

Mekonen KK, Yismaw MB, Abiye AA, Tadesse TA. Adherence to benzathine penicillin G secondary prophylaxis and its determinants in patients with rheumatic heart disease at a cardiac center of an Ethiopian tertiary care teaching hospital. Patient Prefer Adherence. 2020;14:343–52. https://doi.org/10.2147/PPA.S238423.

Meyer JM, Stahl SM. The metabolic syndrome and schizophrenia. Acta Psychiatr Scand. 2009;119(1):4–14. https://doi.org/10.1111/J.1600-0447.2008.01317.X.

Mishra M, Ndisang J. A critical and comprehensive insight on heme oxygenase and related products including carbon monoxide, bilirubin, biliverdin and ferritin in type-1 and type-2 diabetes. Curr Pharm Des. 2014;20(9):1370–91. https://doi.org/10.2174/13816128113199990559.

Molina JM, Cahn P, Grinsztejn B, Lazzarin A, Mills A, Saag M, Supparatpinyo K, Walmsley S, Crauwels H, Rimsky LT, Vanveggel S, Boven K. Rilpivirine versus efavirenz with tenofovir and emtricitabine in treatment-naive adults infected with HIV-1 (ECHO): a phase 3 randomised double-blind active-controlled trial. Lancet. 2011;378(9787):238–46. https://doi.org/10.1016/S0140-6736(11)60936-7.

Mucha GT, Merkel S, Thomas W, Bantle JP. Fasting and insulin glargine in individuals with type 1 diabetes. Diabetes Care. 2004;27(5):1209–10. https://doi.org/10.2337/diacare.27.5.1209.

Müller RH, Gohla S, Keck CM. State of the art of nanocrystals: special features, production, nanotoxicology aspects and intracellular delivery. Eur J Pharm Biopharm. 2011;78(1):1–9. https://doi.org/10.1016/j.ejpb.2011.01.007.

Musoke C, Mondo CK, Okello E, Zhang W, Kakande B, Nyakoojo W, Freers J. Benzathine penicillin adherence for secondary prophylaxis among patients affected with rheumatic heart disease attending Mulago Hospital. Cardiovasc J Afr. 2013;24(4):124–9. https://doi.org/10.5830/CVJA-2013-022.

Nachega JB, Parienti JJ, Uthman OA, Gross R, Dowdy DW, Sax PE, Gallant JE, Mugavero MJ, Mills EJ, Giordano TP. Lower pill burden and once-daily antiretroviral treatment regimens for HIV infection: a meta-analysis of randomized controlled trials. Clin Infect Dis. 2014;58(9):1297–307. https://doi.org/10.1093/cid/ciu046.

Nathan MR, Schmid P. A review of fulvestrant in breast cancer. Oncol Therapy. 2017;5(1):17–29. https://doi.org/10.1007/s40487-017-0046-2.

National Institute of Allergy and Infectious Diseases. (2016). Injectable Cabotegravir compared to TDF/FTC for PrEP in HIV-uninfected men and transgender women who have sex with men. Clinicaltrials.Gov. https://www.clinicaltrials.gov/study/NCT02720094

Ndisang JF, Jadhav A. Hemin therapy improves kidney function in male streptozotocin-induced diabetic rats: role of the heme oxygenase/atrial natriuretic peptide/adiponectin axis. Endocrinology. 2014;155(1):215–29. https://doi.org/10.1210/EN.2013-1050.

Ndisang JF, Jadhav A, Mishra M. The heme oxygenase system suppresses perirenal visceral adiposity, abates renal inflammation and ameliorates diabetic nephropathy in Zucker diabetic fatty rats. PLoS One. 2014;9(1) https://doi.org/10.1371/JOURNAL.PONE.0087936.

Nkanga CI, Fisch A, Rad-Malekshahi M, Romic MD, Kittel B, Ullrich T, Wang J, Krause RWM, Adler S, Lammers T, Hennink WE, Ramazani F. Clinically established biodegradable long acting injectables: an industry perspective. Adv Drug Deliv Rev. 2020; https://doi.org/10.1016/j.addr.2020.11.008.

Ohka F, Natsume A, Wakabayashi T. Current trends in targeted therapies for glioblastoma multiforme. Neurol Res Int. 2012;167:19–46. https://doi.org/10.1155/2012/878425.

Oo WM, Hunter DJ. Repurposed and investigational disease-modifying drugs in osteoarthritis (DMOADs). Ther Adv Musculoskelet Dis. 2022;14:1759720X221090297. https://doi.org/10.1177/1759720X221090297.

Pacheco C, Baião A, Ding T, Cui W, Sarmento B. Recent advances in long-acting drug delivery systems for anticancer drug. Adv Drug Deliv Rev. 2023;194:114724. https://doi.org/10.1016/j.addr.2023.114724.

Padmakumar S, Amiji MM. Long-acting therapeutic delivery systems for the treatment of gliomas. Adv Drug Deliv Rev. 2023;197:114853. https://doi.org/10.1016/j.addr.2023.114853.

Painter NA, Morello CM, Singh RF, McBane SE. An evidence-based and practical approach to using Bydureon™ in patients with type 2 diabetes. J Am Board Fam Med. 2013;26(2):203–10. https://doi.org/10.3122/JABFM.2013.02.120174.

Palumbo MO, Kavan P, Miller WH, Panasci L, Assouline S, Johnson N, Cohen V, Patenaude F, Pollak M, Jagoe RT, Batist G. Systemic cancer therapy: achievements and challenges that lie ahead. Front Pharmacol. 2013;4:1–9. https://doi.org/10.3389/fphar.2013.00057.

Parellada E, Andrezina R, Milanova V, Glue P, Masiak M, Turner MSJ, Medori R, Gaebel W. Patients in the early phases of schizophrenia and schizoaffective disorders effectively treated with risperidone long-acting injectable. J Psychopharmacol. 2005;19(5 Suppl):5–14. https://doi.org/10.1177/0269881105056513.

Parrish R, Sternberg P. Does 'off-label' mean off limits for patient care? Am J Ophthalmol. 2007;143(5):853–5. https://doi.org/10.1016/j.ajo.2007.01.044.

Pedroso MC, Ceravolo GS, Cuman RKN, Bersani-Amado CA, Caparroz-Assef SM. Penicillin G benzathine: characteristic of prescription and use in community pharmacy. Acta Sci. 2001;2001:4.

Peter R, Luzio SD, Dunseath G, Miles A, Hare B, Backx K, Pauvaday V, Owens DR. Effects of exercise on the absorption of insulin glargine in patients with type 1 diabetes. Diabetes Care. 2005;28(3):560–5. https://doi.org/10.2337/diacare.28.3.560.

Petersen MC, Vatner DF, Shulman GI. Regulation of hepatic glucose metabolism in health and disease. Nat Rev Endocrinol. 2017;13(10):572–87. https://doi.org/10.1038/NRENDO.2017.80.

Pettersson B, Varenhorst E, Petas A. Duration of testosterone suppression after a 9.45 mg implant of the GnRH-analogue Buserelin in patients with localised carcinoma of the prostate a 12-month follow-up study. Eur Urol. 2006;50:483–9. https://doi.org/10.1016/j.eururo.2006.03.001.

Pilon D, Muser E, Lefebvre P, Kamstra R, Emond B, Joshi K. Adherence, healthcare resource utilization and Medicaid spending associated with once-monthly paliperidone palmitate versus oral atypical antipsychotic treatment among adults recently diagnosed with schizophrenia. BMC Psychiatry. 2017;17(1):1–18. https://doi.org/10.1186/S12888-017-1358-3/FIGURES/3.

Porcellati F, Rossetti P, Busciantella NR, Marzotti S, Lucidi P, Luzio S, Owens DR, Bolli GB, Fanelli CG. Comparison of pharmacokinetics and dynamics of the Long-acting insulin analogs glargine and Detemir at steady state in type 1 diabetes: a double-blind, randomized, crossover study. Diabetes Care. 2007a;30(10):2447–52. https://doi.org/10.2337/dc07-0002.

Porcellati F, Rossetti P, Ricci NB, Pampanelli S, Torlone E, Campos SH, Andreoli AM, Bolli GB, Fanelli CG. Pharmacokinetics and pharmacodynamics of the long-acting insulin analog glargine after 1 week of use compared with its first administration in subjects with type 1 diabetes. Diabetes Care. 2007b;30(5):1261–3. https://doi.org/10.2337/dc06-2208.

Quiroz JA, Yatham LN, Palumbo JM, Karcher K, Kushner S, Kusumakar V. Risperidone Long-acting injectable monotherapy in the maintenance treatment of bipolar I disorder. Biol Psychiatry. 2010;68(2):156–62. https://doi.org/10.1016/J.BIOPSYCH.2010.01.015.

Ragam A, Kolomeyer AM, Nayak NV, Chu DS. The use of Ozurdex in patients with chronic, recurrent uveitis undergoing anterior segment surgery. Invest Ophthalmol Vis Sci. 2014;55(13):5808.

Rahnfeld L, Luciani P. Injectable lipid-based depot formulations: where do we stand? Pharmaceutics. 2020;12(6) https://doi.org/10.3390/pharmaceutics12060567.

Ramazani F, Van Nostrum CF, Storm G, Kiessling F, Lammers T, Hennink WE, Kok RJ. Locoregional cancer therapy using polymer-based drug depots. Drug Discov Today. 2016;21(4) https://doi.org/10.1016/j.drudis.2016.02.014.

Raskin P, Gylvin T, Weng W, Chaykin L. Comparison of insulin detemir and insulin glargine using a basal-bolus regimen in a randomized, controlled clinical study in patients with type 2 diabetes. Diabetes Metab Res Rev. 2009;25(6):542–8. https://doi.org/10.1002/dmrr.989.

Riddle MC, Rosenstock J, Gerich J. The treat-to-target trial: randomized addition of glargine or human NPH insulin to oral therapy of type 2 diabetic patients. Diabetes Care. 2003;26(11):3080–6. https://doi.org/10.2337/diacare.26.11.3080.

Roberge C, Cros J-M, Serindoux J, Cagnon M-E, Samuel R, Vrlinic T, Berto P, Rech A, Richard J, Lopez-Noriega A. BEPO®: bioresorbable diblock mPEG-PDLLA and triblock PDLLA-PEG-PDLLA based in situ forming depots with flexible drug delivery kinetics modulation. J Control Release. 2020;319 https://doi.org/10.1016/j.jconrel.2020.01.022.

Robinson D, Woerner MG, Alvir JMJ, Bilder R, Goldman R, Geisler S, Koreen A, Sheitman B, Chakos M, Mayerhoff D, Lieberman JA. Predictors of relapse following response from a first episode of schizophrenia or schizoaffective disorder. Arch Gen Psychiatry. 1999;56(3):241–7. https://doi.org/10.1001/ARCHPSYC.56.3.241.

Robinson DG, Woerner MG, Napolitano B, Patel RC, Sevy SM, Gunduz-Bruce H, Soto-Perello JM, Mendelowitz A, Khadivi A, Miller R, McCormack J, Lorell BS, Lesser ML, Schooler NR, Kane JM. Randomized comparison of olanzapine versus risperidone for the treatment of first-episode schizophrenia: 4-month outcomes. Am J Psychiatry. 2006;163(12):2096–102. https://doi.org/10.1176/AJP.2006.163.12.2096.

Rosenstock J, Davies M, Home PD, Larsen J, Koenen C, Schernthaner G. A randomised, 52-week, treat-to-target trial comparing insulin detemir with insulin glargine when administered as add-on to glucose-lowering drugs in insulin-naive people with type 2 diabetes. Diabetologia. 2008;51(3):408–16. https://doi.org/10.1007/s00125-007-0911-x.

Rosenthal S, Decano AG, Bandali A, Lai D, Malat GE, Bias TE. Oritavancin (Orbactiv) a new-generation lipoglycopeptide for the treatment of acute bacterial skin and skin structure infections. Drug Forecast. 2018;43(3):143–7.

Rossetti P, Porcellati F, Ricci NB, Candeloro P, Cioli P, Bolli GB, Fanelli CG. Different brain responses to hypoglycemia induced by equipotent doses of the long-acting insulin analog detemir and human regular insulin in humans. Diabetes. 2008;57(3):746–56. https://doi.org/10.2337/db07-1433.

Rossi G, Frediani S, Rossi R, Rossi A. Long-acting antipsychotic drugs for the treatment of schizophrenia: use in daily practice from naturalistic observations. BMC Psychiatry. 2012;12(1):1–13. https://doi.org/10.1186/1471-244X-12-122/FIGURES/3.

Rosskamp RH, Park G. Long-acting insulin analogs. Diabetes Care. 1999;22(Suppl 2):B109–13.

Rost A, Hantelmann W, Brosig W. Cyproterone acetate, testosterone, LH, FSH, and prolactin levels in plasma after intramuscular application of cyproterone acetate in patients with prostatic cancer. Prostate. 1981;2(3):315–22.

Ruiz-Sancho A, Núñez-Núñez M, Castelo-Corral L, Martínez-Marcos FJ, Lois-Martínez N, Abdul-Aziz MH, Vinuesa-García D. Dalbavancin as suppressive antibiotic therapy in patients with prosthetic infections: efficacy and safety. Front Pharmacol. 2023;14(June):1–7. https://doi.org/10.3389/fphar.2023.1185602.

Saag MS, Benson CA, Gandhi RT, Hoy JF, Landovitz RJ, Mugavero MJ, Sax PE, Smith DM, Thompson MA, Buchbinder SP, Del Rio C, Eron JJ, Fätkenheuer G, Günthard HF, Molina JM, Jacobsen DM, Volberding PA. Antiretroviral drugs for treatment and prevention of HIV infection in adults: 2018 recommendations of the international antiviral society-USA panel. JAMA. 2018;320(4):379–96. https://doi.org/10.1001/jama.2018.8431.

Sajatovic M, Ross R, Legacy SN, Byerly M, Kane JM, Dibiasi F, Fitzgerald H, Correll CU. Initiating/maintaining long-acting injectable antipsychotics in schizophrenia/schizoaffective or bipolar disorder—expert consensus survey part 2. Neuropsychiatr Dis Treat. 2018a;14:1475. https://doi.org/10.2147/NDT.S167485.

Sajatovic M, Ross R, Legacy SN, Correll CU, Kane JM, Dibiasi F, Fitzgerald H, Byerly M. Identifying patients and clinical scenarios for use of long-acting injectable antipsychotics—expert consensus survey part 1. Neuropsychiatr Dis Treat. 2018b;14:1463. https://doi.org/10.2147/NDT.S167394.

Saltzstein D, Shore ND, Moul JW, Chu F, Conception R, et al. Pharmacokinetic and pharmacodynamic comparison of subcutaneous versus intramuscular leuprolide acetate formulations in male subjects. Ther Adv Urol. 2018;10(2):43–50. https://doi.org/10.1177/1756287217738150.

Salvi V, Cerveri G, Aguglia A, Calò S, Corbo M, Martinotti G, Serafini G, Signorelli MS, Aguglia E, Amore M, Biggio G, Di Sciascio G, Mencacci C. Off-label use of second-generation antipsychotics in bipolar disorder: a survey of Italian psychiatrists. J Psychiatr Pract. 2019;25(4):318–27. https://doi.org/10.1097/PRA.0000000000000405.

Schoepf D, Heun R. Bipolar disorder and comorbidity: increased prevalence and increased relevance of comorbidity for hospital-based mortality during a 12.5-year observation period in general hospital admissions. J Affect Disord. 2014;169:170–8. https://doi.org/10.1016/J.JAD.2014.08.025.

Scoble PJ, Reilly J, Tillotson GS. Real-world use of oritavancin for the treatment of osteomyelitis. Drugs Real World Outcomes. 2020;7(Suppl 1):46–54. https://doi.org/10.1007/s40801-020-00194-8.

Seebaluck J, Downes MA, Brown J, Harris K, Isoardi KZ, Chan BS. Case series profile of olanzapine post-injection delirium/sedation syndrome. Br J Clin Pharmacol. 2023;89(2):903–7. https://doi.org/10.1111/BCP.15588.

Shi Y, Lu A, Wang X, Belhadj Z, Wang J, Zhang Q. A review of existing strategies for designing long-acting parenteral formulations: focus on underlying mechanisms, and future perspectives. Acta Pharm Sin B. 2021;11(8):2396–415. https://doi.org/10.1016/j.apsb.2021.05.002.

Smith KA, Kingdom TT, Gray ST, Poetker DM, Orlandi RR. Drug-eluting implants in chronic rhinosinusitis: an evidence-based review with recommendations. Int Forum Allergy Rhinol. 2020;10(7):856–70. https://doi.org/10.1002/ALR.22565.

Srour M, Querques G, Leveziel N, Zerbib J, Tilleul J, Boulanger-Scemama E, Souied EH. Intravitreal dexamethasone implant (Ozurdex) for macular edema secondary to retinitis pigmentosa. Graefes Arch Clin Exp Ophthalmol. 2013;251(6):1501–6. https://doi.org/10.1007/S00417-012-2249-4/FIGURES/2.

Subotnik KL, Casaus LR, Ventura J, Luo JS, Hellemann GS, Gretchen-Doorly D, Marder S, Nuechterlein KH. Long-acting injectable risperidone for relapse prevention and control of breakthrough symptoms after a recent first episode of schizophrenia: a randomized clinical trial. JAMA Psychiatry. 2015;72(8):822–9. https://doi.org/10.1001/JAMAPSYCHIATRY.2015.0270.

Swinnen SG, Dain M-P, Aronson R, Davies M, Gerstein HC, Pfeiffer AF, Snoek FJ, Devries JH, Hoekstra JB, Holleman F. A 24-week, randomized, treat-to-target trial comparing initiation of insulin glargine once-daily with insulin detemir twice-daily in patients with type 2 diabetes inadequately controlled on oral glucose-lowering drugs. Diabetes Care. 2010;33(6):1176–8. https://doi.org/10.2337/dc09-2294.

Syed YY. Rezafungin: first approval. Drugs. 2023;83(9):833–40. https://doi.org/10.1007/s40265-023-01891-8.

Takada R, Yamamuro K, Kishimoto T. Long-lasting extrapyramidal symptoms after multiple injections of paliperidone palmitate to treat schizophrenia. Neuropsychiatr Dis Treat. 2018;14:2541. https://doi.org/10.2147/NDT.S176478.

Texidor WM, Miller MA, Krsak M, Calvert B, Hart C, Storer M, Fish DN. Oritavancin as sequential therapy for gram-positive bloodstream infections. BMC Infect Dis. 2024;24(127):1–6. https://doi.org/10.1186/s12879-023-08725-8.

Thompson GR, Soriano A, Honore PM, Bassetti M, Cornely OA, Kollef M, Kullberg BJ, Pullman J, Hites M, Fortún J, Horcajada JP, Kotanidou A, Das AF, Sandison T, Aram JA, Vazquez JA, Pappas PG. Efficacy and safety of rezafungin and caspofungin in candidaemia and invasive candidiasis: pooled data from two prospective randomised controlled trials. Lancet Infect Dis. 2023;3099(23):1–10. https://doi.org/10.1016/S1473-3099(23)00551-0.

Trezza C, Ford SL, Spreen W, Pan R, Piscitelli S. Formulation and pharmacology of long-acting cabotegravir. Curr Opin HIV AIDS. 2015;10(4):239–45. https://doi.org/10.1097/COH.0000000000000168.

Van Praag HM, Breetveld J, Van Mesdag-Etty H, Westerhuis R, Pen A, Schut T. A controlled comparative study of fluphenazine and fluphenazine enanthate in acute and chronic psychotic patients. Psychiatr Neurol Neurochir. 1970;73(3):165–75.

Verloes R, Deleu S, Niemeijer N, Crauwels H, Meyvisch P, Williams P. Safety, tolerability and pharmacokinetics of rilpivirine following administration of a long-acting formulation in healthy volunteers. HIV Med. 2015;16(8):477–84. https://doi.org/10.1111/hiv.12247.

Wang C, Swerdloff RS. Hormonal approaches to male contraception. Curr Opin Urol. 2010;20(6):520–4. https://doi.org/10.1097/MOU.0b013e32833f1b4a.

Westphal M, Hilt DC, Bortey E, Delavault P, Olivares R, Warnke PC, Whittle IR, Jääskeläinen J, Ram Z. A phase 3 trial of local chemotherapy with biodegradable carmustine (BCNU) wafers

(Gliadel wafers) in patients with primary malignant glioma. Neuro Oncol. 2003;5(2):79–88. https://doi.org/10.1215/15228517-5-2-79.

WHO. WHO guidelines for the treatment of *Treponema pallidum* (syphilis). World Health Organization; 2016. https://doi.org/10.1136/bmj.2.398.180-b.

Williams PE, Crauwels HM, Basstanie ED. Formulation and pharmacology of long-acting rilpivirine. Curr Opin HIV AIDS. 2015;10(4):233–8. https://doi.org/10.1097/COH.0000000000000164.

Winner B, Peipert JF, Zhao Q, Buckel C, Madden T, Allsworth JE, Secura GM. Effectiveness of long-acting reversible contraception. N Engl J Med. 2012;366(21):1998–2007. https://doi.org/10.1056/nejmoa1110855.

Witika BA, Stander JC, Smith VJ, Walker RB. Nano co-crystal embedded stimuli-responsive hydrogels: a potential approach to treat HIV/AIDS. Pharmaceutics. 2021;13(2):127. https://doi.org/10.3390/pharmaceutics13020127.

World Health Organization. Trial results reveal that long-acting injectable cabotegravir as PrEP is highly effective in preventing HIV acquisition in women. 2020.

Wu L, Janagam DR, Mandrell TD, Johnson JR, Lowe TL. Long-acting injectable hormonal dosage forms for contraception. Pharm Res. 2015;32(7):2180–91. https://doi.org/10.1007/S11095-015-1686-2/TABLES/4.

Wyber R, Taubert K, Marko S, Kaplan EL. Benzathine penicillin G for the management of RHD: concerns about quality and access, and opportunities for intervention and improvement. Glob Heart. 2013;8(3):227–34. https://doi.org/10.1016/j.gheart.2013.08.011.

Wyber R, Johnson TD, Patel B. Supply of benzathine penicillin G: the 20-year experience in Australia. Aust N Z J Public Health. 2015;39(6):506–8. https://doi.org/10.1111/1753-6405.12415.

Zhang H, Zhou W, Wang J, Cai Y. Efficacy and safety of oritavancin for the treatment of acute bacterial skin and skin-structure infections: a systematic review and meta-analysis. J Glob Antimicrob Resist. 2021;25:380–9. https://doi.org/10.1016/j.jgar.2021.04.013.

Zhang L, Ye B, Chen Z, Chen ZS. Progress in the studies on the molecular mechanisms associated with multidrug resistance in cancers. Acta Pharm Sin B. 2023;13(3):982–97. https://doi.org/10.1016/j.apsb.2022.10.002.

Chapter 8
Local Tolerability of Biodegradable Long-Acting Injectables

Esther Sutter, Kuno Wuersch, and Helen Booler

List of Abbreviations

ADME	Absorption, distribution, metabolism, and elimination
AMD	Age-related macular degeneration
API	Active pharmaceutical ingredient
AS	Aqueous vehicle-based rotigotine-crystalline suspension
CK	Muscular creatinine kinase
DARPin	Designed ankyrin repeat proteins
EMA	European Medicines Agency
FBGC	Foreign body giant cell
FBR	Foreign body reaction
FDA	Food and Drug Administration
HED	Human equivalent dose
IND	Investigational new drug application
LAI	Long-acting injectable
LOAEL	Lowest observed adverse effect level
MLKL	Mixed lineage kinase-like
MRHD	Maximum recommend human dose
NDA	New drug application
NLPR3	NLR pyrin domain containing 3
OS	Oily vehicle-based rotigotine-crystalline suspension
PEG	Polyethylene glycol
PK	Pharmacokinetic
PLGA	Poly(lactide-co-glycolic) acid
RIPK3	Receptor-interacting protein kinase 3
TAK1	Transforming growth factor beta-activated kinase 1
TCA	Triamcinolone acetonide

E. Sutter (✉) · K. Wuersch · H. Booler
Preclinical Safety, Novartis Pharma AG, Basel, Switzerland
e-mail: esther.sutter@novartis.com

F. Ramazani (ed.), *Biodegradable Long Acting Injectables and Implants*,
https://doi.org/10.1007/978-3-031-72302-5_8

8.1 Introduction

Biodegradable long-acting injectables (LAI) aim to decrease the patient burden of therapeutics by decreasing administration frequency. The key challenge in development of LAIs is the balance of the required pharmacokinetics (PK) for efficacy with the local tolerability in the targeted tissue compartment.

Local tolerability describes the tissue responses of a drug formulation at and around the site of drug administration. Assessment of local tolerability in preclinical toxicity studies aims to identify any potential safety risk for humans either from the active pharmaceutical ingredient (API) or one of the formulation excipients at the site of highest exposure (EMA 2016). The local microscopic observations at the administration site together with the associated in-life observations and gross findings at necropsy are critical parameters to determine if a drug formulation is considered locally tolerated or not tolerated (ICH M3(R2) 2010).

Drug formulations designed as long-acting injectables are intended to stay for an extended period at their site of injection and therefore a local tissue response, like that observed with foreign bodies, is often unavoidable following injection. Implantation of a foreign material into tissue results in a specific inflammatory tissue response described in literature as foreign body reaction, characterized microscopically by the presence of macrophages and multinucleated giant cells, and often including fibrosis. The aim of the response is to degrade and remove the foreign material and to re-establish the integrity of the tissue, with fibrotic responses observed to “wall off” material that is resistant to degradation.

Many parameters, including the scaffold of LAI formulations used, their chemical and physical properties, as well as the API itself can modulate the injection site reaction. In addition, the character and the significance of a local reaction may also differ between routes of administration depending on the sensitivity and regenerative ability of the affected tissue components. For example, a foreign body reaction occurring in the subcutis secondary to the subcutaneous administration of an LAI may be relatively insignificant, whereas if the same reaction occurred in the eye following an intravitreous injection, it may ultimately cause blindness. Further, dependent on the indication, injection into a diseased tissue may be warranted and associated with specific challenges such as the risk for a formulation-related exacerbation of a pro-inflammatory disease state or already present tissue injury. Finally, in addition to the tissue response defining local tolerability of the drug depot, the inflammatory reaction itself, in particular the infiltration of phagocytosing cells, may also contribute to a sustained degradation of the formulation associated with slow drug release over time (Darville et al. 2016; Kim et al. 2022) and therefore be desired.

This chapter provides an overview of the pathophysiology of foreign body reactions, as one of the most frequent reactions to LAIs, and discusses and provides examples how these reactions are influenced by various properties of the formulation. It specifically focuses on the challenges of injecting long-acting injectables into confined spaces and highlights several aspects which may need to be taken into consideration when determining the local tolerability of long-acting injectables.

8.2 Foreign Body Reaction as Injection Site Reaction to LAIs

A foreign body reaction (FBR) describes a distinct local inflammatory reaction associated with relatively inert foreign material injected or implanted into the body and is as such one of the most frequent injection site reactions associated with LAIs. A detailed description of the process has been outlined in multiple publications (Anderson et al. 2008; Carnicer-Lombarte et al. 2021), and is beyond the scope of this chapter and therefore just briefly summarized.

The foreign body reaction is an expected, pathophysiological reaction and its aim is to (1) remove the foreign material by phagocytosis and degradation, initiate wound healing, and re-establish tissue integrity and (2) "wall off" any remaining material and protect the adjacent tissue. Overall, it consists of complex cell-material and cell-cell interactions and can be separated into several sequential steps, including initial tissue injury by the implantation, protein adsorption, inflammatory cell recruitment and adhesion, and fibrosis and/or fibrous capsule formation (Anderson et al. 2008; Vasconcelos et al. 2022). Both the innate and adaptive immune systems are participating, while macrophages as innate immune cells are the central players involved (Mariani et al. 2019). The initial tissue injury is usually associated with immediate extravasation of blood or plasma proteins and infiltration of acute inflammatory cells. Nonspecific proteins (e.g., albumin, fibrinogen, complement, fibronectin) adsorb within minutes to hours to the surface of the foreign body, described in literature as provisional matrix formation, supporting the adherence of inflammatory cells to the material. Neutrophils adhere first and release different cytokines and mediators, which in concert with other chemical signals increase vascular permeability and direct mononuclear cells to the foreign body. Following the differentiation of monocytes into macrophages and with their further proliferation, they become the key players in the subacute to chronic inflammatory phase. Cytoskeleton remodeling and podosomes enable the macrophages to spread over the surface aiming to engulf and phagocytose the material and individual macrophages may fuse to foreign body giant cells (FBGC) which are considered more effective in phagocytosing large and complement-opsonized materials (Ahmadzadeh et al. 2022; Milde et al. 2015). Although FBGC have a distinct morphology to nonfused macrophages, both cell types release various mediators, such as reactive oxygen species, degradative enzymes, and acid into the space between adhered cells and foreign body for material degradation (Ahmadzadeh et al. 2022). However, it is not well understood if FBGC may also have detrimental effects (Milde et al. 2015). While in the acute phase of the foreign body reaction macrophages have a more pro-inflammatory state, a switch to a phenotype with more anti-inflammatory properties occurs in the later stage directing the inflammatory reaction towards wound healing. During tissue repair, granulation tissue—composed of connective tissue fibers, proliferative fibroblasts, and blood vessels—may be formed and further progress to fibrotic tissue, which replaces the lost parenchymal tissue. In case of a slow or even absent degradation of the foreign body material a formed fibrous capsule separates the foreign body from the surrounding tissue. While in the context of LAIs the generation of a fibrotic capsule itself, dependent on the location, may not be harmful and indicate even a settled immune response, it forms a physical barrier and may impact the drug release as the encapsulation can hinder a progressive degradation of the LAI by phagocytosing cells. Dependent on the severity and extension of the initial tissue injury and the degree of inflammatory and fibrotic tissue response a time-dependent resolution of the tissue injury and full tissue regeneration can occur. A complete tissue repair is however strongly dependent on the regeneration capacity of the affected parenchymal tissue. Therefore, in the context of injected LAIs it may be more likely for intramuscular or subcutaneous administrations, while the risk of a nonreversible functional impact related to incomplete regeneration is considered higher for intraocular or intraarticular administrations of LAI.

It is important to mention that the local inflammation and foreign body reaction associated with injected LAIs is not only determining the local tolerability but may be also key in modulating the drug dissolution and absorption. In studies investigating aqueous nano-/microsuspensions, macrophage infiltration with subsequent phagocytosis has been shown to contribute and promote the prodrug dissolution and conversion to active thereby affecting the plasma exposure levels (Darville et al. 2016).

8.3 LAI Properties Modulating the Foreign Body and Injection Site Reaction

All LAIs which hold firm particles (depot-based) are expected to have the potential to induce a foreign body reaction. However, the severity and characteristic can widely differ. The physico-chemical properties of LAIs, the specific technology, and the formulation composition can substantially shape the immunological reaction and modulate the associated foreign body reaction. Protein adsorption to the foreign body material is a central event and surface properties of the LAI can influence this step. During the time course, the composition of adsorbed proteins can change by detachment of early adsorbed proteins and their replacement by different proteins. This dynamic process is known as the Vroman effect and may lead to differences in the foreign body reaction across varied materials (Carnicer-Lombarte et al. 2021; Wilson et al. 2005). Particle properties, such as size, shape, surface texture, charge (hydrophobicity), and rigidity, can change adhesion, fusion, and cytokine secretion of macrophages and other inflammatory cells and result in a pro-inflammatory or anti-inflammatory phenotype. Not only the formulation composition but also the API itself can locally act as an immunomodulator (either as an on- or off-target effect) and thereby enhance or suppress the immunological response in a more beneficial or detrimental way. Finally, the chemistry of the LAI can also determine the sensitivity to degradation and thereby influence the duration and recovery time of a foreign body reaction. Selected properties of LAIs modulating the immune response and thereby the local tolerability are outlined further in the paragraphs below.

Particle Properties (Size, Shape, Surface Texture, Rigidity)

Physical properties such as size, shape, texture, and rigidity of particles are crucial factors modulating phagocytosis and inflammation (Baranov et al. 2021; Rahmati et al. 2020; Vorselen et al. 2020a). A general conclusion on all individual parameters is however challenging to draw as published study results are not always in agreement, which is likely due to differences in the study design, cell types, and in vivo models used. Further, synergistic effects and interdependencies have been observed (Jain et al. 2019).

Smaller sized particles with otherwise similar physicochemical characteristics seem to elicit a stronger acute inflammatory reaction than larger particles exemplified by nanoparticles (1–100 nm range) that often elicit reactive oxygen species and inflammation (Yang et al. 2021). In an in vitro model with bone marrow-derived macrophages as well as an in vivo model following intratracheal administration to mice, amorphous silica particles of 0.030–1 μm in diameter induced a stronger pro-inflammatory cytokine induction than particles of 3–10 μm in diameter (Kusaka et al. 2014). Besides the size, the shape was also reported to be a main property dictating the inflammatory response of hydroxyapatite particles (Lebre et al. 2017). The assessed particles ranged from 0.1 μm to 100 μm in mean size and were needle or spherical shaped and smooth or rough in their surface topography. In dendritic cells, needle-shaped particles showed the most potent IL1β induction followed by the smallest spherical particles assessed, while in macrophages the IL1β induction was even absent for any of the assessed spherical particles. Intraperitoneal injection in mice confirmed needle-shaped hydroxyapatite particles inducing a significantly higher inflammatory response compared to their spherical counterparts.

While the surface topography seemed to have no significant impact on the immune activation for hydroxyapatite particles (Lebre et al. 2017) other studies reported the surface texture being a major determinant for inflammasome activation. Polymeric microparticles with similar overall sizes and surface chemistries but having either a smooth or highly microtextured surface were compared in macrophage assays and mouse peritonitis studies. Both in vitro and in vivo assessments revealed that microparticles with an irregular surface texture are more rapidly phagocytosed and induce a stronger NLPR3 (NLR pyrin domain containing 3) activation and IL1β secretion (Vaine et al. 2013).

Multiple studies have also reported the relevance of rigidity for phagocytosis and immune activation of particles. One of the earliest studies investigated the phagocytosis of IgG (immunoglobulin G)-opsonized polyacrylamide microbeads of different stiffness and reported the preference of macrophages for particles with a higher rigidity (Beningo and Wang 2002). In agreement with these findings, phagocytosing engulfment was shown to be strongly rigidity-dependent for deformable hydrogel microparticles (Vorselen et al. 2020b)

Particle size–dependent differences in the inflammatory reaction and foreign body response are also appreciated in histopathological assessments in vivo. Three different crystal sizes with the same formulation were injected subacromial into rat and assessed at different time points following a single administration (unpublished data) and differences were especially observed in the subacute to chronic phase. Four weeks post-dose uniform, dense, and well-circumscribed macrophage infiltrations with no or only little surrounding fibrosis were observed with a nanosuspension and a microcrystal suspension with crystal size distribution of D90 < 3 μm. In contrast, with a microcrystal suspension with crystal size distribution of D90 < 20 μm, a granulomatous inflammation with numerous multinucleated giant cells, macrophages and fewer lymphocytes associated with fibrosis was present (Fig. 8.1). This morphological difference reflects the expected ability of macrophages to phagocytose smaller particles more easily, which may also result more likely in full reversibility of tissue injury, when compared to larger particles. Notably, the inflammatory reactions were more widespread between the muscular and subcutaneous tissue with the nanosuspension and the smaller crystals than with the larger particles, which may be related to a difference in viscosity between the suspensions. The spread and distribution of drug particles along tissue structure may vary depending on the viscosity and affinity of the delivery vehicle with interstitial fluid and tissues (Kim et al. 2022). Similar observations as described above have also been reported following intramuscular administration of globular shaped entecavir-3-palmitate particles of different median diameters into rats (Ho et al. 2022).

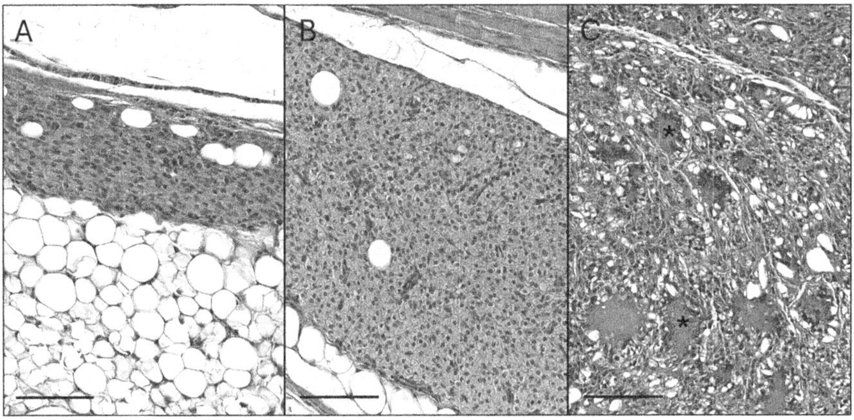

Fig. 8.1 Microscopic differences in the local inflammatory reaction related to crystal size. Four weeks post-dose a single subacromial injection containing a nanosuspension (**a**) and a small microcrystal suspension (**b**) resulted in a dense infiltration of uniform macrophages. A suspension with larger microcrystals (**c**) resulted in a granulomatous inflammation with multinucleated giant cells (*), macrophages, lymphocytes, and associated fibrosis. Scale bar: 100 μm

While for crystal- and biomaterial-induced foreign body responses the NLRP3 inflammasome activation is reported to be key (Franklin et al. 2016; Vasconcelos et al. 2022), the pathways leading to cell death are not fully unravelled. The shape of crystals may be a confounding factor as several studies have reported needle-shaped particles decreasing cell viability compared to other particle shapes (Lebre et al. 2017; Stoehr et al. 2011) and mechanical injury of the cell membrane has been suggested as mechanism. Cytotoxicity may however not be only a consequence of passive, mechanical injury or due to release of reactive oxidative species during phagocytosis, but also a regulated event. Crystal cytotoxicity has been reported to involve receptor-interacting protein kinase 3 (RIPK3) and mixed lineage kinase-like (MLKL) mediated necroptosis as a form of regulated cell death in tubular epithelial cells (Mulay et al. 2016). Further, when investigating a broad range of environmental and metabolic crystals, such as silica, cholesterol, and monosodium urate, in human and murine renal tubular cells, necroptosis was confirmed as one of the pathways leading to cell death, irrespective of crystal size and shape (Honarpisheh et al. 2017).

Soluble LAI Technologies

LAIs do not necessarily need to be depot-based. In some instances, soluble, polymer-based structures have been explored, created through the conjugation of high molecular weight polymer chains to the drug. These are of particular interest for local routes of administration such as the eye (intravitreal), where foreign body-type reactions are considered adverse, and have the potential to cause blindness. In the eye, these formats primarily serve to extend the ocular half-life by increasing the hydrodynamic radius of the molecule (the main driver of vitreal clearance) (Crowell et al. 2019).

Polyethylene glycol (PEG) is the most investigated polymer for drug molecule conjugation, with various molecular weights and formats, including linear, branched and dendrimer, which differentially alter the molecule's PK properties (Li et al. 2013). The first approved intravitreal anti-VEGF drug was a PEG-conjugated aptamer pegaptanib sodium (Macugen®) and was relatively well tolerated clinically but withdrawn from the market, after the approval of more efficacious competitors.

Although soluble formats mitigate the risk of foreign body formation, issues of local tolerability in some cases remain. Abicipar pegol, a PEGylated anti-VEGF designed ankyrin repeat proteins (DARPin) indicated for age-related macular degeneration (AMD), exhibited intraocular inflammation in up to 34% patients in phase I/II clinical trials (Souied et al. 2014). Similarly, in the phase III SEQUOIA and CEDAR trials, the rates of incidence of intraocular inflammation were significantly higher in the abicipar groups compared to ranibizumab (a non-PEGylated antibody fragment, and the standard of care), with approximately 15% patients treated with abicipar pegol developing intraocular inflammation, compared to less than 0.6% in the ranibizumab arms. Over the course of treatment, 56% patients developed systemic anti-drug antibodies, although a definitive relationship between the anti-abicipar antibodies and safety and/or efficacy data was not established (Smith and Kaiser 2014).

The ability of some soluble polymers to serve as adjuvants, increasing the immunogenicity of the therapeutic, is something that is exploited in the development of vaccines. Unfortunately, this can be an undesired and unintended consequence of conjugation for the extension of half-life. Particularly in the eye, where the consequences of immunogenicity and subsequent ocular inflammation are both readily observed, of significant concern, an increase in the propensity for ocular inflammation may remove any potential benefit of extending the treatment interval for patients.

Although challenging, with the right soluble polymer, extension of vitreal half-life without an increase in immunogenicity or ocular inflammation does appear to be possible. KSI-301 is a long-acting anti-VEGF antibody currently undergoing clinical trials for wet-AMD. It is a branched, high molecular weight (800 kDa) phosphorylcholine-containing polymer and was well tolerated in wet AMD patients with no impact on the incidence of ocular inflammation (Chandrasekaran and Madanagopalan 2021).

Excipients of LAI Drug Formulations

Excipients and the type of dispersion medium have shown to significantly determine the local tissue reaction and their tolerability may vary significantly with the route of administration, and with some routes (such as intravitreal) additionally between species (Andaya et al. 2021). Importantly, excipients are also a key factor influencing the distribution in the tissue, aggregation, viscosity, and dissolution of the drug, and with this also the PK profile. As such, solvents are used for lipid-based LAIs to reduce the viscosity of the formulation, by inhibiting crystallization and preventing lipid solidification and to ensure the formulation is injectable (Wilkinson et al. 2022).

Solvents which have been investigated for parental administration of LAIs are for example N-methyl-2-pyrrolidone, dimethyl sulfoxide, and 2-pyrrolidine. In vitro assays and in vivo tolerability studies measuring the release of muscular creatinine kinase (CK) following intramuscular injection have shown a high myotoxic effect when comparing to saline injection (Kranz et al. 2001). However, the toxicity was reduced when the solvents were administered in context of an oil-based formulation (Kranz et al. 2001) emphasizing that the final composition may determine the local tissue response and tolerability of a drug formulation.

For rotigotine, a non-ergot dopamine agonist prescribed to manage early Parkinson's disease, aqueous (AS) and oily (OS) vehicle-based rotigotine-crystalline suspensions were compared in vivo following subcutaneous injection to naive rats (Kim et al. 2022). The physical properties and particle sizes were comparable between the two suspensions and both suspensions exhibited a comparable rectangular or hexagonal crystalline shape. Subcutaneous injection resulted in a noticeable difference in local tissue reaction between these two vehicles. The AS resulted in agglomeration of drug crystals associated with a marked foreign body reaction characterized by a time-dependent infiltration of neutrophils and macrophages and a fibrotic encapsulation of the material. In contrast, with the OS there was no agglomeration of drug crystals, and no inflammatory or fibrotic reaction have been described. Importantly, the two vehicles had also differed PK profiles with the OS being associated with a rapid decrease of the plasma level and the AS exhibiting a more sustained plasma level concentration while both had a similar AUC. With the intramacrophage site having been previously suggested to be a primary site of drug dissolution and release (Darville et al. 2016), it is reasonable to argue that the foreign body reaction associated with the AS modified the degradation and PK properties resulting in an extended systemic exposure, while the oily vehicle maintained a uniform crystal distribution at the injection site resulting in a faster drug dissolution and absorption into the blood stream (Kim et al. 2022). This example illustrates the interplay between the local tissue reaction and PK of LAIs and the potential challenge to balance the requirements for both characteristics.

A modulation of local inflammatory changes has also been reported following in vivo assessment of different nanosuspension stabilizing excipients (Chamanza et al. 2018). Following single intramuscular administration of an antiviral drug formulated with five different stabilizing excipients in rat, all five stabilizers resulted in a similar local inflammatory reaction. Only subtle differences were described but supported the conclusion that nature and dynamics of local inflammatory changes may be modulated by the type of stabilizing excipients (Chamanza et al. 2018).

Active Pharmaceutical Ingredients

Although the physicochemical properties of long-acting injectable drug delivery formulations are crucial parameters driving the local tissue response, the active pharmaceutical ingredient (API) and its pharmacological effects can also modulate the local tissue reaction and tolerability.

Preclinical data supporting an API-associated modulation of the local tolerability was observed for an intraarticular LAI inhibiting transforming growth factor beta-activated kinase 1 (TAK1) (publication in progress). The TAK1 inhibitor was initially formulated as an aqueous suspension with needle-shaped microcrystals and was shown to significantly reduce synovial fluid cytokine levels in an acute joint inflammation rat model. When subsequently assessing local tolerability in a single-dose intraarticular tolerability study in healthy rats, a pronounced inflammatory response of the synovial tissue was observed. This was characterized by an acute infiltration of neutrophils, monocytes, and synovial necrosis one day post-dose, while progressing to a granulomatous inflammation eight days post-dose, as evidenced by aggregates of macrophages, single multinucleated giant cells, and peripheral infiltrating lymphocytes. Although the morphology of the inflammation was consistent with the expected foreign body reaction related to microcrystals, the high severity grade and the additional strong proliferation of undefined stromal cells arranged in bundles and whorls was unexpected. The crystal shape was considered a possible confounding factor as studies have reported pronounced activation of the NLRP3 inflammasome by needle-shaped particles (Lebre et al. 2017). When comparing similar aqueous suspensions with needle or rhomboid shaped crystals, no significant improvement in the pattern or the severity grade of the local inflammatory response was observed, suggesting that the crystal shape had no significant impact. A partial improvement was achieved when the needle-shaped microcrystals were suspended in a lipid in situ forming gel, which reduced the synovial inflammation at both timepoints, one- and eight-days post-dose. However, with the drug-loaded lipid gel formulation, a marked exacerbation of synovial hyperplasia was observed as compared to the placebo lipid gel, suggestive of a proliferative response to the injected formulation. A possible explanation for the reduction in inflammation might be that the lipid gel acted as a lubricant, enveloping the suspended microcrystals, and thereby dampening the mechanical irritation and injury of the synovia. This limited the translocation of microcrystals into the tissue and resulted in a less severe foreign body related inflammatory response. However, the strong proliferative aspects observed with both formulations suggested this was not a typical foreign body reaction, but rather an additional API-related on- or off-target effect by the TAK1 inhibitor molecule. Overall, the intraarticular tolerability profile for this molecule was considered unacceptable for either of the tested long-acting formulations.

Zilretta™, triamcinolone acetonide (TCA) formulated as extended-release formulation in poly(lactide-co-glycolic) acid (PLGA) microspheres, was the first approved long-acting intra-articularly administered analgesic for the management of osteoarthritis knee pain. Intra-articularly administered corticosteroids typically have a short residence time, thereby limiting the duration of any analgesic and/or anti-inflammatory effects and may lead to systemic metabolic reactions. Zilretta™ enables slow drug release and sustained joint exposure, thereby also lowering the systemic exposure and mitigating potential systemic adverse reactions. However, the API of Zilretta™ has been shown to have a detrimental effect on cartilage proteoglycan content as evidenced by a loss of Sarafin O staining (Bodick et al. 2018). In terms of cartilage homeostasis, corticosteroids are a double-edged sword, with both concentration-dependent beneficial effects and detrimental and degenerative effects following repeated intraarticular injection (Fubini et al. 2001; Wernecke et al. 2015). Proteoglycans are crucial for the cartilage biomechanical properties, homeostasis, and repair (Alcaide-Ruggiero et al. 2023) and a decrease in their concentration is an undesired on-target effect of TCA. Observed also with the approved crystalline-based formulation Kenalog®-40, this finding appeared to be more severe with the PLGA-based formulation Zilretta™ which was considered related to the substantial cumulative API exposure (Bodick et al. 2018). Further preclinical findings and studies conducted for approval of Zilretta™ are provided in more detail in Sect. 8.4 on intraarticular administration.

8.4 Administration Routes: Regulatory Requirements and Challenges

This section focuses on the regulatory and toxicologic requirements related to the route of administration and exemplifies on intraarticular administration. Delivery of a drug to a confined space such as a joint cavity or the ocular globe request different considerations regarding the toxicologic assessment compared to oral or intravenous administration. A review of the literature on available cross-species information of knee joint parameters is presented and the nonclinical program for the approval of Zilretta™ discussed as an example for a long-acting corticosteroid formulation administered intraarticularly.

The Food and Drug Administration (FDA) Drug Nomenclature Monograph "Route of Administration" (C-DRG-00301) lists more than 100 routes of administration from auricular (otic) to vaginal. However, based on a 2018 publication (Zhong et al. 2018), oral administration takes the lion share with about 62% of all FDA-approved drugs, followed by injection (22%), cutaneous (9%), mucosal (5%), inhalation (1%), and other (less than 1%) administrations.

While a regulatory guidance specific for locally administered LAIs is lacking several other guidelines provide recommendations for local tolerability testing in general which are applicable for LAI drug development as well.

The International Council for Harmonisation of Technical Requirements for Registration of Pharmaceuticals for Human Use (ICH) M3 guideline recommends testing the local tolerability by the intended therapeutic route of administration as part of the general toxicity study and not as a stand-alone study (ICH M3(R2) 2010). This paradigm in the spirit of 3R is also supported by the European Medicines Agency (EMA) guideline on nonclinical local tolerance testing of medicinal products (EMA 2016).

The ICH M3 guideline also covers the assessment of unintended injection sites such as paravenous administration for IV drugs and states that this should be assessed prior to exposure of large numbers of patients (phase 3 clinical trials). But there are regional differences and FDA usually does not require such studies while PMDA and EMA recommend single paravenous administration assessment for intended IV route of administration.

The FDA good review practice "nonclinical safety evaluation of reformulated drug products and products intended for administration by an alternate route" (FDA 2015) is a guideline for the nonclinical evaluation of previously approved drug substances when a new formulation or a new route of administration is proposed. In other words, if the drug substance has not yet been approved, the nonclinical safety assessment should follow the general recommendations in ICH M3. Oftentimes reformulation and a new route of administration are coinciding and need to be tested in nonclinical safety studies for potential new toxicities.

All routes of administration can lead to systemic exposure and therefore, it is important to consider systemic toxicity and existing data should be evaluated. It is also important to evaluate differences in pharmacokinetics and absorption, distribution, metabolism, and elimination (PK/ADME) of the drug substance related to the route of administration as well as the formulation. Reformulations can also have new toxic effects even if the route of administration remains the same because of new combination of active and inactive ingredients or due to higher local exposures. If systemic exposure by a new route of administration is equivalent or lower compared to the approved route, then the histological evaluation may be limited to the locally exposed tissue.

Finally, the FDA guideline also lists extended release injected or implanted formulations as a specific category and the length of the study should cover the expected duration of the extended release. Furthermore, the fate of the material associated with the formulation (solid material from an implant for example) should also be determined.

There is also an excellent review on less common administration routes including the advantages and disadvantages in different species regarding efficacy a toxicology assessment (Emami et al. 2018).

In contrast to intramuscular or subcutaneous administration, confined tissue spaces such as the synovial cavity or the globe of the eye have different requirements to appropriately calculate dose volume and allometric scaling is important for nonclinical tolerability assessment such as safety margin or exposure multiple calculations. For test articles injected intravitreally into the eye, vitreous volume is used to scale margin/exposure (Pruimboom-Brees et al. 2023). The knee joint is usually selected preclinically for intraarticular injections and several good publications on joint parameters exist. However, there is a general lack of a standardized methodology for morphometry, and publications differ in their completeness of the assessed endpoints and anatomic structures. Furthermore, cross-species comparison is impeded by the heterogeneity of publications with regard to the number of subjects studied, differences in age or disease status adding to the cross-species differences related to anatomic and physiological differences such as tissue composition, range of motion, and weight bearing and moving patterns, and an absence of correlation between body weight and cartilage volume in human in contrast to dogs (Dhollander et al. 2017). Many different experimental methods for the determination of synovial fluid volume have been reported using instead of radioactive tracers for instance albumin (Geborek et al. 1988), hydroxyethyl starch (Heilmann et al. 2008), calcium (Matsuzaka et al. 2002), or urea (Kraus et al. 2007). Most data have been published on human, dog, and rabbit knee joints and selected data is presented in below table (Table 8.1). As an example, synovial fluid volume has been reported for human and rabbits but is less clearly defined for dogs. For this species, there is often reference made to general veterinary practice or publications extrapolating the volume based on scaling factors derived from other endpoints, such as cartilage or synovial membrane surface areas. While for the initial concentration of a test article in the synovial cavity, the synovial fluid volume is important, it is less clear whether the synovial membrane as active tissue in phagocytosing and processing of the test article should be used for dose administration scaling and comparison. This consideration is even more important for LAIs, which are likely to be phagocytosed by the synoviocytes.

Table 8.1 Overview on selected published knee joint parameters across species

Species	Synovial fluid volume	Synovial membrane volume	Synovial membrane area	Cartilage surface	Cartilage thickness	Cartilage volume
Human	131 ± 53 mL (40–290 mL)[a*]	8327 ± 2079 mm^3[b]	20314 ± 522 mm^2[b]	121 ± 14.1 cm^2[c] (total cartilage)	1.95 ± 0.26 mm[c] (total cartilage)	23.3 ± 3.8 mL[c] (total cartilage)
	6.7 ± 2.3 mL (4.3–12.3 mL)[d]			14.2 ± 2 cm^2[c] (tibial cartilage)	1.62 ± 0.23 mm[c] (tibial cartilage)	2.48 ± 0.54 mL[c] (tibial cartilage)
	13.6 ± 7.4 mL[d*]			1215 ± 170 mm^2[e] (tibial cartilage)	2.1 ± 0.25 mm[e] (tibial cartilage)	2500 ± 480 mm^3[e] (tibial cartilage)
	3.05 mL (0.56–71.7 mL)[f*]					
Beagle dog		5585 ± 620 mm^3[b]	9552 ± 953 mm^2[b]	427–757 mm^2[g] (total cartilage)	1.75–2.5 mm[g] (total cartilage)	320–647 mm^3[g] (total cartilage)
						1080 ± 70 mm^3[h*] (total cartilage)
Mongrel dog				196 ± 18 mm^2[e] (tibial cartilage)	0.77 ± 0.04 mm[e] (tibial cartilage)	142 ± 15 mm^3[e] (tibial cartilage)
Minipig		5445 ± 1162 mm^3[b]	9389 ± 1213 mm^2[b]			
Rabbit	133 ± 20 μL[i]			69 ± 4.9 mm^2[e] (tibial cartilage)	0.37 ± 0.05 mm[e] (tibial cartilage)	23.5 ± 3.9 mm^3[e] (tibial cartilage)
	390 ± 86 μL[j]					
Rat				6.2 ± 0.9 mm^2[e] (tibial cartilage)	0.14 ± 0.12 mm[e] (tibial cartilage)	0.81 ± 0.17 mm^3[e] (tibial cartilage)

[a]Matziolis et al. (2015), [b]Hurtig et al. (2016), [c]Eckstein et al. (2001), [d]Heilmann et al. (2008), [e]Wehr et al. (2007), [f]Kraus et al. (2007), [g]Dhollander et al. (2017), [h]Boileau et al. (2008), [i]Delecrin et al. (1992), [j]Heilmann et al. unpublished (referenced in Heilmann et al. 2008), *Values of diseased joints

Related to the clinical importance of the cartilage condition in human health, data on cartilage surface area, volume, and thickness have been published for beagle dogs including correlations with body weight (Dhollander et al. 2017). On the other hand, other reports compare the synovial membrane area and volume across species and explicitly point out its importance for uptake and degradation of biomaterials (Hurtig et al. 2016). Complicating fact is that published measurements in the public domain can differ significantly, which can be related to various aspects as mentioned above including the methodology, age, breed, or disease status. As an example for knee joints, Wehr et al. report for mongrel dogs of about 25 kg a mean cartilage thickness of 0.77 mm at the medial tibia (Wehr et al. 2007); in contrast, Dhollander et al. report for beagle dogs (weight: 7.2–17.1 kg) a thickness ranging from 1.75 to 2.5 mm over the entire joint cartilage (Dhollander et al. 2017). The availability of high-quality data on the various structures of the knee joint is of paramount interest not only for PK modeling but also for toxicity risk assessment and determination of exposure multiples and safety margins between the nonclinical species and humans. This challenge is also exemplified by the various exposure multiple calculations for Zilretta comparing dog versus human PLGA-formulated corticosteroid (Bodick et al. 2018).

For the calculated exposure multiples between canine and human equivalent dose (HED), Bodick et al. referenced a dog knee joint surface area of 6300 mm^2 (Bodick et al. 2018), which was derived by allometric scaling by body surface area from rabbit data published by Davies and Lond (1946). This seems significantly higher compared to the reported data by Dhollander et al. (2017). Some of this difference may be constituted in the fact that joint surface reported by Davies and Lond (1946) may also include the surface of the synovia in contrast to the more limited surface of the cartilage. The appropriateness of allometric scaling from rabbit to dog has not been demonstrated to the best of our knowledge. The limited regulatory and toxicological experience in the development of intraarticular medicines make the assessment even more challenging for LAIs with their extended local concentration. However, approved intraarticular drugs such as Zilretta™ can provide some guidance on the nonclinical data that is required and how to best assess exposure relationship to humans. Furthermore, they can exemplify potential issues related to local drug administration as well as extended-release formulation and provide mitigation strategies.

The preclinical package used for the approval of Zilretta™ has been published (Bodick et al. 2018) and is also available on the FDA drug database (NDA 208845 2017) including comments by the FDA reviewers. The corticosteroid triamcinolone acetonide (TCA) had been previously approved by the FDA for intramuscular and intraarticular use as a crystalline suspension releasing TCA for several days (Kenalog®-40). PLGA microspheres had been used for extended-release formulations for intramuscular or subcutaneous administration. However, as Zilretta™ was using PLGA intraarticular, this was a new route of administration and required a nonclinical safety assessment to qualify the excipient and to conduct clinical trials and seek market approval. For the new drug application (NDA), a total of three nonclinical toxicity studies in dogs have been submitted—two single dose and one repeat-dose study. Dosing groups were administered with PLGA-based extended-release formulation (FX006), the crystalline suspension (active comparator—Kenalog®-40), or PLGA-blank vehicle. The contralateral knees were used as negative controls. Although repeat-dose studies have been conducted, the application sought approval for single-use indication and, therefore, the FDA considered the single dose studies as the pivotal studies. For the previous IND submission (IND 111325), only the two single dose studies have been submitted. All studies included as the highest technical feasible dose 75 mg FX006 corresponding to 18.75 mg of

TCA. In the first study with the last necropsy on day 43, granulomatous inflammation in the synovial membrane including macrophages and single multinucleated giant cells with cytoplasmic microspheres had been observed up to 6 weeks post injection. A second single dose study assessed recovery at 3, 4, 6, and 9 months. While multinucleated giant cells were still present at the 3- and 4-month recovery time points, these cells were not present anymore at the 6-month time point suggesting reversibility of the granulomatous inflammation. During the IND submission, the latter study was reviewed by FDA based on a draft report including data up to 6 months but lacking 9-month and toxicokinetic data. Based on the additional recovery data, a low safety ratio could be defined. The ratio was calculated based on a human knee joint volume of 1–3 mL and a canine knee joint volume of 0.25–1 mL. The maximum dose of FX006 at TCA doses was assumed to be 60 mg in humans and 18.75 mg in dogs resulting in a concentration range in the joint of 20–60 mg/mL and 18.75–75 mg/mL in humans and dogs, respectively. At the time of the IND review, a final safety ratio range of about 1 to 3-fold has been derived. During the review of NDA, the local safety margin was calculated as provided (Table 8.2). The human dose of TCA was defined as the maximum recommend human dose (MRHD) of 32 mg while the maximum canine dose of 18.75 mg was defined as the lowest observed adverse effect level (LOAEL). Zilretta™ did not exceed the systemic exposure compared to Kenalog®-40. It is of interest to note that both the human and canine synovial volumes were defined differently during IND versus NDA review. The human synovial volume used during the NDA review corresponds well with the one reported by Heilmann (Heilmann et al. 2008). It is less clear how the slightly larger canine synovial volume compared to the IND review had been determined. On the other hand, it is important to distinguish between the synovial volume and the injected volume. For the toxicity studies, 1 mL has been used for injection based on veterinary best practice and as reported by Getgood (Getgood et al. 2019). The synovial volume for concentration calculations needs to be the sum of the injected volume in addition to the residual volume. Therefore, the volume 1.4 mL is a more reasonable estimation for the immediate concentration after administration compared to the lower volumes used during the IND review.

The microscopic assessment of the knee joint was done based on sagittal sections and included a modified Mankin score. As for systemic nonclinical toxicity studies sagittal orientation is preferred for joints of intraarticular studies. Processing is usually enhanced by multiple sections covering lateral and medial aspects of the joint and including the recessus for examination, as LAI formulations may accumulate at

Table 8.2 Calculation of the safety margin for Zilretta™ (NDA)

Species	Injected dose of TCA (mg)	Synovial volume (mL)	Synovial concentration (mg/mL)
Human	32	7	4.6
Dog	18.75	1.4	13.4
Safety margin	n.a.	n.a.	13.4/4.6 = 2.9×

this specific location. The nonclinical studies also investigated if microspheres would be detectable distal of the administration site via lymphoid transportation (popliteal and other lymph nodes); however, this has not been reported. The studies have also included assessment of systemic exposure by blood TK as well as microscopic evaluation of a standard tissue list.

8.5 Considerations for Determining Local Tolerability of LAIs

In nonclinical in vivo studies, the in-life observations and the macroscopic and microscopic findings at the administration sites are key when assessing local tolerability for long-acting drug formulations. According to the EMA guideline on local tolerance testing the overall evaluation of the local tolerability results should include the significance of the findings for the clinical use of the product (EMA 2016). This is a challenge and requires a careful assessment of the benefit–risk ratio. For the interpretation of the significance of nonclinical findings multiple aspects may need to be taken into consideration, of which some are highlighted as follows.

Adversity and Route of Administration

The two terms "adverse" and "nonadverse" are key in the interpretation of findings in nonclinical toxicology studies. In general, "adverse" is a term indicating harm to the test species while "nonadverse" indicates the lack of harm (Kerlin et al. 2016). However, there are variable opinions and recommendations on the more detailed criteria for definition of these two terms and even a reversible change can be considered adverse. In nonclinical toxicity studies, the presence or absence of adverse findings determine the nonadverse-effect level (NOAEL), which is an important threshold level for the estimation of the first dose in humans (ICH M3(R2) 2010). Microscopic findings often associated with adversity include inflammation, necrosis, degeneration, and hemorrhage, and these are as well findings which are frequently observed at the local administration site of LAIs. The clinical significance of such findings depends not only on the severity grade but also on the location and the affected cells or tissue structures, and any potential impairment of an organ function. The adversity of a specific finding will vary by the location at which it is observed, for example, significant ocular inflammation or foreign body reactions within the eye would be considered an adverse finding for intraocular LAIs, as there would be impairment or loss of vision. However, similar findings in the subcutis may be acceptable (causing no functional impact). Tolerability can also depend on specific anatomic structures and may be subject to species differences. Intravitreal toxicity studies with PLGA microspheres of different diameters (20, 50, or 100 μm) all induced a similar foreign body responses in rabbits. However, the smallest microspheres tested were the least tolerated particles inducing the greatest toxicity related to the easier penetration into the retina resulting in retinal degeneration (Thackaberry et al. 2017). However, with particulate formulations in the rabbit, the particles do generally stay within vitreous. Similar formulation, when administered to non-human primates, exhibit migration of particles into the anterior chamber—a finding that can be adverse not only because of its direct impact on vision, but additionally because microparticles can obstruct the drainage of aqueous leading increased intraocular pressure (Adamson et al. 2016). In joints, such as the knee-joint, articular cartilage is a key component and has fundamental roles in the joint function and body movement. In rodents as well as rabbits, articular cartilage has a certain endogenous regenerative potential, while like humans, dogs, pigs and non-human primates have only limited or no endogenous repair of chondral defects. Any degenerative change of the articular cartilage, as a crucial component for joint movement, may therefore be as well considered adverse for intraarticular LAIs. In addition, inflammatory reactions by the synovial membrane or ligaments related to LAI may have a higher clinical significance compared to the same LAI administered intramuscular or subcutaneous. Following subcutaneous and intramuscular LAI administration, the overall affected area of the entire skin or skeletal muscle tissue is rather small, the tissue reactions are focal, and the tissue itself has a good regenerative capacity. Although local inflammation, initial necrosis, degeneration, and possible fibrosis are frequent observations with LAIs, the function of the skin, or the skeletal muscle, may not be impacted and such findings should therefore not be considered per se adverse.

Acute Local Tissue Reactions of LAIs: Comparison to Vaccines

For intramuscular vaccines, the threshold for considering local tissue reactions adverse is high, which may support to benchmark acute local reactions to intramuscular or subcutaneous LAIs in a similar way. Most vaccines are administered intramuscular, besides other less commonly used administration routes. As for LAIs, intramuscular administered vaccines can result in marked acute tissue responses. Most vaccines result in some degree of inflammation, muscle degeneration or necrosis, edema, hemorrhage, and/or fibrosis dependent on the timing of necropsy after administration, and the inflammation can be very prominent, especially when formulated with immune activators (Sellers et al. 2020). Microscopic findings with marked to severe gradings are generally considered nonadverse, when unaccompanied by clinical signs (other than transient elevations in body temperature and redness/swelling at the injection site) and with evidence of reversibility (Sellers et al. 2020). However, ulceration of the skin, severe edema or abscessation, extensive necrosis, and lesion progression are considered findings which may need additional efforts and discussions to determine adversity (Sellers et al. 2020). Strong injection-site inflammation and edema have also been reported preclinically with

Pfizer-BioNTech COVID-19 vaccines, but which were considered as well tolerable and nonadverse (Rohde et al. 2023). While the acute tissue responses may have similarities between vaccines and LAIs, vaccines are not designed to form a local depot and show therefore usually a fast resolution, which is a major difference to LAIs.

Reversibility of Local Tissue Reactions

Long-acting injectable drug formulations are designed to remain at the injection site for an extended duration, resulting in local tissue reactions that may last for weeks or even several months. In line with the ICH M3 guideline, assessment of reversibility should be included for LAIs (ICH M3(R2) 2010). Importantly, the foreign body reaction, as one of the most frequent tissue responses to LAIs, is a normal, expected, and time-dependent biological process with the goal to remove the foreign material and to re-establishment tissue integrity. As such, a progression of the acute LAI-related tissue response towards reversibility and tissue repair can be expected. However, microscopic observations indicating an unusual progression of the tissue response and/or and impaired tissue repair can be of concern and should be put further into perspective to the administration site as well as the human risk assessment. For instance, a persistent acute inflammatory phase failing to progress into a macrophage driven chronic response or an extensive fibrosis replacing vast areas of parenchymal tissue may be considered a concern, independent of the administration site. In case of depots of poorly degradable LAIs the material may become encapsulated forming a fibrotic nodule. While this may be less of a concern in the subcutaneous space when dependent on the nodule size and thickness of the fibrotic capsule, it can be considered not acceptable for the intravitreal or intraarticular space. Remaining minimal, focal fibrosis in the subcutaneous or intramuscular tissue will likely have no functional impact, and therefore may be regarded as of no concern. In contrast, fibrosis of the synovia in joints has been shown to contribute to joint stiffness and pain within osteoarthritis patients (Zhang et al. 2021) and may therefore be considered a nontolerable finding.

Targeted Tissue Compartment: Diseased Versus Healthy Tissue

LAIs intended for intraocular or intraarticular administration present an additional layer of complexity as they are injected into diseased tissue. In both the eye and the joint, there are typically preexisting degenerative and/or inflammatory disease-related changes specific to the indication, and local reactions related to the LAI may exacerbate these conditions. Of note, not only the tolerability profile but also the local drug exposure may change, as diseased tissue compared to healthy tissue may for instance have a different volume (joint) or a differing cellular composition influencing the pharmacokinetics. From a regulatory perspective, safety and tolerability assessment of LAIs is only required in healthy animals. While animal disease models can have certain limitations and may not precisely represent the human disease state, including and assessing tolerability endpoints in preclinical efficacy studies can provide valuable information for the overall human risk assessment. Consequently, a higher tolerability profile may be required for LAIs injected into diseased tissue compared to LAIs intended for subcutaneous or intramuscular administration.

8.6 Conclusion

Despite technical and regulatory challenges, there is a great interest in LAI in general and for confined tissue spaces in particular. There are multiple reasons for this including patient adherence to therapy, cost-effective treatments, and a reduction of harmful events by repeated administration to sensitive tissues.

Long-acting injectables are drug formulations specifically designed to remain at the injection site for an extended duration, leading to a prolonged local tissue reaction. The overall tolerability assessment of LAI-associated local tissue reactions is challenging and there is limited published literature and lack of specific regulatory guidelines. To determine the clinical significance of nonclinical findings and assess the overall local tolerability, it is crucial to put these findings in perspective to other aspects. These include for instance the route of administration, potential organ function impairment, the reversibility of local tissue responses, as well as a comprehensive human risk assessment that considers the indication, patient population, and benefit–risk ratio. Dependent on the route of administration additional challenges arise especially if the route of administration is a confined tissue space such as the intraarticular or intravitreal. While it is a pivotal aim of nonclinical toxicity studies to define toxicity thresholds and derive exposure multiples or safety ratios, the cross-species scaling for these tissue compartments is less well defined compared to intramuscular or subcutaneous routes. Similarly, the interpretation of tissue reactions toward the LAI is more challenging as articular and ocular structures are composited of different cell types manifesting in the specific tissue structures such as articular cartilage or retina. Finally, the difference in their regenerative potential is vastly different from the subcutaneous or intramuscular space and preexisting disease-related degenerative or inflammatory changes in the ocular or articular tissue compartment add an additional layer of complexity.

Independent of the route of administration, the physicochemical properties of the LAI formulation and the scaffold have the potential to significantly impact the local tissue response. It may be therefore essential to explore multiple formulations in preclinical in vivo and/or in vitro studies to identify the optimal balance between the necessary pharmacokinetics for efficacy and the local tolerability within the targeted tissue compartment. Furthermore, approved long-acting drug formulations may offer valuable insights into the local tolerability, potential issues, and mitigation strategies related to local drug administration. Leveraging the nonclinical and clinical experiences gained from these drugs can guide and facilitate the determination of local tolerability for new long-acting drug formulations early on in drug development.

References

Adamson P, Wilde T, Dobrzynski E, Sychterz C, Polsky R, Haworth R, Tang C-M, Korczynska J, Cook F, Papanicolaou I, Tsikna L, Roberts C, Hughes-Thomas Z, Walford J, Gibson D, Warrack J, Smal J, Verrijk R, Miller PE, et al. Single ocular injection of a sustained-release anti-VEGF delivers 6 months pharmacokinetics and efficacy in a primate laser CNV model HHS Public Access. J Control Release. 2016;244:1–13. https://doi.org/10.1016/j.jconrel.2016.10.026.

Ahmadzadeh K, Vanoppen M, Rose CD, Matthys P, Wouters CH. Multinucleated giant cells: current insights in phenotype, biological activities, and mechanism of formation. Front Cell Dev Biol. 2022;10:873226. https://doi.org/10.3389/fcell.2022.873226.

Alcaide-Ruggiero L, Cugat R, Domínguez JM. Proteoglycans in articular cartilage and their contribution to chondral injury and repair mechanisms. Int J Mol Sci. 2023;24(13):10824. https://doi.org/10.3390/ijms241310824.

Andaya R, Booler H, de Almeida Nagata D, Lawson C, Vogt J, Schuetz C, Chang DP, Bantseev V. Intravitreal administration of acetyl triethyl citrate and benzyl benzoate is retinotoxic in rabbits but not in cynomolgus monkeys. Toxicol Pathol. 2021;49(3):621–33. https://doi.org/10.1177/0192623320971571.

Anderson JM, Rodriguez A, Chang DT. Foreign body reaction to biomaterials. Semin Immunol. 2008;20(2):86–100. https://doi.org/10.1016/j.smim.2007.11.004.

Baranov MV, Kumar M, Sacanna S, Thutupalli S, van den Bogaart G. Modulation of immune responses by particle size and shape. Front Immunol. 2021;11:607945. https://doi.org/10.3389/fimmu.2020.607945.

Beningo KA, Wang Y. Fc-receptor-mediated phagocytosis is regulated by mechanical properties of the target. J Cell Sci. 2002;115(4):849–56. https://doi.org/10.1242/jcs.115.4.849.

Bodick N, Williamson T, Strand V, Senter B, Kelley S, Boyce R, Lightfoot-Dunn R. Local effects following single and repeat intra-articular injections of triamcinolone acetonide extended-release: results from three nonclinical toxicity studies in dogs. Rheumatol Ther. 2018;5(2):475–98. https://doi.org/10.6084/m9.figshare.7016654.

Boileau C, Martel-Pelletier J, Abram F, Raynauld JP, Troncy É, D'Anjou MA, Moreau M, Pelletier JP. Magnetic resonance imaging can accurately assess the long-term progression of knee structural changes in experimental dog osteoarthritis. Ann Rheum Dis. 2008;67(7):926–32. https://doi.org/10.1136/ard.2007.077297.

Carnicer-Lombarte A, Chen ST, Malliaras GG, Barone DG. Foreign body reaction to implanted biomaterials and its impact in nerve neuroprosthetics. Front Bioeng Biotechnol. 2021;9:622524. https://doi.org/10.3389/fbioe.2021.622524.

Chamanza R, Darville N, van Heerden M, De Jonghe S. Comparison of the local tolerability to 5 long-acting drug nanosuspensions with different stabilizing excipients, following a single intramuscular administration in the rat. Toxicol Pathol. 2018;46(1):85–100. https://doi.org/10.1177/0192623317737295.

Chandrasekaran PR, Madanagopalan VG. KSI-301: antibody biopolymer conjugate in retinal disorders. Ther Adv Ophthalmol. 2021;13:25158414211027708. https://doi.org/10.1177/25158414211027708.

Crowell SR, Wang K, Famili A, Shatz W, Loyet KM, Chang V, Liu Y, Prabhu S, Kamath AV, Kelley RF. Influence of charge, hydrophobicity, and size on vitreous pharmacokinetics of large molecules. Transl Vis Sci Technol. 2019;8(6) https://doi.org/10.1167/tvst.8.6.1.

Darville N, Van Heerden M, Mariën D, De Meulder M, Rossenu S, Vermeulen A, Vynckier A, De Jonghe S, Sterkens P, Annaert P, Van Den Mooter G. The effect of macrophage and angiogenesis inhibition on the drug release and absorption from an intramuscular sustained-release paliperidone palmitate suspension. J Control Release. 2016;230:95–108. https://doi.org/10.1016/j.jconrel.2016.03.041.

Davies DV, Lond MB. Synovial membrane and synovial fluid of joints. Lancet. 1946;248(6432):815–9. https://doi.org/10.1016/S0140-6736(46)91133-6.

Delecrin J, Oka M, Kumar P, Takahashi S, Kotoura Y, Yamamuro T, Daculsi G. Measurement of synovial fluid volume: a new dilution method adapted to fluid permeation from the synovial cavity. J Rheumatol. 1992;19(11):1746–52.

Dhollander A, Malone A, Price J, Getgood A. Determination of knee cartilage volume and surface area in beagle dogs: a pilot study. J Exp Orthop. 2017;4(1) https://doi.org/10.1186/s40634-017-0109-1.

Eckstein F, Winzheimer M, Hohe J, Englmeier KH, Reiser M. Interindividual variability and correlation among morphological parameters of knee joint cartilage plates: analysis with three-dimensional MR imaging. Osteoarthritis and Cartilage. 2001;9(2):101–11. https://doi.org/10.1053/joca.2000.0365.

EMA. Guideline on non-clinical local tolerance testing of medicinal products. 2016.

Emami A, Tepper J, Short B, Yaksh TL, Bendele AM, Ramani T, Cisternas AF, Chang JH, Mellon RD. Toxicology evaluation of drugs administered via uncommon routes: intranasal, intraocular, intrathecal/intraspinal, and intra-articular. Int J Toxicol. 2018;37(1):4–27. https://doi.org/10.1177/1091581817741840.

FDA. Nonclinical safety evaluation of reformulated drug products and products intended for administration by an alternate route. U.S. Food & Drug Administration; 2015.

Franklin BS, Mangan MS, Latz E. Crystal formation in inflammation. Annu Rev Immunol. 2016;34:173–202. https://doi.org/10.1146/annurev-immunol-041015-055539.

Fubini SL, Todhunter RJ, Burton-Wurster N, Vernier-Singer M, MacLeod JN. Corticosteroids alter the differentiated phenotype of articular chondrocytes. J Orthop Res. 2001;19(4):688–95. https://doi.org/10.1016/S0736-0266(00)00060-7.

Geborek P, Saxne T, Heinegård D, Wollheim FA. Measurement of synovial fluid volume using albumin dilution upon intraarticular saline injection. J Rheumatol. 1988;15(1):91–4.

Getgood A, Dhollander A, Malone A, Price J, Helliwell J. Pharmacokinetic profile of intra-articular fluticasone propionate microparticles in beagle dog knees. Cartilage. 2019;10(2):139–47. https://doi.org/10.1177/1947603517723687.

Heilmann H-H, Lindenhayn K, Walther H-U. Das Synovia-Volumen gesunder und arthrotischer menschlicher Kniegelenke. Zeitschrift Für Orthopädie Und Ihre Grenzgebiete. 2008;134(02):144–8. https://doi.org/10.1055/s-2008-1039786.

Ho MJ, Jeong MY, Jeong HT, Kim MS, Park HJ, Kim DY, Lee HC, Song WH, Kim CH, Lee CH, Choi YW, Choi YS, Han YT, Kang MJ. Effect of particle size on in vivo performances of long-acting injectable drug suspension. J Control Release. 2022;341:533–47. https://doi.org/10.1016/j.jconrel.2021.12.011.

Honarpisheh M, Foresto-Neto O, Desai J, Steiger S, Gómez LA, Popper B, Boor P, Anders HJ, Mulay SR. Phagocytosis of environmental or metabolic crystalline particles induces cytotoxicity by triggering necroptosis across a broad range of particle size and shape. Sci Rep. 2017;7(1) https://doi.org/10.1038/s41598-017-15804-9.

Hurtig M, Drangova M, Holdsworth D. Allometric scaling for implementation of sustained release intra-articular medications – a pilot study. Osteoarthritis Cartilage. 2016;24:S529–30. https://doi.org/10.1016/j.joca.2016.01.968.

ICH M3(R2). M3(R2) Nonclinical safety studies for the conduct of human clinical trials and marketing authorization for pharmaceuticals. 2010.

Jain N, Moeller J, Vogel V. Mechanobiology of macrophages: how physical factors coregulate macrophage plasticity and phagocytosis. Annu Rev Biomed Eng. 2019;21:267–97. https://doi.org/10.1146/annurev-bioeng-062117.

Kerlin R, Bolon B, Burkhardt J, Francke S, Greaves P, Meador V, Popp J. Scientific and regulatory policy committee: recommended ('best') practices for determining, communicating, and using adverse effect data from nonclinical studies. Toxicol Pathol. 2016;44(2):147–62. https://doi.org/10.1177/0192623315623265.

Kim MS, Ho MJ, Joung MY, Choi YS, Kang MJ. Effect of dispersion medium on pharmacokinetic profile of rotigotine crystalline suspension following subcutaneous injection. Pharmaceutics. 2022;14(12) https://doi.org/10.3390/pharmaceutics14122630.

Kranz H, Brazeau GA, Napaporn J, Martin RL, Millard W, Bodmeier R. Myotoxicity studies of injectable biodegradable in-situ forming drug delivery systems. Int J Pharm. 2001;212(1):11–8. www.elsevier.com/locate/ijpharm

Kraus VB, Stabler TV, Kong SY, Varju G, McDaniel G. Measurement of synovial fluid volume using urea. Osteoarthritis Cartilage. 2007;15(10):1217–20. https://doi.org/10.1016/j.joca.2007.03.017.

Kusaka T, Nakayama M, Nakamura K, Ishimiya M, Furusawa E, Ogasawara K. Effect of silica particle size on macrophage inflammatory responses. PLoS One. 2014;9(3) https://doi.org/10.1371/journal.pone.0092634.

Lebre F, Sridharan R, Sawkins MJ, Kelly DJ, O'Brien FJ, Lavelle EC. The shape and size of hydroxyapatite particles dictate inflammatory responses following implantation. Sci Rep. 2017;7(1) https://doi.org/10.1038/s41598-017-03086-0.

Li W, Zhan P, De Clercq E, Lou H, Liu X. Current drug research on PEGylation with small molecular agents. Prog Polym Sci. 2013;38(3–4):421–44. https://doi.org/10.1016/j.progpolymsci.2012.07.006.

Mariani E, Lisignoli G, Borzì RM, Pulsatelli L. Biomaterials: foreign bodies or tuners for the immune response? Int J Mol Sci. 2019;20(3):636. https://doi.org/10.3390/ijms20030636.

Matsuzaka S, Sato S, Miyauchi S. Estimation of joint fluid volume in the knee joint of rabbits by measuring the endogenous calcium concentration. Clin Exp Rheumatol. 2002;20(4):531–4.

Matziolis G, Roehner E, Windisch C, Wagner A. The volume of the human knee joint. Arch Orthop Trauma Surg. 2015;135(10):1401–3. https://doi.org/10.1007/s00402-015-2272-0.

Milde R, Ritter J, Tennent GA, Loesch A, Martinez FO, Gordon S, Pepys MB, Verschoor A, Helming L. Multinucleated giant cells are specialized for complement-mediated phagocytosis and large target destruction. Cell Rep. 2015;13(9):1937–48. https://doi.org/10.1016/j.celrep.2015.10.065.

Mulay SR, Desai J, Kumar SV, Eberhard JN, Thomasova D, Romoli S, Grigorescu M, Kulkarni OP, Popper B, Vielhauer V, Zuchtriegel G, Reichel C, Bräsen JH, Romagnani P, Bilyy R, Munoz LE, Herrmann M, Liapis H, Krautwald S, et al. Cytotoxicity of crystals involves RIPK3-MLKL-mediated necroptosis. Nat Commun. 2016;7 https://doi.org/10.1038/ncomms10274.

NDA 208845. 2017. https://www.fda.gov/drugs/development-approval-process-drugs/drug-approvals-and-databases

Pruimboom-Brees IM, Gupta S, Chemuturi N, Booler HS, Nimz E, Ferrell Ramos M, Caruso A, Maass KF, Bantseev V, Huang Q, Choules MP, Nussbaum JC, Kanodia J, Thompson C, Durairaj C. International consortium for innovation and quality: an industry perspective on the nonclinical and early clinical development of intravitreal drugs. Clin Transl Sci. 2023;16(5):723–41. https://doi.org/10.1111/cts.13480.

Rahmati M, Silva EA, Reseland JE, Heyward CA, Haugen HJ. Biological responses to physicochemical properties of biomaterial surface. Chem Soc Rev. 2020;49(15):5178–224. https://doi.org/10.1039/d0cs00103a.

Rohde CM, Lindemann C, Giovanelli M, Sellers RS, Diekmann J, Choudhary S, Ramaiah L, Vogel AB, Chervona Y, Muik A, Sahin U. Toxicological assessments of a pandemic COVID-19 vaccine—demonstrating the suitability of a platform approach for mRNA vaccines. Vaccines. 2023;11(2) https://doi.org/10.3390/vaccines11020417.

Sellers RS, Nelson K, Bennet B, Wolf J, Tripathi N, Chamanza R, Perron Lepage MF, Adkins K, Laurent S, Troth SP. Scientific and regulatory policy committee points to consider*: approaches to the conduct and interpretation of vaccine safety studies for clinical and anatomic pathologists. Toxicol Pathol. 2020;48(2):257–76. https://doi.org/10.1177/0192623319875085.

Smith AG, Kaiser PK. Emerging treatments for wet age-related macular degeneration. Expert Opin Emerg Drugs. 2014;19(1):157–64. https://doi.org/10.1517/14728214.2014.884559.
Souied EH, Devin F, Mauget-Faÿsse M, Kolár P, Wolf-Schnurrbusch U, Framme C, Gaucher D, Querques G, Stumpp MT, Wolf S. Treatment of exudative age-related macular degeneration with a designed ankyrin repeat protein that binds vascular endothelial growth factor: a phase I/II study. Am J Ophthalmol. 2014;158(4):724–732.e2. https://doi.org/10.1016/j.ajo.2014.05.037.
Stoehr LC, Gonzalez E, Stampfl A, Casals E, Duschl A, Puntes V, Oostingh GJ. Shape matters: effects of silver nanospheres and wires on human alveolar epithelial cells. Part Fibre Toxicol. 2011;8 https://doi.org/10.1186/1743-8977-8-36.
Thackaberry EA, Farman C, Zhong F, Lorget F, Staflin K, Cercillieux A, Miller PE, Schuetz C, Chang D, Famili A, Daugherty AL, Rajagopal K, Bantseev V. Evaluation of the toxicity of intravitreally injected PLGA microspheres and rods in monkeys and rabbits: effects of depot size on inflammatory response. Investig Ophthalmol Vis Sci. 2017;58(10):4274–85. https://doi.org/10.1167/iovs.16-21334.
Vaine CA, Patel MK, Zhu J, Lee E, Finberg RW, Hayward RC, Kurt-Jones EA. Tuning innate immune activation by surface texturing of polymer microparticles: the role of shape in inflammasome activation. J Immunol. 2013;190(7):3525–32. https://doi.org/10.4049/jimmunol.1200492.
Vasconcelos DP, Águas AP, Barbosa JN. The inflammasome in biomaterial-driven immunomodulation. J Tissue Eng Regen Med. 2022;16(12):1109–20. https://doi.org/10.1002/term.3361.
Vorselen D, Labitigan RLD, Theriot JA. A mechanical perspective on phagocytic cup formation. Curr Opin Cell Biol. 2020a;66:112–22. https://doi.org/10.1016/j.ceb.2020.05.011.
Vorselen D, Wang Y, de Jesus MM, Shah PK, Footer MJ, Huse M, Cai W, Theriot JA. Microparticle traction force microscopy reveals subcellular force exertion patterns in immune cell–target interactions. Nat Commun. 2020b;11(1) https://doi.org/10.1038/s41467-019-13804-z.
Wehr B, Grams A, Hudelmaier M, Kotyk J, Wachsmuth L, Eckstein F. 79 Tibial cartilage surface area, thickness and volume in various animal species and in humans. Osteoarthritis Cartilage. 2007;15:C54–5. https://doi.org/10.1016/s1063-4584(07)61712-9.
Wernecke C, Braun HJ, Dragoo JL. The effect of intra-articular corticosteroids on articular cartilage: a systematic review. Orthop J Sports Med. 2015;3(5):1–7. https://doi.org/10.1177/2325967115581163.
Wilkinson J, Ajulo D, Tamburrini V, Le Gall G, Kimpe K, Holm R, Belton P, Qi S. Lipid based intramuscular long-acting injectables: current state of the art. Eur J Pharm Sci. 2022;178:106253. https://doi.org/10.1016/j.ejps.2022.106253.
Wilson CJ, Clegg RE, Leavesley DI, Pearcy MJ. Mediation of biomaterial–cell interactions by adsorbed proteins: a review. Tissue Eng. 2005;11(1–2):1–18. https://doi.org/10.1089/ten.2005.11.1.
Yang W, Wang L, Mettenbrink EM, Deangelis PL, Wilhelm S. Nanoparticle toxicology. Annu Rev Pharmacol Toxicol. 2021;61:269–89. https://doi.org/10.1146/annurev-pharmtox-032320.
Zhang L, Xing R, Huang Z, Ding L, Zhang L, Li M, Li X, Wang P, Mao J. Synovial fibrosis involvement in osteoarthritis. Front Med. 2021;8:684389. https://doi.org/10.3389/fmed.2021.684389.
Zhong H, Chan G, Hu Y, Hu H, Ouyang D. A comprehensive map of FDA-approved pharmaceutical products. Pharmaceutics. 2018;10(4):263. https://doi.org/10.3390/pharmaceutics10040263.

Chapter 9
Envisaging the Future of Long-Acting Injectables and Implants

Farshad Ramazani, Andreas Fisch, and Bernd Riebesehl

List of Abbreviations

FDA	Food and Drug Administration
IM	Intramuscular
IVIVC	In-vitro–in-vivo correlation
LAIs	Long-acting injectables
NMEs	New molecular entities
NMP	N-Methylpyrrolidone
PAT	Process analytical technology
PK	Pharmacokinetics
PK/PD	Pharmacokinetics/pharmacodynamic
PLGA	Polylactic-co-glycolic acid
QTPP	Quality target product profile
REMS	Risk evaluation and mitigation strategy
SC	Subcutaneous

9.1 Introduction

Long-acting injectables (LAIs) are known to enhance patient adherence and benefit–risk ratio in the management of chronic diseases. This technology is emerging across various therapeutic areas, including cancer, central nervous system disorders, endocrine/metabolism conditions, women's health, blood disorders, and infectious diseases. Until now, several LAI platform technologies including polylactic-co-glycolic acid (PLGA) microparticles, solid implants, in situ forming gel/implants, crystal suspensions, and oil-based formulations of lipophilic (pro)drugs have been

F. Ramazani (✉) · A. Fisch · B. Riebesehl
Technical Research and Development, Novartis Pharma AG, Basel, Switzerland
e-mail: farshad.ramazani@novartis.com

F. Ramazani (ed.), *Biodegradable Long Acting Injectables and Implants*,
https://doi.org/10.1007/978-3-031-72302-5_9

established in the pharmaceutical market (Nkanga et al. 2020). Each year, several novel molecules or already marketed oral or immediate-release molecules, as life cycle management, are formulated in these platform technologies and enter clinical trials. In addition, the recent surge in clinical trials for LAIs demonstrates their versatility and potential to address unmet medical needs from treating osteoarthritis and tendon disease to managing chronic pain (Bursulaya et al. 2018; Kou et al. 2023; Mekhail 2020). Nevertheless, the development of LAI products is complex and challenging, but it potentially generates value beyond the loss of patent exclusivity. One key challenge is minimizing the time associated with LAI development. To shorten development timelines, it is important to anticipate and predict the feasibility limiting factors, utilizing modeling and simulation tools (Siemons et al. 2023). For example, for having a simple LAI based on drug crystal suspension, a discovery chemist may design drug candidates for high potency, long plasma half-life, and poor water solubility to enable long dosing intervals (Bassand et al. 2022; T'jollyn and Ackaert 2023). Understanding the factors that impact the in vivo performance of LAI is of great importance; for instance, factors such as drug crystal lipophilicity, particle size, excipient, body mass index, gender, and animal species have an impact on the performance of the formulations and make it difficult to extrapolate the pharmacokinetics (PK) from animal to human (Nguyen et al. 2022).

The development of LAIs requires consideration of biomaterials (excipients) that need to be biocompatible, stable, capable of high drug loading, and suitable for long-term drug release. Several biodegradable polymers have already been established in the marketed products showing high biocompatibility when injected into different body tissues. However, depending on the site of injection, drug, and formulation types, some local tolerability issues have been identified which may halt the development of LAIs. Understanding the mechanism of body reaction to foreign material and designing novel material with better tolerability would increase the chance of success in the development of new LAI products.

At the end, a patient-centric approach is essential in the development of LAIs. This involves considering patient preferences, needs, and behaviors throughout the development process. The chapter envisions future strategies to facilitate the development of LAIs by minimizing development timelines and considerations in drug substance design, manufacturing and scale-up, biomaterial, and analytics.

9.2 How to Minimize LAIs Development Timeline?

LAI development is seen as a chance to extend the life of an existing oral drug. To enhance patient experience, existing drug molecules are repurposed for novel uses as LAI, utilizing the wealth of information on clinical PK and preclinical and clinical safety as a foundation for development (Pizzorno et al. 2019; Selmin et al. 2020). Nevertheless, the analysis of the development pipeline reveals a trend toward developing new molecular entities (NMEs) tailored for LAIs, emphasizing characteristics such as poor solubility, long in vivo half-life, and high potency (Bassand et al.

2022). As compared to life cycle management of existing drugs, developing NMEs as LAI is more complicated. For both cases, safety and efficacy have to be investigated through preclinical and clinical trials. However, since there are some data already available for life cycle management of the oral drug, the chance of success is higher as compared to NME. There are several ways to minimize the development timeline and risks associated with the development of LAI. Since there is an inherent, potential toxicological risk of drugs never tested in humans, either locally or systemically, the first clinical study should be performed with an immediate release formulation, typically in a dose escalation mode. Only in the second step should a dose escalation study be performed with the LAI formulation. This sequential approach is also preferable since the LAI formulation typically cannot be removed once injected.

Modeling and simulation can support the development timelines by integrating in vitro, preclinical, and clinical studies into mathematical modeling. Numerous advancements in the field are leading to an increasing reliance on modeling. These tools are, e.g., minimizing and substituting animal studies, aiding in formulation design, predicting formulation complexity (e.g., drug–polymer interaction, stability), simulating initial projections of LAI doses and duration of drug release, modeling drug release based on route of administration, establishment of in-vitro–in-vivo correlation (IVIVC) and deepening our understanding of the mechanistic aspects of physiological conditions that influence PK profiles of LAI (Kotla et al. 2023b; Siemons et al. 2023). Furthermore, simulation can shorten the scale-up process and minimize the number of technical batches needed for scaling up, e.g., by simulating reactor design and how energy is distributed within the reactor and manufacturing equipment.

Designing a road map for development of complex formulation is very important as it can promote communication across diverse teams and departments who are involved in such complex discovery projects.

9.3 What to Consider in Developing LAI to Minimize the Risk of Failure?

LAI for Systemic Versus Local Drug Action

Most LAIs are locally dosed but distributed by blood circulation and exhibit their pharmacological action in the target tissues. Thus, the drug plasma concentration can act as surrogate for the concentration in the target tissue, which is more difficult to measure. This scenario simplifies optimization of LAI drug release based on in-vivo plasma concentration curves. By contrast, local tissue target concentrations after local LAI dosing can typically not be measured easily and drug plasma concentrations are diluted by body distribution and cannot directly serve as surrogate for the local drug concentrations that enable efficacy.

If biomarkers supporting clinical efficacy or safety endpoints are known, then plasma concentrations can be used to optimize formulations and dosing schemes. In the early stage of any drug development especially LAI, it is important to identify biomarkers to understand drug's mechanism of action, study toxicity and efficacy of the molecules and even classify patients who may respond to the treatment (Huber et al. 2022).

Drug Substance Considerations

For designing drug substances that are suitable for LAI development, poor water solubility is a key metric for optimization (Nkanga et al. 2020). For example, the molecules need to be often highly insoluble in water or body fluids. In addition, metabolic stability is vital in influencing their elimination speed or half-life in circulation. Prodrug approaches add complexity, requiring consideration of cleavage kinetics during preclinical characterization for predicting pharmacokinetics in patients with various diseases. A fundamental principle across all LAI approaches is selecting highly potent drug candidate in relevant biological assays to reduce the required dose and injection volume, enhancing the final product's safety and tolerability due to lower volume of injection.

Manufacturing and Scalability

Complex manufacturing processes pose challenges in LAI formulation development. To ensure successful scale-up manufacturing of a drug product, several key steps should be followed:

(a) Formulation screening: Design the formulation based on quality target product profile (QTPP). It is important to perform a thorough exercise to define the QTPP and based on the criteria design a suitable formulation that meets the needs of patients. For example, if the route of administration is an inflamed tissue, it is wise to not use an organic solvent in the formulation that may worsen the inflammation in the local tissues. See Table 9.1 for an example of QTPP for an LAI. For selecting the best formulation technology, it is important to analyze all the existing technology and rank them against QTPP criteria. After selecting top technologies, the formulation screening can be performed to investigate which formulation has the best pharmacokinetics/pharmacodynamic (PK/PD) profile, stability, tolerability, and manufacturability.
(b) Process development: Evaluate the manufacturing process and find critical process parameters and define an acceptable range based on critical quality attributes of your products. Check the reproducibility of manufacturing process.
(c) Manufacturing equipment validation: Design and validate manufacturing equipment from pilot to large-scale manufacturing.

Table 9.1 Example of quality target product profile of an implant

QTPP elements	Target
Dosage form	Solid implant
Dosage design	Long-acting ≥6 months drug release
Route of administration	Intraocular injection
Therapeutic indication	Age-related macular degeneration
Population	Adults
Posology	0.46 mm in diameter and 6 mm in length
Dosage strength	To be determined (projected 50 mg)
Shelf life and storage	Stable at 2–8 °C for 2 years
Preferred primary packaging	Applicator ready for administration
API source and quality	Company name
Excipients	Biodegradable and biocompatible, commercially available in parenteral Good Manufacturing Practice (GMP) quality
Sterility	Terminally sterile or aseptically manufactured
Target marketing territories	Worldwide

(d) Process validation: Perform thorough process validation to verify that the large-scale process always produces drug products meeting quality standards that are predetermined on a small scale. Perform thorough process validation to verify that the large-scale process always produces drug products meeting quality standards that are predetermined in small scale, for example, by performing validation studies, data analysis, and process controls establishment.

(e) Supply chain management: Find a reliable excipient supply chain to ensure the availability of raw materials for large-scale manufacturing. This is especially important for biodegradable polymers as their quality greatly impacts the performance of the final drug product.

(f) Technology transfer: If the scale-up is being performed at a different manufacturing site, make sure to well plan the technology transfer process. The transfer involves effective communication, clear documentation, and thorough training to transfer knowledge and skills smoothly from one site to the other. All the manufacturing equipment and their size and dimension with details need to be equivalent to the pilot plant and identical as some small changes may have a huge impact on the final product.

By following these steps, drug manufacturers can increase the likelihood of successful scale-up manufacturing, thereby achieving consistent quality, meeting market demands and ensuring patient safety. For instance, developing PLGA microparticles involves massive process development and several scale-ups from the lab and pilot plant to the manufacturing plant. Successful scale-up requires rigorous cooperation between formulation scientists working in the lab and pharmaceutical engineers in pilot plant. The small-scale manufacturing equipment, e.g., the emulsification mixer or solvent evaporation vessels need to be scalable and must already be established as GMP equipment in the pilot plant. Process analytical technology (PAT) tools must be implemented for monitoring drug product manufacturing; for instance, in this case, monitoring particle formation, solidification, and solvent removal using online tools is beneficial for scale-up. In addition, all critical process parameters, e.g., reactor size and dimension, solvent evaporation speed, solidification time and temperature, mixing speed, and drying, must be identified for successful scale-up manufacturing (Nkanga et al. 2020).

Novel Biomaterials and Drug Molecules

Since the excipient quality is important in defining the shelf life and performance of polymer-based LAI, it is recommended to use high-quality polymers. The development scientist needs to engage closely with their polymer suppliers to select the desired polymer and increase the probability of success during development. PLGA-based polymers have been used for formulating small molecules and peptides, but they have failed to encapsulate larger molecules such as protein drugs or sensitive peptides. This is in part due to their acidity build-up over drug-release time and during shelf life. The stability issues are not only related to the polymers but also due to manufacturing methods which often involve organic/water solvent, high temperature, and harsh terminal sterilization (Van De Weert et al. 2000). Therefore, novel (e.g., non-PLGA or modified PLGA) polymers and manufacturing methods are needed to answer those questions and make LAI delivery possible for sensitive biotherapeutics. From emerging RNA therapeutics and other novel drug modalities, it is expected to see the LAI formulation concept to be implemented for such molecules. Although most of the lessons learned from small molecule development are translatable to novel modalities, they may have their own unique challenges to tackle.

In Vivo Behavior and Tolerability of LAI Formulations

Local tolerability refers to tissue reactions at the site of drug administration. Evaluating local tolerability in preclinical studies is crucial for identifying potential clinical safety risks. Key factors in this evaluation include histopathological observations at the administration site, along with in-life observations and gross findings at necropsy. These elements help determine if a drug formulation is well tolerated locally or not. LAIs are designed to remain at the injection site for extended periods, generally causing a local foreign body reaction. This response aims to degrade and remove foreign material and restore tissue integrity. The reaction can be influenced by the formulation's matrix, its chemical and physical properties, and the drug product composition itself. The reaction may also vary between different administration sites due to differences in tissue type and mechanical forces (Darville et al. 2016; Kim et al. 2022). Injecting into diseased tissue presents additional challenges, such as exacerbating a pro-inflammatory state or preexisting tissue injury. An inflammatory response might also aid in the slow degradation of the formulation and sustained drug release. The challenge in developing LAIs is to create a formulation that provides extended drug exposure while minimizing local tissue reactions. Mechanistic understanding of phagocytosis and immunologic events is helpful for the development of LAI. Novel biocompatible polymer with better tolerability or engineering the surface of particle with material that can enhance their stealth properties from immune system may improve biocompatibility of the LAI (Kotla et al. 2023a). A regulatory drug approval can occur if the clinical benefit–risk assessment supports a favorable balance of clinical patient benefits over risks. A grading of injection site reactions supports a more rational assessment. For some LAIs, a risk evaluation and mitigation strategy (REMS) has been requested by the Food and Drug Administration (FDA) to ensure that benefits outweigh the risks. For detailed information about local tolerability of LAIs, refer to Chap. 9 of this book.

Analytical Prospective

From an analytical standpoint, the comprehension of drug-release mechanisms and the connection between in vitro and in vivo drug release in development of LAI presents a notable challenge. The choice of experimental conditions, release mechanism, and selection of release medium, which are different from lab to lab, all greatly influence in vitro drug release (Nkanga et al. 2020). Various techniques, including continuous flow and sample-and-separate methods, are employed to measure drug release in vitro for batch-to-batch consistency, quality assurance and formulation development. As most of LAI drugs are highly lipophilic, making sink condition for in vitro drug release can be obtained by adusting factors such as pH, temperature, and addition of surfactants to the release medium. These additives may negatively impact drug-release kinetics. Moreover, as mentioned above, local tissue response significantly impacts drug release in the body. As a result, developing an in vitro method that represents in vivo conditions with body response to foreign material is not possible. Although some IVIVCs have been established for specific LAI formulations, during the late phase or after commercialization, effectively translating data between animals and humans remains challenging. Further research in this field, particularly utilizing modeling and simulation, is necessary to enhance our understanding of drug-release mechanisms and enhance the ability to predict in-vivo release for LAI formulations (Shah and Hong 2022).

9.4 Patient-Centric Drug Development

Patient-focused drug development has been implemented by FDA that places patients at the center of healthcare decision-making. The goal is to ensure that the experiences, perspectives, and needs of patients are not only considered but are also actively incorporated into all aspects of their care (Algorri et al. 2023; Timpe et al. 2020). The FDA has also issued a detailed guideline on the concept (FDA 2020). Some examples of enabling patient-centric concept for LAI design and development are discussed.

Patient-centric approach can be implemented at any stage of development from early to late phase and even after commercialization. For a patient-centric drug development, a close collaboration between development team and patient representative at any stage of development is important to design an LAI drug product with its administration device. For example, patients might prefer SC over other routes of injection as they can self-administer the dose at home versus requiring a healthcare professional with potential special training to perform the injection. As shown in Fig. 9.1, there is a notable transition from intramuscular (IM) to subcutaneous (SC) delivery for systemic LAI treatments, driven by the convenience of at-home self-administration, the psychological impact of smaller needles on pain perception, and the reduction of costs and burdens on healthcare systems.

It is of great importance to design drug delivery systems that can be SC-injected using small needle size, e.g., 27G. This can be achieved, for example, by controlling particle size and size distribution using novel method such as microfluidics, using novel excipients or cosolvents that can improve rheology properties of in situ forming gels and developing novel self-administration devices that can inject the formulation with minimum pain. One of the issues with SC injection is the limited volume

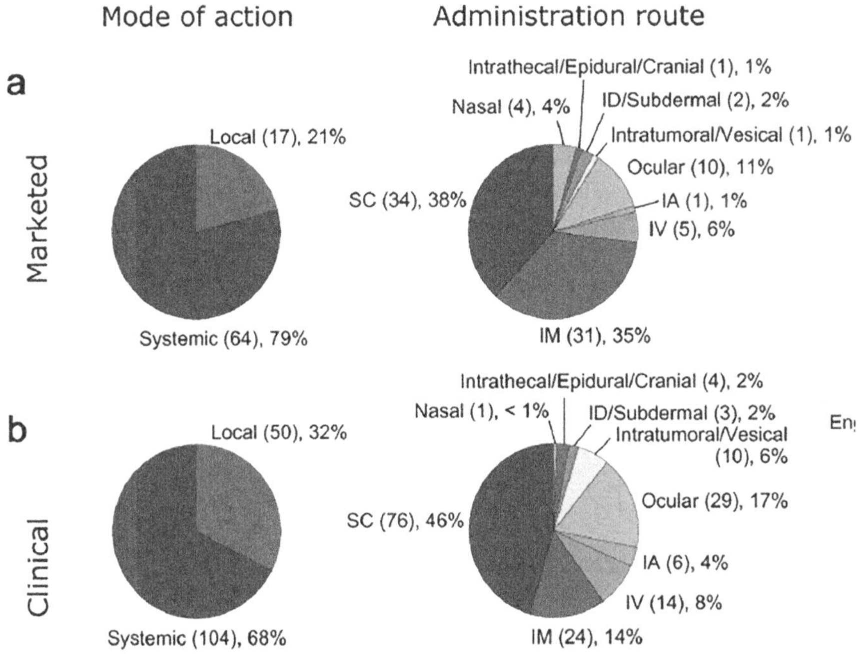

Fig. 9.1 LAIs marketed (**a**) and recent clinical trials (**b**) showing a shift toward SC injection. (With permission from Bassand et al. (2022))

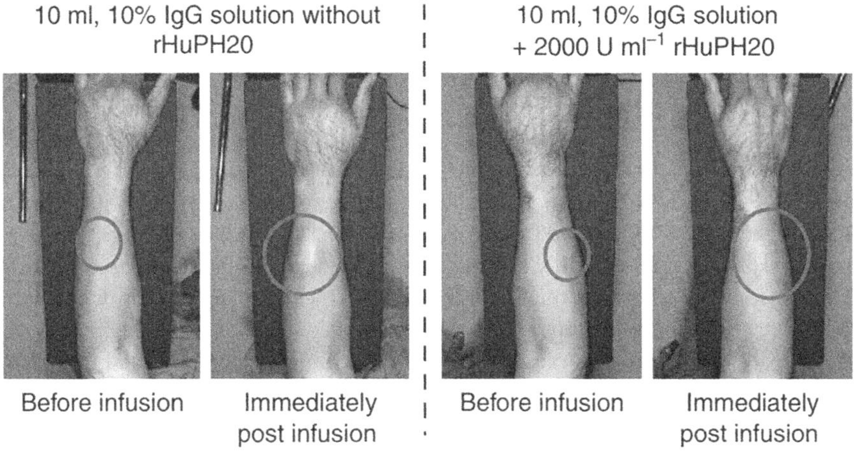

Fig. 9.2 Injection site before and immediately after rituximab SC injection co-formulated without (left panel) and with (right panel) recombinant human hyaluronidase (rHuPH20) (Shpilberg and Jackisch 2013)

of injection. This issue can be circumvented by combination therapy using hyaluronidase to disrupt the hyaluronan structure under the skin (Shpilberg and Jackisch 2013), enabling larger injection volumes and/or better spreading of the injection solution in the SC tissue and less injection pain. SC administration of MabThera® (combination of hyaluronidase and Rituximab) was much faster compared to IV infusion of Rituximab (7 min vs. 90 min). Approximately, 80% of patients preferred SC injection of MabThera® as compared to infusion. The anticipated outcome of SC administration is a substantial reduction in both caregiver and patient burden, aiming to provide added value to the healthcare system (Harvey et al. 2022). Figure 9.2 shows how hyaluronidase enables SC injection of large volume drug solution.

9.5 Recent Clinical Approvals

In recent clinical trials, novel biodegradable polymers such as BEPO® technology technology which is based on an amphiphilic copolymer of lactic acid and polyethylene glycol have been employed for control release of drugs. In situ forming gel can be prepared by dissolving this polymer in N-methylpyrrolidone (NMP) (Roberge et al. 2020). Compared to PLGA microparticles or implants, in situ forming gels are simpler to develop without the need for scale-up manufacture and extensive process development. In addition, their administration is easy and can be performed using an autoinjector. However, it is worth mentioning that NMP may cause toxicity in higher doses, e.g., > 5.3 mg/day, which may limit the development of in situ forming gels based on NMP (Nkanga et al. 2020). In addition, controlling drug release, especially the initial burst, is difficult with such systems as compared to PLGA microparticles and implants.

Nanocrystal suspensions are also paving the way for future of LAI development. Several new drug products with extremely long duration of drug release in the body have already been developed including the very potent, lipophilic, and poorly water-soluble prodrug Paliperidone palmitate (Invega Sustenna, Trinza and Hafyera, J&J). In theory, nanocrystal formulations release drugs faster than microcrystals due to their higher surface-to-volume ratio. However, once injected, the drug release from nanocrystals is influenced by the local environment, including factors such as phagocytosis, granuloma formation, lymph node uptake, and enzymatic degradation (Darville et al. 2014). In addition, injecting a high concentration of drug into a confined space can lead to saturation solubility, which may delay drug dissolution and prolong the drug release duration. Despite the various factors affecting drug release from crystal suspensions, recently marketed drugs formulated as nanosuspensions demonstrate ultra-long drug-release durations. This suggests that nanocrystals will likely become a key technology in the development of LAIs in the future.

To achieve self-administration, even PLGA microparticle formulations were developed in a dual-chamber autoinjector. Bydureon Bcise©, which consists of microparticles loaded with Exenatide—an antidiabetic peptide that is used for lowering blood glucose in patients with type 2 diabetes—is supplied with an autoinjector (Li et al. 2021). The manufacturer developed an autoinjector that contains both microparticles and the vehicle for suspension preparation. Unlike other marketed microparticles which are supplied using an aqueous-based vehicle, Bydureon Bcise© vehicle is based on medium-chain triglycerides (Tice 2022). Upon shaking for few seconds, it forms a homogenous single-dose suspension that can be immediately injected into subcutaneous space. This autoinjector and vehicle system can be a platform for self-administering microparticles formulations in the future.

9.6 Conclusion

The advancement of LAIs holds significant promise in pharmaceutical development for the treatment of chronic diseases. Gaining a deep understanding of key factors involved in the formulation and production processes and utilizing modeling and simulation tools provides avenues to advance LAI development while mitigating risks of failure. By adopting a patient-centric approach, we can forge a pathway for better implementation of these promising therapeutic platforms.

References

Algorri M, Cauchon NS, Christian T, O'Connell C, Vaidya P. Patient-centric product development: a summary of select regulatory CMC and device considerations. J Pharm Sci. 2023;112(4):922–36. https://doi.org/10.1016/j.xphs.2023.01.029.

Bassand C, Villois A, Gianola L, Laue G, Ramazani F, Riebesehl B, Sanchez-Felix M, Sedo K, Ullrich T, Duvnjak Romic M. Smart design of patient-centric long-acting products: from preclinical to marketed pipeline trends and opportunities. Expert Opin Drug Deliv. 2022;19(10) https://doi.org/10.1080/17425247.2022.2106213.

Bursulaya B, Fisch A, Lajiness JP, Lajiness P, Machauer R, Malekar S, Petrassi HMJ, Ramazani F, Remond A-C, Ullrich T, Usselmann P, Vangrevelinghe E. Substituted indazoles for treating tendon and/or ligament injuries. 2018.

Darville N, van Heerden M, Vynckier A, De Meulder M, Sterkens P, Annaert P, Van den Mooter G. Intramuscular administration of paliperidone palmitate extended-release injectable microsuspension induces a subclinical inflammatory reaction modulating the pharmacokinetics in rats. J Pharm Sci. 2014;103(7):2072–87. https://doi.org/10.1002/jps.24014.

Darville N, van Heerden M, Mariën D, De Meulder M, Rossenu S, Vermeulen A, Vynckier A, De Jonghe S, Sterkens P, Annaert P, Van den Mooter G. The effect of macrophage and angiogenesis inhibition on the drug release and absorption from an intramuscular sustained-release paliperidone palmitate suspension. J Control Release. 2016;230:95–108. https://doi.org/10.1016/j.jconrel.2016.03.041.

FDA. Patient-focused drug development: collecting comprehensive and representative input guidance for industry, Food and Drug Administration staff, and other stakeholders. Silver Springs: Center for Drug Evaluation and Research; 2020.

Harvey MJ, Zhong Y, Morris E, Beverage JN, Epstein RS, Chawla AJ. Assessing the transition from intravenous to subcutaneous delivery of rituximab: benefits for payers, health care professionals, and patients with lymphoma. PLoS One. 2022;17(1):e0261336. https://doi.org/10.1371/journal.pone.0261336.

Huber C, Friede T, Stingl J, Benda N. Classification of companion diagnostics: a new framework for biomarker-driven patient selection. Ther Innov Regul Sci. 2022;56(2):244–54. https://doi.org/10.1007/s43441-021-00352-2.

Kim MS, Ho MJ, Joung MY, Choi YS, Kang MJ. Effect of dispersion medium on pharmacokinetic profile of Rotigotine crystalline suspension following subcutaneous injection. Pharmaceutics. 2022;14(12):2630. https://doi.org/10.3390/pharmaceutics14122630.

Kotla NG, Langlois J-B, Fisch A, Kramer I, Halleux C. Surface modified cationic PLGA microparticles as long-acting injectable carriers for intra-articular small molecule drug delivery. Eur J Pharm Biopharm. 2023a;193 https://doi.org/10.1016/j.ejpb.2023.10.020.

Kotla NG, Pandey A, Vijaya Kumar Y, Ramazani F, Fisch A. Polyester-based long acting injectables: advancements in molecular dynamics simulation and technological insights. Drug Discov Today. 2023b;28(2):103463. https://doi.org/10.1016/j.drudis.2022.103463.

Kou H, Qing Z, Zhao G, Sun X, Zhi L, Wang J, Chen X, Guo H, Zhang R, Ma J. Effect of lorecivivint on osteoarthritis: a systematic review and meta-analysis. Heliyon. 2023;9(8) https://doi.org/10.1016/j.heliyon.2023.e18682.

Li T, Chandrashekar A, Beig A, Walker J, Hong JKY, Benet A, Kang J, Ackermann R, Wang Y, Qin B, Schwendeman AS, Schwendeman SP. Characterization of attributes and in vitro performance of exenatide-loaded PLGA long-acting release microspheres. Eur J Pharm Biopharm. 2021;158:401–9. https://doi.org/10.1016/j.ejpb.2020.10.008.

Mekhail NA. Safety and efficacy of novel epidural clonidine micropellets for lumbosacral radiculopathy. Clin Neurol Neurosurg. 2020;3:2. https://doi.org/10.37421/jcnn.2020.3.109.

Nguyen VTT, Darville N, Vermeulen A. Pharmacokinetics of long-acting aqueous nano-/microsuspensions after intramuscular administration in different animal species and humans—a review. AAPS J. 2022;25(1):4. https://doi.org/10.1208/s12248-022-00771-5.

Nkanga CI, Fisch A, Rad-Malekshahi M, Romic MD, Kittel B, Ullrich T, Wang J, Krause RWM, Adler S, Lammers T, Hennink WE, Ramazani F. Clinically established biodegradable long acting injectables: an industry perspective. Adv Drug Deliv Rev. 2020;167:19–46. https://doi.org/10.1016/J.ADDR.2020.11.008.

Roberge, Christophe et al. BEPO®: Bioresorbable Diblock MPEG-PDLLA and Triblock PDLLA-PEG-PDLLA Based in Situ Forming Depots with Flexible Drug Delivery Kinetics Modulation. J Control Release 2020;319:416–27.

Pizzorno A, Padey B, Terrier O, Rosa-Calatrava M. Drug repurposing approaches for the treatment of influenza viral infection: reviving old drugs to fight against a long-lived enemy. Front Immunol. 2019;10(MAR):531. https://doi.org/10.3389/FIMMU.2019.00531.

Selmin F, Musazzi UM, Magri G, Rocco P, Cilurzo F, Minghetti P. Regulatory aspects and quality controls of polymer-based parenteral long-acting drug products: the challenge of approving copies. Drug Discov Today. 2020;25(2):321–9. https://doi.org/10.1016/J.DRUDIS.2019.12.008.

Shah JC, Hong J. Model for long acting injectables (depot formulation) based on pharmacokinetics and physical chemical properties. AAPS J. 2022;24(3) https://doi.org/10.1208/s12248-022-00695-0.

Shpilberg O, Jackisch C. Subcutaneous administration of rituximab (MabThera) and trastuzumab (Herceptin) using hyaluronidase. Br J Cancer. 2013;109(6):1556–61. https://doi.org/10.1038/bjc.2013.371.

Siemons M, Schroyen B, Darville N, Goyal N. Role of modeling and simulation in preclinical and clinical long-acting injectable drug development. AAPS J. 2023;25(6) https://doi.org/10.1208/s12248-023-00864-9.

T'jollyn H, Ackaert O. The AAPS Journal Theme Issue: "Perspectives on clinical drug development of long-acting injectables". AAPS J. 2023;25(6):104. https://doi.org/10.1208/s12248-023-00871-w.

Tice T. Inspirational chemistry of bioabsorbable long-acting injectables. Drug Deliv Transl Res. 2022;13(2):371–7. https://doi.org/10.1007/S13346-022-01199-2/METRICS.

Timpe C, Stegemann S, Barrett A, Mujumdar S. Challenges and opportunities to include patient-centric product design in industrial medicines development to improve therapeutic goals. Br J Clin Pharmacol. 2020;86(10):2020–7. https://doi.org/10.1111/bcp.14388.

Van De Weert M, Hennink WE, Jiskoot W. Protein instability in poly(lactic-co-glycolic acid) microparticles. Pharm Res. 2000;17(10):1159–67. https://doi.org/10.1023/A:1026498209874.

The manufacturer's authorised representative in the EU is Springer Nature Customer Service Centre GmbH, Europaplatz 3, 69115 Heidelberg, Germany. If you have any concerns regarding our products, please contact ProductSafety@springernature.com

Printed and bound by CPI Group (UK) Ltd, Croydon, CR0 4YY

30/09/2026

02225558-0010